New Frontiers in Regional Science: Asian Perspectives

Volume 4

More information about this series at http://www.springer.com/series/13039

New Frontiers in Regional Science: Asian Perspectives

This series is a constellation of works by scholars in the field of regional science and in related disciplines specifically focusing on dynamism in Asia.

Asia is the most dynamic part of the world. Japan, Korea, Taiwan, and Singapore experienced rapid and miracle economic growth in the 1970s. Malaysia, Indonesia, and Thailand followed in the 1980s. China, India, and Vietnam are now rising countries in Asia and are even leading the world economy. Due to their rapid economic development and growth, Asian countries continue to face a variety of urgent issues including regional and institutional unbalanced growth, environmental problems, poverty amidst prosperity, an ageing society, the collapse of the bubble economy, and deflation, among others.

Asian countries are diversified as they have their own cultural, historical, and geographical as well as political conditions. Due to this fact, scholars specializing in regional science as an inter- and multi-discipline have taken leading roles in providing mitigating policy proposals based on robust interdisciplinary analysis of multifaceted regional issues and subjects in Asia. This series not only will present unique research results from Asia that are unfamiliar in other parts of the world because of language barriers, but also will publish advanced research results from those regions that have focused on regional and urban issues in Asia from different perspectives.

The series aims to expand the frontiers of regional science through diffusion of intrinsically developed and advanced modern regional science methodologies in Asia and other areas of the world. Readers will be inspired to realize that regional and urban issues in the world are so vast that their established methodologies still have space for development and refinement, and to understand the importance of the interdisciplinary and multidisciplinary approach that is inherent in regional science for analyzing and resolving urgent regional and urban issues in Asia.

Topics under consideration in this series include the theory of social cost and benefit analysis and criteria of public investments, socio-economic vulnerability against disasters, food security and policy, agro-food systems in China, industrial clustering in Asia, comprehensive management of water environment and resources in a river basin, the international trade bloc and food security, migration and labor market in Asia, land policy and local property tax, Information and Communication Technology planning, consumer "shop-around" movements, and regeneration of downtowns, among others.

Kiyoko Hagihara • Chisato Asahi
Editors

Coping with Regional Vulnerability

Preventing and Mitigating Damages from Environmental Disasters

Springer

Editors
Kiyoko Hagihara
School of Sociology
Bukkyo University
Kyoto, Kyoto
Japan

Chisato Asahi
Division of Urban Policy
Tokyo Metropolitan University
Hachioji, Tokyo
Japan

ISSN 2199-5974 ISSN 2199-5982 (electronic)
New Frontiers in Regional Science: Asian Perspectives
ISBN 978-4-431-55168-3 ISBN 978-4-431-55169-0 (eBook)
DOI 10.1007/978-4-431-55169-0

Library of Congress Control Number: 2015949226

Springer Tokyo Heidelberg New York Dordrecht London

Printed on acid-free paper

Springer Japan KK is part of Springer Science+Business Media (www.springer.com)

Preface

Although Japan has developed a relatively affluent society since the end of World War II from an economic perspective, it has also experienced several negative events including serious environmental pollution and disasters such as floods and earthquakes. In particular the recent Great East Japan Earthquake and Tsunami and the Fukushima nuclear power plant disaster have taught us to always be prepared to cope with such events. This is not limited to Japan. In many countries throughout Asia, people confront various kinds of natural events, including disasters, especially those caused by global climate change. Whether a natural event turns into a disaster depends on the severity of the hazard as well as the condition of the social sphere of its potential victims, that is, vulnerability.

In this book we focus on regional vulnerability considering the fact that regional socioeconomic conditions determine the nature of the damage and thus the risk management policy. We provide the theory and methodology to understand and cope with regional vulnerability through an interdisciplinary approach. The fields mainly included in this work are welfare and environmental economics, the planning and management area of civil engineering, risk management and conflict management. In particular, we focus on the hazards and vulnerability surrounding water issues and provide readers with an understanding of how regional analysis is incorporated into vulnerability analysis. Also considered is what risk management should be when the diversified regional background of the vulnerability is taken into account.

An important feature of this book is that it provides contrastive regional coverage: the vulnerability of a developed country—urban and regional areas of Japan—and that of a developing country, Bangladesh. The contents consist of three parts: (1) Socioeconomic Vulnerability in a Regional Perspective, (2) Evaluation of Regional Vulnerability and (3) Risk Management of Regional Vulnerability.

In order to accomplish the above-mentioned undertakings, we asked specialists who have been studying these issues in a range of related fields to contribute to this book. In particular, we are grateful towards Professor Keith W. Hipel, who kindly accepted writing Chap. 17 and discussing his leading-edge ideas in the area of conflict resolution.

We would like to thank all our contributors for their patience and dedication; they revised their contributions several times to implement our editorial guidance. Without the hard work of all of the authors, this book would have never seen the light of day.

We would also like to thank the editorial board at the Japan Section of the Regional Science Association for compiling this volume as one of the epochal series of *New Frontiers in Regional Science: Asian Perspectives*, in particular, Professor Yoshiro Higano, who kindly reviewed the contents of this book as editor in chief. It is our hope that this book contributes to an improvement of human well-being in areas suffering from the impacts of various kinds of disasters and vulnerabilities.

Kyoto, Japan
Hachioji, Japan
April 2015

Kiyoko Hagihara
Chisato Asahi

Contents

Contributors

Tofayel Ahmed is a researcher at the Japan Association of Drainage and Environment Bangladesh Office, Dhaka, Bangladesh.

Chisato Asahi is associate professor in the Division of Urban Policy, Faculty of Urban Liberal Arts, Tokyo Metropolitan University, Tokyo, Japan

Yosuke Fukushima is a government official of the Ministry of Land, Infrastructure, Transport and Tourism, Tokyo, Japan, and received his master's degree in engineering from Kyoto University.

Kiyoko Hagihara is professor in the Department of Public Policy, School of Sociology, Bukkyo University, Kyoto, Japan, and professor emeritus in Tokyo Metropolitan University, Tokyo, Japan.

Yoshimi Hagihara is an emeritus professor in the Disaster Prevention Research Institute, Kyoto University, Kyoto, Japan.

Keith W. Hipel is university professor of Systems Design Engineering at the University of Waterloo, senior fellow of the Centre for International Governance Innovation, and president of the Academy of Science, Royal Society of Canada.

Daisuke Kamiya is associate professor in the Department of Civil Engineering and Architecture, Faculty of Engineering, University of the Ryukyu, Okinawa, Japan.

Masanori Kawano is a government official of the Ministry of Land, Infrastructure, Transport and Tourism, Tokyo, Japan, and received his master's degree in engineering from Kyoto University.

Akira Sakai is professor in the Faculty of Economics, University of Marketing and Distribution Sciences, Kobe, Japan.

Maiko Sakamoto is associate professor, University of Tokyo, Chiba, Japan.

Sho Shibata is a staff of the International Operations, NEWJEC INC., Osaka, Japan, and received his master's degree in engineering from Kyoto University.

Kunio Takahashi is a senior researcher at the Japan Association of Drainage and Environment, Tokyo, Japan, and received his Ph.D. degree in engineering from Kyoto University.

Sotaro Tsuboi is a researcher at the Disaster Reduction and Human Renovation Institution, Kobe, Japan, and received his Ph.D. degree in urban science from Tokyo Metropolitan University.

List of Figures

List of Tables

Part I
Overview and Context

Chapter 1
Sustainability and Human Well-Being

Kiyoko Hagihara and Chisato Asahi

Abstract Subsequent to the publication of Brundtland's report on sustainable development, the number of books and papers which include the words 'sustainable' in their title has grown enormously. However, the very elasticity of the concept has given rise to questions about what it is supposed to mean: the sustainability of what, for whom, for how long, and why? First, the concept of well-being is defined as a multi-level structure in this book. The first level of well-being consists of the sustainability of people's basic living conditions such as income, health and so on. When we consider the improvement of people's well-being, we should give precedence to securing the first level of well-being. Second, the equity problem is considered in the framework of welfare economics theory. Multiple criteria analysis is considered to be a promising method which takes into account not only both efficiency and equity but also many non-monetary items which constitute well-being. Concepts in sustainability and vulnerability are complementary and closely related; mitigating the vulnerability of the human-environment system can increase its resilience or sustainability. Vulnerability is not only an issue for developing countries such as Bangladesh, which is one of the poorest nations, but is also an issue for developed countries such as Japan.

Keywords Sustainability • Well-being • Multiple criteria analysis • Vulnerability

---

K. Hagihara (✉)
School of Sociology, Bukkyo University, 96 Kita-hananobo, Murasakino, Kita-ku, Kyoto 603-8301, Japan
e-mail: hagihara@bukkyo-u.ac.jp

C. Asahi
Division of Urban Policy, Faculty of Urban Liberal Arts, Tokyo Metropolitan University, 1-1 Minami-osawa, Hachioji, Tokyo 192-0397, Japan
e-mail: asahi@tmu.ac.jp

K. Hagihara, C. Asahi (eds.), *Coping with Regional Vulnerability*, New Frontiers in Regional Science: Asian Perspectives 4, DOI 10.1007/978-4-431-55169-0_1

1.1 Introduction

In the last decade, and following the publication of Brundtland's report on sustainable development, the number of books and papers with the word "sustainable" in their title has grown enormously. However, the word is used in many different ways (Jordan 2008).

The Brundtland formulation, taken from the 1987 report of the World Commission on Environment and Development, states: "Sustainable development is development that meets the needs of the present without compromising the ability of future generations to meet their own needs" (World Commission on Environment and Development 1987). It is quite difficult to define the word "development" exactly and to determine the needs of future generations. In addition, developed countries and developing countries might hold to different interpretations of the above-quoted words. Consequently, a consensus has not been reached regarding the meaning of the concept of sustainability.

The elasticity of the concept has given rise to questions about what it is supposed to mean: the sustainability of what and for whom, and how is sustainability put into effect (Adger and Jordan 2009)?

1.2 Sustainability of What?

Recently, the word "well-being" has frequently been used not only in academic papers but also in public documents. However, even in public documents such as the United Nations' Millennium Ecosystem Assessment and the 2000 Local Government Act in the United Kingdom, the concept of well-being is used in different ways (Atkinson and Joyce 2011).

In the United Nation's Millennium Ecosystem Assessment (MA), sustainability is regarded as sustainability of both well-being, which means human welfare, and ecological quality, which affects human well-being. The MA's conceptual framework analysed ecosystem services and the impacts of change on multiple dimensions of human well-being. It showed how provisioning, regulating, supporting and cultural services lead to security, basic materials for a good life, health, and good social relations. The framework drew on the work of Amartya Sen (e.g. Sen 1984, 1999, 2010) in defining human well-being as a set of opportunities and capabilities, rather than simply assets and income. Hence, the MA defined the constituents of well-being as underpinning freedoms and choice (Millennium Ecosystem Assessment, 2003:5) (Brown 2009).

Dasgupta (2001) also contended that well-being is a pluralistic concept and includes human ability in terms of carrying out various kinds of freedom. Stiglitz et al. (2009) likewise demonstrated that well-being is a pluralistic concept and its various aspects should be taken into account simultaneously.

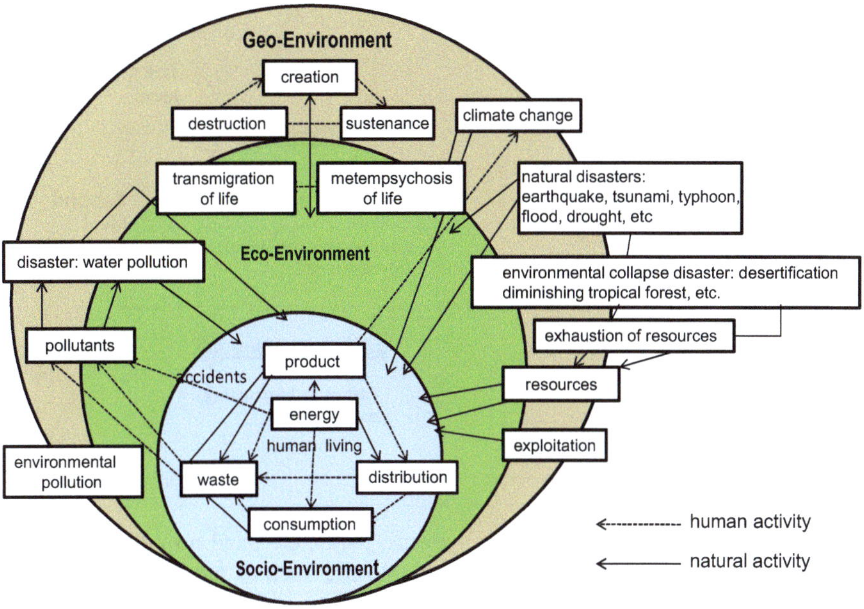

Fig. 1.1 GES (Geo-, Eco- and Socio-) environment

It is argued that sustainable development necessarily requires the integration of social, economic and environmental goals. Integration begins with the recognition that environment and society are closely linked, as shown in Fig. 1.1 (Hagihara et al. 1995).

There are three layers of environment that are relevant in this book (Fig. 1.1): the Geo-environment where the law of geophysics governs, e.g. climate change and natural disasters such as earthquakes, floods, drought and typhoons; the Eco-environment where the law of ecology governs, e.g. ecosystem functions and services; and the Socio-environment where rules of society govern, e.g. human well-being (Hagihara et al. 1998; Hagihara and Hagihara 2004; Hagihara 2008). If the geological space where the Geo-environment exists is the entire earth, the Eco-environment cannot exist without the Geo-environment, and the Socio-environment cannot exist without the Geo- and Eco-environments. Humans are members of the Geo-, Eco- and Socio-environments. Conceptions such as resilience and vulnerability are also important aspects in every layer.

As described above, well-being consists of multiple factors. Moreover, well-being is regarded as hierarchical in nature in this book. That is, human welfare consists of three levels: the first level includes income, health and a safe environment; the second level comprises a convenient and comfortable environment; the third level includes realization of a meaningful life, fruitful communication, taking enjoyment in pastimes and a flourishing natural environment (see Fig. 1.2). The following is a more detailed description of the structure of well-being: at the first level as the base of human well-being, people can enjoy safety and a secure life being supplied with various kinds of infrastructure including transportation, energy,

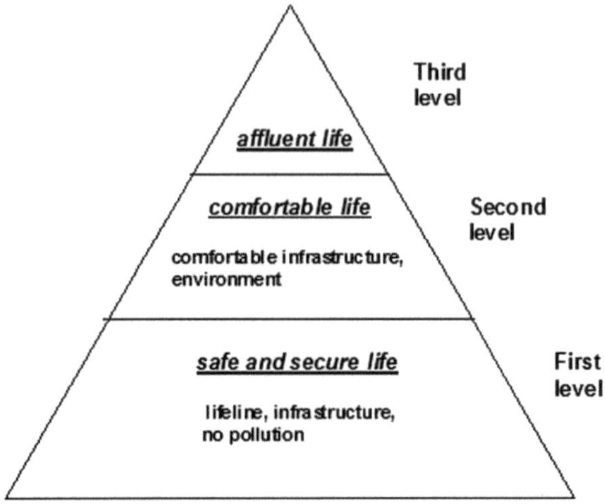

Fig. 1.2 Structure of human well-being

housing, municipal water, sewerage systems, parks, etc., and no pollution; at the second level, people can enjoy a comfortable life such as having a comfortable traffic system, comfortable housing, comfortable surroundings including green vegetation, clean water and healthy soil; at the third level, people can enjoy an affluent lifestyle, that is, people can enjoy rich surroundings, be in close contact with each other, participate in building up their community, and so on. The first level is always the first priority in a society whether in developed countries or in developing countries. Consequently, 'sustainability of what' means human well-being by securing the first level of welfare.

1.3 Sustainability for Whom?

The sustainability agenda is large and radical. It covers intergenerational justice, resource use, pollution, urban and rural planning, participation, poverty and social inclusion (Adger and Jordan 2009). In order to consider some of those issues, welfare economics is briefly outlined below.

Welfare economics is the framework within which the normative significance of economic events is evaluated. It aims to determine the desirability of a particular policy in terms of some explicitly stated ethical criteria. Welfare economics can be described and employed as an investigation of methods of obtaining a social ordering over alternative possible states of the world.

A social state refers to a particular allocation of resources. The ranking of a social state is inevitably a normative procedure; that is, it involves making value judgements. Two main value judgements are involved. The first is that the social ordering ought to be based on individual orderings of alternative social states, that is, on individual preferences, where it is implicitly assumed that each individual is the best judge of his or her own preferences. This assumption is called individualism. The second widely accepted value judgement is the Pareto principle.

The normative analysis of economic efficiency proceeds with only individualism and the Pareto principle as value judgements. There is an important relationship between resource allocations generated by the market economy and those ranked highly by the Pareto criterion. This relationship is summarized in what are referred to as the two basic theorems of welfare economics.

Cost-benefit Analysis (CBA) is a common method used to aid decision-making. It has been widely used as an evaluation tool for public policymaking by governments in Japan and around the world.

The main strength of CBA as an appraisal technique is as a test for economic efficiency in resource allocation. The main criticism of CBA is the acceptability of the ethical framework underpinning it. CBA is primarily based on welfare economics. One of the welfare criteria is Pareto Optimality, which requires that no one is made worse off and at least one individual believes he/she is better off after a policy decision. The problem with this is that most policy changes make some people better off and some people worse off simultaneously. So, modern welfare economics is now based upon the Kaldor-Hicks principle of potential compensation, called the potential Pareto principle. That is, if the gainers from a policy gain sufficiently to compensate the losers, the policy is an improvement regardless of whether compensation is actually paid.

In applying the potential Pareto principle, however, it is possible that a policy could actually lower the sum of utilities if people with different levels of wealth have different marginal utilities of money. If the low-wealth person's marginal utility of money is higher than that of the high-wealth person, then it is possible that the utility loss of the low-wealth person could outweigh the utility gain of the high-wealth person. Thus, whereas the Pareto principle allows us to avoid interpersonal utility comparisons by guaranteeing increases in aggregate utility for policies with positive net benefits, the potential Pareto principle does not do so. The potential compensation criterion is useful in separating efficiency and equity, but has meant that discussions of actual compensation have been avoided on the grounds that equity issues are outside the economists' realm (see, for example, Hanley and Spash 1993; Hanley 2001; Hanley and Barbier 2009).

According to the economic literature, sustainable development is economic and social development that maintains a certain minimum level of human welfare for present and future generation of humans, in the sense of either maximizing welfare over time or meeting the demands of distributional justice between generations. However, if sustainability is also about equity in distribution over generations, then it raises the same question as to what it is we are supposed to be distributing equally. The question "Equity of what?" is directly related to the common question among environmental activists and advocates, "Sustainability of what?" (O'Neill and Holland 2008; Munda 2005). Much of the environmental valuation literature is concerned with the issue of sustainability; however, income distribution is ignored (see, for example, Hanley and Spash 1993; Hanley 2001; Hanley and Barbier 2009).

Yet, a sense of fairness is a critical factor in economic decisions (Gowdy 2004; Fleurbaey and Maniquet (2011)).

Although CBA is adequate for evaluating the efficiency of the policy in question, it does not definitively take into account the equity issues and sustainability aspects of the policy. Therefore, multicriteria analysis (MCA), which includes participatory multicriteria analysis, has been proposed to evaluate policies (Munda 2005; Kallis et al. 2006). Sustainability raises a set of issues based on the civil rights of current and future generations as well as respect for ecological systems (Messner 2006). MCA takes into account a much wider variety of methods than CBA (see, for example, Nijkamp 1977; Figueira et al. 2005; Getzner et al. 2005; Vincke 1992).

1.4 Attaining Sustainability

An essential characteristic of MCA is its explicit consideration of various evaluation criteria explicitly, which are weighted in the course of the analysis. In MCA approaches, diverse quantitative, qualitative, and fuzzy criteria can be defined to reflect different kinds of effects as well as trade-offs and synergies. Even CBA results can be included in MCA.

Recently, there has been a push for greater public participation and the inclusion of non-governmental stakeholders in project appraisals. There are different methods for including participation, such as scenario workshops, mediated modelling, and social multicriteria evaluation (Munda 2005). However, some difficulties have been identified for participatory MCA. Several methodological issues and questions are still subject to debate, such as: What MCA method and which participatory approach should be selected for a particular evaluation problem? Who should determine the criteria? Who decides on the weightings? Who is to be included in the participation process? How objective results be can attained (Messner 2006)?

With regard to sustainability, a process for realizing it is carefully considered. In the process, equity, vulnerability, social marginalization and well-being must be taken into account. If we ignore social marginalization, vulnerability and the uneven distribution of benefits from human development, then we risk sowing the seeds of future conflict and witnessing the breakdown of collective responsibility. In other words, the processes of decision making directly affect the sustainability of their outcomes (Adger and Jordan 2009).

Public participation is widely and sometimes rather uncritically embraced as a "good thing", but more is needed to be known about how best to incorporate it into the management of sustainability (Jordan 2008), and ultimately, what it actually delivers in terms of residents' and their communities' survivability "on the ground". By survivability we mean, for individuals, continued life, and for communities, their ability to continue to function as a community. If the first level of the above-defined well-being is vulnerable, not only sustainability but also survivability might not be attained.

Sustainability is not a matter of simply maintaining some aggregate level of human and natural capital. It requires the maintenance of specific resources and goods that meet different human needs and capacities (O'Neill 2009).

Dam construction in Japan can provide an example of these issues (Hagihara and Sakamoto 2006). In the recent conflict over the construction of a weir in Yoshino River, a referendum was carried out in Tokushima city, which is located in the downstream area. As a result, the construction of the weir was stopped. However, the people residing mid- and upstream of Yoshino River, who wanted the weir because they are at risk of being flooded, were not included in the referendum, so their views were not reflected in the decision to stop construction. Rather, the result only reflected the opinion of the people in Tokushima city and other environmentalist groups. Sustainability has been secured in Tokushima city, but both sustainability and survivability of communities and people residing mid- and upstream of Yoshino River were not secured at all.

1.5 Regional Vulnerability

Concepts in sustainability and vulnerability are complementary and closely related; mitigating the vulnerability of the human-environment system can increase its resilience or sustainability. However, whereas sustainability science research has to date largely focused on the environmental and economic dimensions, empirical vulnerability research has tended to emphasize the social dimension (Pearsall 2010).

Before considering the vulnerability, the concept of risk is shown in Fig. 1.3.

Vulnerability refers to the relationship between poverty, risk, and efforts to manage risk. Some general principles related to vulnerability as a concept include the following (Alwang et al. 2001):

(a) it is forward-looking and defined as the probability of experiencing a loss in the future relative to some benchmark of welfare,
(b) a household can be said to be vulnerable to future loss of welfare and this vulnerability is caused by uncertain events,
(c) the degree of vulnerability depends on the characteristics of the risk and the household's ability to respond to the risk,
(d) vulnerability depends on the time horizon, in that a household may be vulnerable to risks over the next month, year, etc. and responses to risk take place over time, and
(e) the poor and near-poor tend to be vulnerable because of their limited access to assets (broadly defined) and limited abilities to respond to risk.

The extent of vulnerability depends on hazard (see Fig. 1.3). Thus, vulnerability is defined as follows:

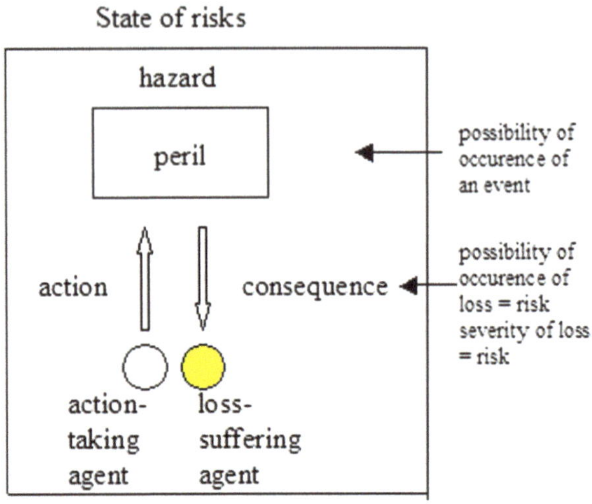

Fig. 1.3 Concept of risk

$$\text{vulnerability} = f(\textit{peril}, \text{coping} : \textit{hazard})$$

where coping means both characteristics of the risk and the household's ability to respond to the risk.

For instance, Bangladesh is one of the poorest nations in the world and the above-mentioned first level of well-being is not fully established in the country. In particular, most people living in villages drink groundwater as drinking water, but the groundwater is highly contaminated with arsenic. They are vulnerable with respect to accessing secure drinking water. The reasons why they are vulnerable are that most of the people are poor, they have little knowledge about the arsenic contamination of some wells and the cancer risk posed by drinking arsenic-contaminated water, and they do not have enough funds to obtain secure drinking water by constructing water systems including secure wells nearby or buying arsenic-free bottled water or by other means.

Vulnerability is also an issue for developed countries such as Japan. Japan is composed of both urban and rural areas. The rural areas play important roles including supplying food as well as conserving national land, the head sources of streams and the natural environment. However, the rural areas have lost population, and it has become difficult for the rural communities to fulfil these roles. The remaining rural residents may be in danger of not only losing their way of life, but also their very lives, due to floods and ecological destruction along the rivers (Hagihara et al. 2009). The urban areas are likely to be affected as well (Hagihara and Hagihara 2010). Therefore, as Hagihara and Hagihara (1991) have suggested, the depopulation problem in rural areas must be regarded as a problem for the sustainability of the whole region including both urban and rural areas.

Furthermore, there are issues with the quality of drinking water in Japan, including cancer risk. A recent serious problem has been various suspensions of the water supply system not only due to disasters such as earthquakes and the 2011

Fukushima nuclear power plant accident at a nuclear power plant but also on account of superannuated water supply systems, most of which were constructed in the mid-twentieth century, when Japan was experiencing high economic growth.

Japan is not a poor country like Bangladesh, but the financial situation has become quite severe since the crash of the so-called stock market bubble in the 1990s. The maintenance and replacement of various social infrastructures, including roads, bridges and, of course, water supply systems and sewerage systems, are financially quite difficult for both national and local governments.

Water supply systems are constructed by the local governments such as those of cities and towns. So, households in cities and towns are vulnerable with respect to the maintenance and replacement of water supply systems.

1.6 Coping with Regional Vulnerability: The Scope of this Book

Our aim is to manage the regional vulnerability and to bring not only sustainable but also survival life to people who live in both rural and urban areas. Part I consists of Chaps. 1 and 2. Chapter 2 analyses water resources conflicts in Japan and abroad – especially the conflict between India and Bangladesh – the circumstances in these areas, and the reasons for the strife that afflicts them. The global social risks of managing water resources are swiftly increasing. Managing water resources conflicts is the most essential and serious problem for the survival of human beings. With the aim of ensuring regional sustainability, the author tries to answer the question: 'What is Sustainability?', and shows the need for a paradigm shift in water resources management amidst conditions of social inequality. Chapter 2 also examines a meta-methodology for managing water resources conflicts, with a focus on the GES (Geo-, Eco- and Socio-) environment, a time-based circulation system of mid- and long-term environmental change, and an adaptive water resources planning. Finally, some examples of research are briefly introduced from three phases of meta-methodology, including 'producing alternatives', 'multiple evaluation' and 'conflict management'.

1.6.1 Part II: Socioeconomic Vulnerability in a Regional Perspective

The remainder of the chapters in this book are grouped into three broad parts, which address the same core issues raised in this chapter but from different theoretical and empirical perspectives. Part II is broadly concerned with regional vulnerability of a developed country – urban and regional areas of Japan – and that of a developing country, Bangladesh.

In Chap. 3, the depopulation problem in Japan is considered from the viewpoint of local public goods equilibrium. If in moving from one region to another, migrants do not account for the effect of their moving on the other residents, then one region may be overpopulated and the other underpopulated. In the framework of a simple model, it is suggested that the central government may be justified in using a system of intergovernmental grants to overcome these inefficiencies. In order to confirm the role of intergovernmental grants, the model is applied to a village which is designated as a depopulated area. Furthermore, the effects of the countermeasures taken over the 40 years since the first depopulation law was implemented are investigated. Finally, it is pointed out that there is a need to take into account another situation: there are areas which are similar to depopulated areas, but are not designated as such under the law.

In Chap. 4, regional vulnerability with regard to flood and earthquake risk in the Tokyo ward area is considered. Until the 1970s, the majority of flood disasters occurred in the lowlands to the east of Tokyo. In recent years, however, with the increasing frequency of localized torrential downpours – referred to as 'guerrilla rainstorms' – locally concentrated, devastating damage has occurred. Because of its densely concentrated urban structure, the Tokyo Metropolitan Area is also at high risk of fires and building collapse during earthquakes. Furthermore, a large-scale earthquake is expected to occur there in the near future. Tsuboi and Asahi examine these two types of disasters and show the features of urban-specific vulnerabilities and countermeasures. Although a variety of disaster prevention measures for floods and earthquakes are being conducted in the Tokyo Metropolitan Area, lowland areas could still suffer serious damage because of their densely concentrated urban structure. Since disasters affect society as a whole, in addition to urban structure and mitigation measures, the pre-disaster welfare state of individual households, the capacity of administrative bodies, and the level of cooperation within them should be used to define the vulnerability of these areas. Thus, residents' self-help and cooperative activities and their administrative promotion are emphasized.

The depopulation problem in isolated islands in the Ryukyu archipelagos in Okinawa Prefecture is presented in Chap. 5. The economic base is very poor in Okinawa, so tourism is a relatively important policy among the active economic measures. However, in addition to the ordinary water scarcity problems, the increase in water demand from the tourism sector has become serious. Daisuke Kamiya describes and analyses these issues. Remote islands in Okinawa Prefecture are roughly classified according to which islands contain tourist resorts and which do not. Both types have some severe problems. Resort islands have problems concerning water shortages, waste disposal, and the succession of tradition. Non-resort islands face problems concerning depopulation and aging. Firstly, the effect of tourism promotion on water shortages was analysed, which showed the available water resources of the small islands. Then, vulnerability to tsunami disasters in depopulated and aging remote islands was examined. Residents' daily life activities and their association with their neighbours were investigated, revealing that problems exist concerning the structure of the social network in terms of the support people required at the time of disaster.

Ensuring safe water supply and proper sanitation is a basic requirement for maintaining public health and safety, as well as the foundation of sustainable development. Although arsenic contamination mitigation from tube wells is an urgent issue related to public health in Bangladesh, a significant portion of the population still subsists on arsenic-contaminated water. One of the primary reasons for this is the difficulty involved in accessing safe water, even in areas where arsenic mitigation facilities exist. As for sanitation, although coverage of sanitary facilities has been increasing in the country, existing toilets have various disadvantages, one of which is that improper human excreta management is causing surface water pollution, hindering the use of this water as an alternative drinking water source. To maintain soil quality and food productivity in the future, it is expected that human excreta will be used to fertilize agricultural land. In Chap. 6, socio-environmental problems and the risks related to water supply and sanitation will be identified and their interrelationships will be discussed. Additionally, the factors considered during the selection of technical options aimed at reducing risks will be discussed based on Sakai et al.'s sanitation improvement-related field activities.

Safe drinking water and sanitation are closely related to the environmental and health risks faced by villagers in the rural areas of Bangladesh where tube-well arsenic contamination is an issue of particular concern. Since a significant percentage of the people affected by this problem have poor access to safe drinking water, the risk of increasing numbers of people suffering from arsenicosis is considered to be high. Regarding access to sanitation facilities, while the coverage ratio of improved facilities is increasing, improper management of existing facilities is often observed in rural areas. The authors conducted questionnaire surveys related to safe drinking water and sanitation awareness in two rural Bangladeshi villages with differing socio-environments, the results of which are presented in Chap. 7. On the basis of these results the current status and local awareness levels of issues concerning drinking water and sanitation in their villages will be compared while taking socio-environmental differences into consideration. Sakai et al. also discuss how the questionnaire survey results were used to help the villagers discuss alternatives that will improve their living environments and reduce health risks through the mitigation of arsenic contamination and sanitation improvements.

1.6.2 Part III: Evaluation of Regional Vulnerability

Vulnerability must be measured and evaluated to be integrated into social decision-making. Part III provides some perspectives and methods to describe vulnerabilities in an evaluable way.

The first three chapters of this part present the evaluation model from the welfare economics perspective. In Chap. 8, marginal willingness to pay for public investment under urban environmental risk is considered. In particular, Hagihara et al. show a model that takes into account a bounded rationality on the amount of risk perception, knowledge of information, and people's threshold for acceptance of risk. Then, based on a case study of urban environmental risk, the evaluation of risk in municipal water use is shown. In order to investigate the knowledge of information and risk perception of people, a survey is conducted using questionnaires. The survey reveals that risk awareness is a factor in drinking water, and that people change their choice based on risk information. Then, the effects of information on risk and public investment are considered. Based on some numerical examples, marginal willingness to pay is found to be low in the case of high risk, because of consumers' self-defensive activities. In other words, marginal willingness to pay for public investment is high when there is no or little averting behaviour. Moreover, it is shown that consumers' perception of risk is largely dependent on information on risk, countermeasures taken by public authorities, and overconfidence in private averting goods.

The reliability of a system to provide municipal water under water quality and quantity risks and its evaluation for efficiency are considered in Chap. 9. First, the quality and quantity risks and the current institutional measures for the provision of municipal water are examined. Second, the present supply system is studied from the perspective of who is in charge of the risk management. Third, "the reliability evaluation model", which shows that the marginal benefit of an increase in reliability is equivalent to the marginal cost of the project for risk management, is presented. The model can show the optimal reliability level or the optimal investment level to achieve a certain reliability level. The welfare measurement of the reliability benefit is classified into several types. The proper welfare measure for a change in risk (reliability) for risk-averse individuals depends on three specific conditions: the availability of conditional markets, whether the project cost is certain or not, and whether the risks for economic agents are collective or individual. Asahi and Hagihara classified some of the measures for water supply risks into each welfare measure in line with these conditions and showed the fundamental perspectives for the provision and evaluation of reliability in water supply systems.

In Chap. 10, the effects of welfare and the measurement of social overhead capital for a household are examined. In Japan, a decline in the population and the dire situation of public finance have given rise to difficulties in the maintenance and replacement of social overhead capital that was invested during the period of high economic growth and the subsequent period of fiscal stimulus that began in the 1990s. The importance of the basic services provided by infrastructure services such as public utility services or flood control has increased because of transformation in regional structures caused by a decrease in population and continual concern about disasters. To solve the trade-offs between the capacity limits of new public investments and the increasing need for safety provided by social overhead capital, Asahi and Hagihara present a model for welfare evaluation of the vulnerability handled by the infrastructure and examine the provision of municipal water. First, the welfare evaluation model for the vulnerability that households face is shown. Second, the welfare measures of vulnerability are developed especially from the viewpoint of the theoretical appropriateness of the welfare measures. Third, preliminary inspections show the validity of the application of the model and the conditions required for evaluation.

The latter part of Part III shows the social environmental perspective of the vulnerability evaluation. In recent years, arsenic contamination of drinking water has become a serious problem in Bangladesh. Many foreign institutions have provided support to Bangladesh in terms of constructing arsenic-free wells, providing arsenic removal equipment, and so forth. However, most of these initiatives are not accepted by local residents because they cannot understand how to maintain the equipment or their effectiveness for reducing arsenic contamination. Furthermore, they find certain equipment to be too inconvenient to use in their daily lives. In Chap. 11, social environment analysis is considered with the aid of a social survey. A survey was conducted in two villages in Bangladesh in order to define the relationship between arsenic contamination in drinking water and their social environment. First, Fukushima et al. attempted to analyse residents' satisfaction regarding the drinking water available to them. Second, the authors introduced the unhappiness function in their model. Finally, they identified alternatives acceptable to the residents by devising a structural model addressing distrust of external support.

Bangladesh is a densely populated agricultural nation characterized by a tropical monsoon climate. Most of the country consists of flat lowlands and a vast delta. These weather and geographical conditions in combination with vulnerable facilities, such as embankments, lead to severe flooding about once every ten years. On the other hand, the spread of 'low-cost Pit Latrine' toilets by the government and international organizations has increased. However, since they are poorly managed, the Pit Latrine cannot be considered a sustainable and sanitary facility, and they have an adverse environmental impact. Appropriate technology for toilets should contribute to improved sanitation, the use of human excreta as a resource and the conservation of the quality of water. The ecological sanitation (henceforth, Eco San) toilet meets the above needs. Takahashi et al. examined the contribution to sanitary improvement and evaluated the benefits of the Eco San toilet, the findings of which are presented in Chap. 12. The Eco San toilet is found to be an improved toilet, and it reduces households' expenditure through its mitigation of waterborne diseases. Considering the average medical expenditures, the benefits will be about 2000 BDT/year/household. However, it cannot be overstated that such a result can be achieved only through proper management.

1.6.3 Part IV: Risk Management of Regional Vulnerability

Part IV covers the management of risk characterized by vulnerability.

In Chap. 13, the system of the provision of municipal water through the market mechanism, including some forms of "privatization", is considered in terms of the normative aspects of resource allocation and welfare under the uncertainty of water quality. Asahi and Hagihara investigate the conditions under which market mechanisms work effectively for the fair allocation of municipal water, as well as public goods and environmental risk, under the assumption of non-convexity in production and preference. This is particularly noteworthy in that considerably more studies have considered the supply side of municipal water than the demand side. Asahi and Hagihara outline three steps for considering the effects of the uncertainty of quality on the municipal water provision system. First, the first and second fundamental theorems of welfare economics are reviewed, and the cases in which the market mechanism fails to satisfy the optimal allocation are shown. Second, among those cases, the problems relevant to municipal water provision under environmental risk and asymmetric information between the water supplier and regional consumers are considered. The authors review the analytical framework of credence goods and show the advantages of using it when the uncertainty of the water quality is at issue. Third, they assume that some private firms have the chance to contract to provide municipal water and examine how the credence of municipal water (i.e., quality regulation) affects the price and water quality level using an analytical model. A regulator of municipal water supply should control the municipal water supplier by means of penalties and inspections, based on the risk attitudes of the supplier and the probabilities of detection under informational asymmetry.

Recently, arsenic contamination of drinking water has become a serious problem in rural areas of Bangladesh. Arsenic can cause skin disease, cancers, and death. Bangladesh has suffered many kinds of disasters, including cyclones, floods, droughts, and salt damage. Therefore, Bangladesh is considered a fragile area for arsenic problems. In 1993, arsenic contamination was first found in Bangladesh in the groundwater from shallow tube-wells (STWs). Roughly 70 % of Bangladesh is contaminated with arsenic, and 30 % of the STWs used for drinking water exceed the arsenic concentration standard. Since the revelation of arsenic contamination, various water supply options have been installed to supply drinking water. However, some of them have already been abandoned. These options are not acceptable for local residents without consideration of their social environments and without support after their introduction. Chapter 14 presents appropriate plans to install water supply options in areas with arsenic-contaminated drinking water given Bangladesh's local social environment.

To cope with groundwater arsenic contamination, which is widespread throughout Bangladesh, securing alternative drinking water sources is an urgent issue. However, the current prevalence of inappropriate sanitary practices in many parts of the country has resulted in extensive surface water pollution, which limits the safe use of such water for drinking. To facilitate improved water quality and water conservation, water supply and sanitation improvements should be integrated. In Chap. 15, a project aimed at improving the living environment through water supply safety and sanitation upgrades is introduced. Sakai et al. focus on one of the two study-area villages discussed in Chap. 7 that are located in arsenic-affected areas. However, in this village, due to the limited support provided previously, most villagers have been forced to subsist on arsenic-contaminated groundwater, and have been mostly unwilling to take steps that would improve the situation. Herein, Sakai et al. present a scheme aimed at overcoming the problems resulting from the previous failed projects, and discuss locally appropriate technology options that consider continuous community-based management practices. This scheme focuses on pond sand filter (PSF) systems and ecological sanitation, as it is expected that the use of PSF systems would increase villager awareness regarding their drinking water source, the quality of which must be maintained by sound ecological sanitation measures. Finally, after describing the current situation, issues related to the community-based management of the installed facilities will be discussed.

In Chap. 16, the authors aim to demonstrate an adaptive waterside environmental management process that incorporates sustainability, survivability and participation. The concept of sustainability has given rise to questions about what it is supposed to mean: the sustainability of what, for whom, for how long, and why? For instance, there is a case in which the sustainability of a city and the region as a whole is achieved, but on the other hand, some people in certain parts of the city and region may be in danger of not only risking their sustainability but also their very lives due to floods and ecological destruction at the waterside. In this chapter, Hagihara et al. focus on differences among residents both in the same area and in different areas and consider waterside management in urban areas taking into account sustainability, survivability and participation in the adaptive waterside environmental management process. First, cost-benefit analysis is reviewed critically from the viewpoint of sustainability and survivability. It is suggested that the adaptive waterside environmental management process which uses a systems analysis methodology is a promising method to aid management decisions. The methodology is applied in three areas along the Kamo River in Kyoto city, two of which are located in the upper river area and one in the downstream area. Based on a social survey of residents, the environmental characteristics of each area are determined. The environmental valuation function for each area is then defined, and possible priorities for waterside environmental management are presented. Waterside environmental management incorporating sustainability and survivability is thus demonstrated.

Upstream-downstream conflicts constitute an important class of water resource disputes in which external guidance may be required in order to reach an enhanced resolution. At a bilateral level, Bangladesh and India are currently enmeshed in a bitter dispute over the control of the flow of the Ganges River by India to downstream Bangladesh. Because the river constitutes an important source of water for both countries within a region lacking sufficient supplies of fresh water, many conflicts have arisen over the utilization of this valuable water resource both between and within these nations. As the downstream country, Bangladesh is situated in a disadvantageous position. To satisfy water demand and avoid using groundwater polluted by natural arsenic, Bangladesh would like to use as much water as possible from the Ganges River, especially during the dry season.

In the final chapter of this book, Hipel et al. present a comprehensive approach to modelling Third Party intervention within the framework of the Graph Model for Conflict Resolution (GMCR) and demonstrate its usefulness by applying it to the conflict existing between India and Bangladesh over the regulation of the Ganges River by India at the Farakka Barrage, located just upstream in India before it flows into Bangladesh. A general system of systems engineering approach to Third Party Intervention within the GMCR structure is designed to reflect a range of ways in which it can be implemented in practice. Having an insightful and powerful tool like Third Party GMCR permits one to ascertain how a Third Party can guide a serious conflict to a more reasonable resolution which may be mutually beneficial to all concerned parties.

Effective governance is required by society to cope with a broad range of interconnected issues including regional vulnerability for which society would like to prevent and mitigate damage from disasters caused by humans and natural causes such as flooding, earthquakes and tsunamis. Within Bangladesh, factors pertaining to disasters can be classified as either natural or socio-environmental ones. Socio-environmental factors can be categorized more specifically as socio-environmental destruction, socio-environmental pollution, and socio-environmental cultural factors. The water resources conflict between Bangladesh and India studied in Chap. 17 falls within the domain of socio-environmental cultural factors. Whatever the exact situation, conflict, which will inevitably arise when there are competing interests, can be systematically addressed employing the Third Party GMCR methodology put forward in Chap. 17.

Acknowledgements We are grateful to Mr. Conrad W. Hipel for editorial comments and suggestions on earlier drafts.

References

Adger, W. N., & Jordan, A. (2009). Sustainability: Exploring the processes and outcomes of governance. In W. N. Adger & A. Jordan (Eds.), *Governing sustainability* (pp. 3–31). Cambridge: Cambridge University Press.

Alwang, J., Siegel, P. B., & Jorgensen, S. L. (2001). Vulnerability: A view from different disciplines. *Social protection discussion paper series*, No. 0115, The World Bank.

Atkinson, S., & Joyce, K. E. (2011). The place and practices of well-being in local governance. *Environment and Planning C: Government and Policy, 29*, 133–148.

Brown, K. (2009). Human development and environmental governance: A reality check. In W. N. Adger & A. Jordan (Eds.), *Governing sustainability* (pp. 32–51). Cambridge: Cambridge University Press.

Dasgupta, P. (2001). *Human well-being and the natural environment*. Oxford: Oxford University Press.

Figueira, J., Greco, S., & Ehrgott, M. (Eds.). (2005). *Multiple criteria decision analysis: State of the art surveys*. New York: Springer.

Fleurbaey, M., & Maniquet, F. (2011). *A theory of fairness and social welfare*. Cambridge: Cambridge University Press.

Getzner, M., Spash, C. L., & Stagl, S. (Eds.). (2005). *Alternatives for environmental valuation*. Oxford: Routledge.

Gowdy, J. M. (2004). The revolution in welfare economics and its implications for environmental valuation and policy: *Land Economics, 80*(2), 239–257.

Hagihara, Y. (2008). *Adaptive system planning methodology for environment risk management*. Kyoto: Kyoto University Press (in Japanese).

Hagihara, K., & Hagihara, Y. (1991). The role of intergovernmental grants in under populated regions. *Regional Studies, 25*, 163–172.

Hagihara, K., & Hagihara, Y. (2004). The role of environmental valuation in public policymaking: The case of urban waterside area in Japan. *Environment and Planning C: Government and Policy, 22*, 3–13.

Hagihara, K., & Hagihara, Y. (2010). Waterside environmental management based on a consideration of sustainability and survivability. In WATER 2010*:* Tenth symposium on stochastic hydraulics and fifth international conference on water resources and environment research, Quebec city, Canada, 2010.

Hagihara, Y., & Sakamoto, M. (2006). *Conflict management: Social risk in water resources' problems*. Tokyo: Keiso Shobo (in Japanese).

Hagihara, Y., Takahashi, K., & Hagihara, K. (1995). A methodology of spatial planning for waterside area. *Papers in Regional Science, 25*(2), 19–45.

Hagihara, Y., Hagihara, K., & Takahashi, K. (1998). *Urban environment and waterside planning*. Tokyo: Keiso Shobo (in Japanese).

Hagihara, Y., Hagihara, K., & Kawano, M. (2009). A study on environmental management through characteristics of up and down stream regions. *Annals of Disaster Prevention Research Institute of Kyoto University, 52*(B), 885–897 (in Japanese).

Hanley, N. (2001). Cost-benefit analysis and environmental policymaking. *Environment and Planning C: Government and Policy, 19*, 103–118.

Hanley, N., & Barbier, E. B. (2009). *Pricing nature: Cost-benefit analysis and environmental policy*. Cheltenham: Edward Elgar.

Hanley, N., & Spash, C. L. (1993). *Cost-benefit analysis and the environment*. Cheltenham: Edward Elgar.

Jordan, A. (2008). The governance of sustainable development: Taking stock and looking forwards. *Environment and Planning C: Government and Policy, 26*, 17–33.

Kallis, G., Videira, N., Antunes, P., Guimaraes Pereira, A., Spash, C. L., Coccossis, H., Corral Quintana, S., del Moral, L., Hatzilacou, D., Lobo, G., Mexa, A., Paneque, P., Pedregal Mateos, B., & Santos, R. (2006). Participatory methods for water resources planning. *Environment and Planning C: Government and Policy, 24*, 215–234.

Messner, F. (2006). Applying participatory multicriteria methods to river basin management: Improving the implementation of the water framework directive. *Environment and Planning C: Government and Policy, 24*, 159–167.

Millennium Ecosystem Assessment (2003). *Ecosystem and human well-being: a framework for assessment*. Washington DC: Island Press.

Munda, G. (2005). Multiple criteria decision analysis and sustainable development. In J. Figueira, S. Greco, & M. Ehrgott (Eds.), *Multiple criteria decision analysis: State of the art surveys* (pp. 953–986). New York: Springer.

Nijkamp, P. (1977). *Theory and application of environmental economics* (North-Holland, Amsterdam); Translated by Fujioka, A., Hagihara, K., & Kanesawa, T. et al. (1985). *Kankyo-keizaigaku no Riron to Ouyo*. Tokyo: Keiso Shobo.

O'Neill, J. H. (2009). Sustainability, welfare and value over time. In W. N. Adger & A. Jordan (Eds.), *Governing sustainability* (pp. 283–304). Cambridge, UK: Cambridge University Press.

O'Neill, J., Holland, A., & Light, A. (2008). *Environmental values*, Routledge.

Pearsall, H. (2010). From brown to green? Assessing social vulnerability to environmental gentrification in New York city. *Environment and Planning C: Government and Policy, 28*, 872–886.

Sen, A. (1984). *Commodities and capabilities*. Oxford India Paperbacks: Oxford University Press.

Sen, A. (1999). *Development as freedom*. Oxford: Oxford University Press.

Sen, A. (2010). *The idea of justice*. London: Penguin Books.

Stiglitz, J., Sen, A., & Fitoussi, J.-P. (2009). Report by the commission on the measurement of economic performance and social progress. www.stiglitz-sen-fitoussi.fr

Vincke, P. (1992). *Multicriteria decision-aid*. Chichester: John Wiley & Sons.

WCED. (1987). *Our common future World Commission on Environment and Development*, Oxford: Oxford University Press.

Chapter 2
Water Resources Conflict Management: Social Risk Management

Yoshimi Hagihara and Kiyoko Hagihara

Abstract The global social risks of managing water resources are swiftly increasing. Managing water resources conflicts is the most essential and serious problem for the survival of human beings. Firstly, this chapter analyses water resources conflicts in Japan and abroad – especially the conflict between India and Bangladesh – the circumstances in these areas, and the reasons for the strife that afflicts them. With the aim of ensuring regional sustainability, the authors try to answer the question: 'What is Sustainability?', and show the need for a paradigm shift in water resources management amidst conditions of social inequality. Next, the chapter examines a meta-methodology for managing water resources conflicts, with a focus on the GES (Geo-, Eco- and Socio-) environment, a time-based circulation system of mid- and long-term environmental change, and an adaptive water resources planning. Finally, some examples of research are briefly introduced from three phases of meta-methodology, including 'producing alternatives', a multiple evaluation' and 'conflict management'.

Keywords Water resources conflict management • Social risk management • Sustainability • Survivability • GES environment

2.1 Introduction

The twenty-first century is called the century of water. Most people know that living organisms are made of this substance, to one degree or another. Certain social risks come with managing water resources, and our global social institutions are not sustainable. Successfully managing water resources conflicts is the most essential and serious problem for the survival of human beings. In most cases, water

Y. Hagihara (✉)
Kyoto University, 264 Suemarucho, Nakagyou-ku, Kyoto 604-0901, Japan
e-mail: y-hagi@gf7.so-net.ne.jp

K. Hagihara
School of Sociology, Bukkyo University, 96 Kita-hananobo, Murasakino, Kita-ku, Kyoto 603-8301, Japan
e-mail: hagihara@bukkyo-u.ac.jp

K. Hagihara, C. Asahi (eds.), *Coping with Regional Vulnerability*, New Frontiers in Regional Science: Asian Perspectives 4, DOI 10.1007/978-4-431-55169-0_2

resources conflicts have historically been caused by problems involving economic power and weapons, and are driven by those same factors today.

Some authors describe the grave situation of real water conflicts (e.g. Pearce 1992; Ohlsson 1996; Postel 1999; Barrow and Clarke 2002) and mathematical models and algorithms have been developed for meta-game theory (e.g. Fraser and Hipel 1984; Fang et al. 1993).

Following in their footsteps, this chapter first discusses the circumstances of and reasons for water resources conflicts in Japan and abroad. International water resources conflicts mainly result from water shortages that impact survival. Water resources conflicts in Japan result from debates on 'development versus the environment', and actions that express residents' concerns.

Next, the chapter focuses on the question, 'What is Sustainability?' This section explores the potential for contaminating the environment, and examines a paradigm shift in the management of water resources amidst conditions of social inequality, in order to ensure regional sustainability.

Thirdly, meta-methodology for water resources conflict management is discussed with a focus on the following three figures.

1. The relationship between cause and effect in regards to global disasters, with the recognition of a GES environment from a philosophical perspective.
2. A time-based circulation system of mid- and long- term environmental change, and water resources planning for preventing and mitigating disasters, shown from a conceptual standpoint.
3. An adaptive water resources planning meta-methodology for managing social conflicts is proposed, based on systems analysis.

Finally, some examples of research are briefly introduced from three phases of meta-methodology, including 'producing alternatives', 'a multiple evaluation' and 'conflict management'.

2.2 Water Resources Conflicts Happen Everywhere

2.2.1 Global Water Resources Problems Converge When Identifying Conflict

The global social risks of managing water resources are swiftly increasing (Hagihara and Sakamoto 2006). A paradigm shift in the handling of such problems in this area is needed to make the leap 'from sustainability to survival for humankind'. Social risks mainly consist of the following:

(a) Uneven distribution and variation of rainfall;
(b) Rapid increase in the world's population;
(c) A rise in the demand for water per capita;

(d) Damage from salt and desertification via large-scale irrigation, which depletes underground water supplies;
(e) A decrease in water resources via contamination from industry and agriculture;
(f) Great damage to societal structures from large floods and earthquakes;
(g) Limiting water resources within international borders; and
(h) A lack of fairness in managing regional water resources;

Recent global climate change must be considered in light of risk (a). Moreover, risk (f) has grown rapidly.

Thus, the global social risks of managing water resources are summarized as follows: Available water makes up only 0.01 % of the earth's fresh water. The absolute amount is constant. Risk (a) restricts the use of water in dry countries and regions. Risks (b) and (c) cause water shortages and contaminate fresh water. The decrease in available water and the way it is allocated, combined with risks (d) and (e), lead to new international challenges to survival. In addition, if risk (f) occurs, any region in the world could collapse, and that regions' social system of water resources could very well disintegrate. Risks (g) and (h) are also possible (Fig. 2.1a and b).

The aforementioned context tells us that various problems of water resources management converge at points of conflict, not for reasons of sustainability, but for survival. In other words, social risks of managing water resources shall result in water resources conflict management. A typical example of a water resources conflict is the Ganges River between India and Bangladesh (Sakamoto and Hagihara 2005).

The Ganges has been an important source of water in the Bengal Region since ancient times, but today, Bangladesh and India have fought over it. The situation changed drastically in 1975 when India constructed the Farakka Barrage over the Ganges River near the border with Bangladesh. The two countries did not agree on this construction at all. India had chosen to build the barrage unilaterally. Both sides signed treaties regarding use of the Ganges (after the barrage was built) in 1975, 1977, and 1996. The treaties are sometimes in effect and sometimes not, depending on the political background of the two countries.

These treaties are different in some respects. However, they generally benefit India, which is upstream on the river. Bangladesh is more vulnerable to floods and drought than India because of the Farakka Barrage. The Bangladeshi government cannot plan to utilize its national water resources without considering the effects of the Farakka Barrage. From this angle, floods and drought in Bangladesh might be described as man-made disasters. It is difficult for Bangladesh to offer a new, effective option to solve the conflict because Bangladesh is at an economic and topographic disadvantage. It is difficult to think that India would change its preference; therefore, this conflict will continue.

Fig. 2.1 (**a**) The Ganges River in the dry season (Nov. 2002), downstream of Farakka Barrage in Bangladesh (Photo by Y. Hagihara). (**b**) The Ganges River in the rainy season (Mar. 2007), downstream of Farakka Barrage in Bangladesh (Photo by Y. Hagihara)

2.2.2 *Water Resources Conflicts in Japan*

Water resources conflicts in Japan arise from debates on 'development versus the environment,' and actions that express residents' concerns. While international water resources conflicts mainly result from water shortages that impact survival. Water development projects are being debated globally from a 'development versus environment' perspective; such projects include the development of water resources in the Narmada River in India, and the South-to-North Water Transfer Project in China, which is a plan of withdrawing water from the Yangtze River in particular to Beijing. Large countries, such as China and India, have vast control over their water resources, and often neglect indigenous minorities in the process of managing them; for example, the U.S. has discriminated against Native Americans, and Japan has treated its indigenous Ainu people unfairly (Hagihara 2009).

Water resources conflicts in Japan are shown in Fig. 2.2 (Sato 2008) and explained briefly as follows.

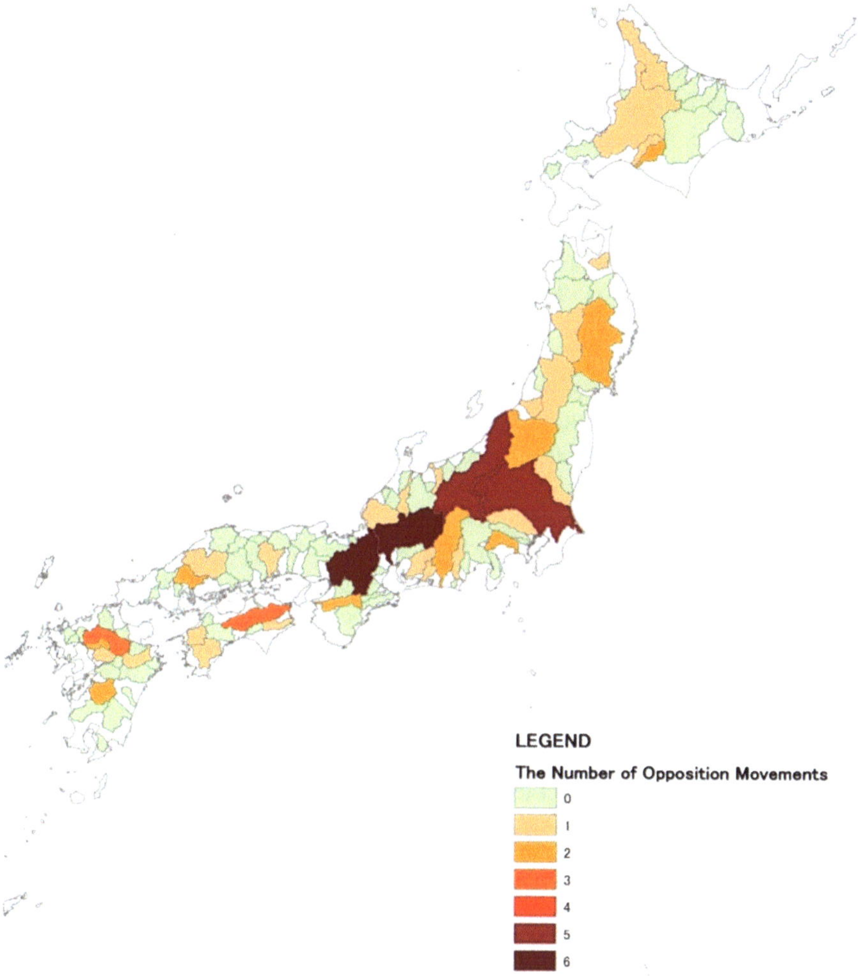

Fig. 2.2 Numbers of opposition movements against water resources development projects in Japan

1. 1945–1959: A movement emerged against dam construction projects on the ground that the projects would have caused residential areas to be submerged in the early stage of the period; there was also a movement that advocated compensating residents for moving to new locations in the latter half of the period.
2. 1960–1969: There were movements against dam construction projects; the circumstances were the same as before (1945–1959). Some people doubted the legitimacy and public nature of the development projects in question.
3. 1970–1979: In addition to the same movements as in the period of 1960–1969, the reasons for opposing development projects became diversified. Concerns

included preserving nature, negative effects on the fishing industry, local life and culture, and over-estimating the demand for water.

4. 1980–1992: No opposition movements occurred because government policy changed. A large number of compensation funds were prepared for residents whose homes would be submerged. Opposition movements gained momentum via mass media. For instance, the Nagara River Estuary Weir Development (NREWC) Project gained a lot of national attention in the media, and led to the creation of further opposition movements supported by various actors, including stakeholders and negotiators. Locals become involved in one of the opposition movements. The NREWC Project began in 1988 and finished in 1995, when Japanese had a relatively strong economy.
5. 1992–2014: The Nagara project became a turning point. Opposition movements changed due to more detailed criteria required for planning and designing projects. Also, 1994 was the year that large amounts of people began using the Internet. Some of the more prominent opposition movements were against projects in the Kawabe River Reservoir of the Kuma River in Kyushu Region, Daijyu Weir of the Yoshino River in Shikoku Region, and Daido River Reservoir of the Yodo River in Kinki Region. After the Japanese bubble economy collapsed in 1992, the country entered a recession. The Japanese government changed nearly all of its policies in 1998 to decrease investment for public sector development. Accordingly, the projects at the Kawabe River Reservoir and Daijyuu Weir were stopped in 2008, despite the serious risk of flooding in those regions.

Many disputes occur in Japan among residents, local governments, and other concerned parties with respect to development and the environment when it comes to water resource development projects. For instance, in the case of the Daijyu Weir of the Yoshino River, a referendum was held, and the opposition won. As a result, construction was halted.

2.3 What Is Sustainability?

2.3.1 Conducting Reasonable and Fair Human Activities for Sustainability: A Paradigm Shift for Planning How to Manage Water Resources

The evaluation process for determining whether human activities will contaminate the environment depends on the probability of rivers, lakes and seashores to become polluted. Much discussion arose in 1975 on the potential for contamination in Japan. Human activities developed without regard to anticipating dangers; assessing development projects allows us to ascertain the risks to public health and the ecosystem, which could lead to water shortages or crop failure.

Problems in using water result from excessive human consumption without considering the impact on regional and domestic resources; the effects of human activity on social and ecological processes have caused floods, water shortages, and environmental pollution. By examining natural water circulation systems and taking into account estimates based on recent global climate change in (international and domestic) regions and valleys, we see that human activity has exacerbated and complicated current problems. Nature cannot be conquered as René Descartes, the French philosopher, mathematician, and writer desired; rather human beings can only be part of nature. According to Asian philosophy, water resources planning aims to create harmony between human desires and natural water resources, mainly for the sake of sustainability and humanity's survival. Planning how to use water should be comprehensive and include elements such as institutions, socio-economics, social and natural (geophysical and ecological) environments, and technology. Challenges in managing water not only result from hydrological issues; hydrology is just one aspect of the solution.

Historically, the major goals of water resource development in Japan have changed based on irrigation, flood control, the dynamics of power, industrial activity, and the urban usage of water. Different factors existed in other times; for example, the use of the environment has recently been considered.

Japan's problems stem from the fact that various ministries and agencies traditionally involved in water resources planning have their own ways of operating, which is common in other countries as well. Each ministry involved in water resources development strove to become self-sufficient in the 50 years following World War II. Thus, water resources infrastructures are relatively stable due to the effort of each ministry. However, many issues remain, such as coordinating each ministry's risk management system. There are many institutional barriers to a comprehensive water resources management system.

Nevertheless, from the viewpoint of multilevel planning systems, each ministry has worked to resolve problems. As of now, there are four major challenges in regional water resources planning:

- The integrated use of water – it is necessary to connect rivers, reservoirs, waterworks, sewage systems, and irrigation systems;
- The environment – including regional scales of integrated risk management based on ecology (fauna and flora);
- The coordination of upstream and downstream factors (Hagihara and Hagihara 1991), and the relationship between basins in a region;
- Conflicts regarding goals and alternatives.

The solution to these problems requires a systems analysis approach. A framework for dealing with these challenges would have basic inputs consisting of natural phenomena, such as rainfall, and social and economic phenomena generated by human activities. It would have variables of how water functions, such as storage, transportation, and water quality transformation, and control variables such as hydraulic structures with soft technology. It would also have outputs for regional land use planning and non-harmful human activity.

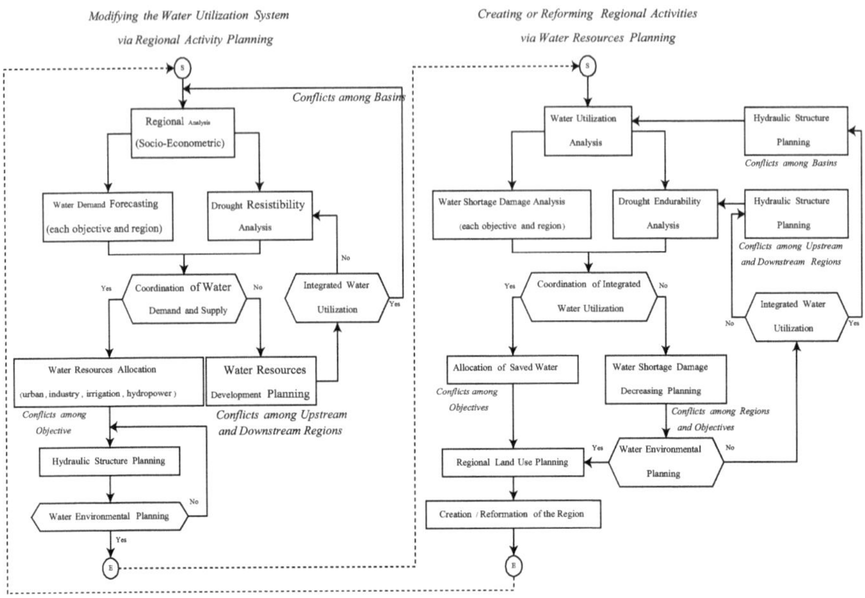

Fig. 2.3 Systems analysis of regional water resources problems for purposes of sustainability (Watanabe and Hagihara 1993)

A planning process for the circulation of water resources, which consists of two different views, is shown in Fig. 2.3. The 'modification' of a water use system (by the planning of regional activities) and the 'creation or reformation' of regional activities (by water resources planning) should be combined. The former regards water as an environmental element, whereas the latter treats water as a scarce resource.

2.3.2 *Social Inequality Is Not to Be Forgotten*

From an economic standpoint, water resources planning is based on the concept of cost-benefit analysis (CBA). In most countries (including Japan), CBA tends to neglect various types of social inequality such as income, gender, and race, and also ecological factors such as forests and wild animals.

Further, current models of sustainability sometimes ignore social inequality, both in Japan and abroad. A 2006 OECD report revealed that social inequality in Japan had grown since 1985, and that it also widened in the U.S. and the U.K., but not in France or Germany. This report used the Gini coefficient to measure social inequality.

For example, in the above society, where social inequality is prevailing, the most important concept in water resources planning for social equality is taking into account the followings:

$$\min |\mu_1 - \mu_2| \, and \min \sigma_2 \tag{2.1}$$

where μ_1 and μ_2 is the mean value for a rich group (1) and a poor group (2), respectively, and σ_2 is a standard deviation of income for the group (2).

2.4 Meta-methodology for Water Resources Conflict Management

2.4.1 Recognising the Relationship to the GES Environment

A sect of Hinduism in India conceives of the Deity as a Trinity whereby destruction (Siva), sustenance and prosperity (Vishnu) and creation (Brahma) are unified into one being as shown in Fig. 2.4. Buddhism also has the philosophy that sees life as *samsara* (transmigration and metempsychosis of the soul). While the Trinity deals with the universe, the *samsara* deals with the lives of living beings, including humans.

If we regard the targeted regional areas as a universe when planning how to handle water resources, we should first look at what destroys, creates, and sustains the elements of the geo-system in connection with the eco-system. Then we can proceed to the socio-system, where we choose active or passive elements from the viewpoint of social conflicts of water resources planning (Hagihara et al. 1995).

Figure 2.5 shows the cause-and-effect relationships between global disasters based on the geo-system (controlled by geophysical laws), and the eco-system (controlled by ecological laws), and the socio-system (controlled by man-made social rules).

This meta-figure could be applied to a regional scale; from a social science perspective, the regional level lies at the root of the global level. From the standpoint of meteorology, the change in perspective from the macro to the micro level is significant, and is important from the angle of sociology and ecology. The research must combine these aspects to examine water resources.

As mentioned previously, the philosophy governing water resources planning should connote the objectives described in (Eq. 2.1). Further, this philosophy and the spatial design of waterside environments aim to:

- Create spaces that bring the five senses together;
- Combine water in waterside environments with hard and soft technology (which is composed of geo-, eco- and socio-systems, as described in Fig. 2.5)

In the waterside environments, people will be able to understand (consciously or unconsciously) the cycle of destruction, creation, and sustenance, and the transmigration and metempsychosis of life (including the lives of humans), which include humans sensory impressions.

Fig. 2.4 The Cave Temple of at Elephanta Island near Mumbai in western India (Photo by Y. Hagihara in 2005)

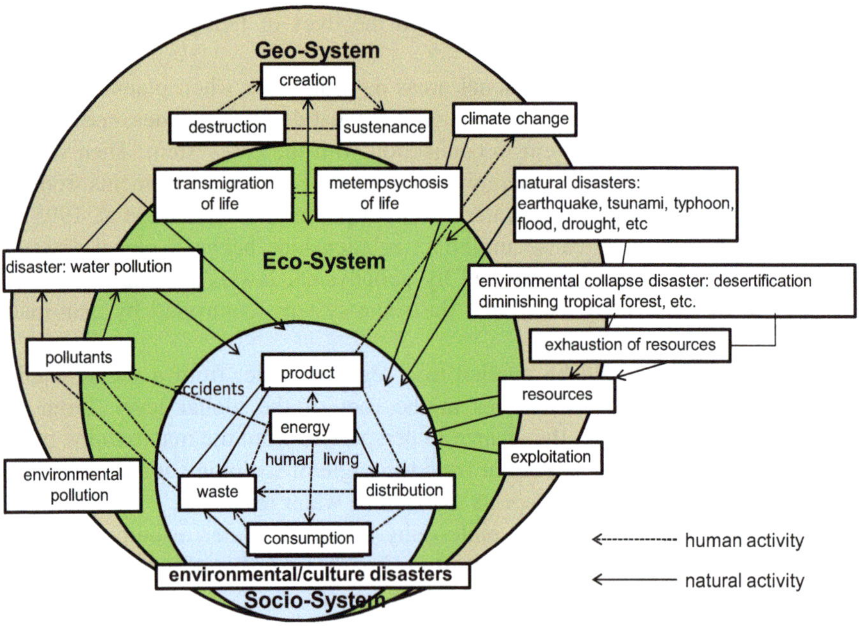

Fig. 2.5 The cause-and-effect relationships of global disasters

There are five aspects to a comprehensive system of water resources planning: (1) disaster prevention, (2) playing, (3) feeling, (4) ecology, and (5) culture (Hagihara et al. 1998).

The concept of earth seems to be dominated today by Cartesian philosophy, as shown on the left side of Fig. 2.6. Typical phenomena include urban structures,

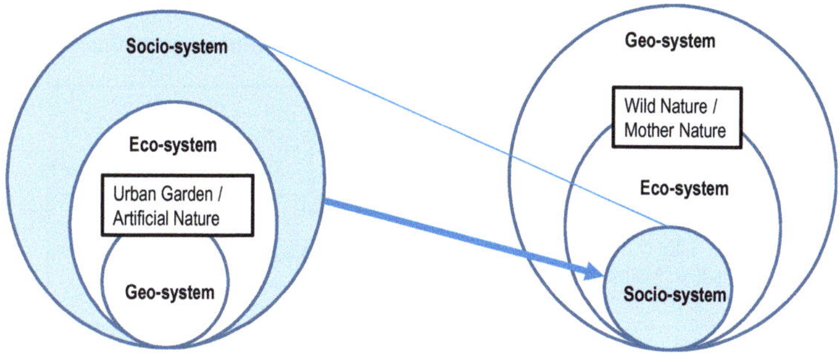

Fig. 2.6 Changing the concept of the earth

especially in large cities, and the gardens in these cities. European modernization is generally based on the logic of *logos*, from Greek philosophy, whereas Asian philosophy is based on the logic of *lemma* from Buddhism.

Figure 2.6 shows the European concept of the earth integrating into the Asian concept. The six phases shall be considered as the basis for our research on water resources. In this sense, any scientists who studies water resources problems must determine in which phase his or her research falls; otherwise, scientists might not be able to collaborate.

The European concept is:

- *A* is *A* and not ^{not}A".

The Asian concept means:

- *A* is not *A* but also not ^{not}A, and
- *A* is *A*, but also ^{not}A".

The Japanese Zen garden, which is based on hills and streams without water, shows the concept of *lemma*.

2.4.2 *The Circulation System for Mid- and Long-Term Water Resources Planning*

Time varies in Figs. 2.5 and 2.6; thus, a length of time should be selected (Hagihara 2008). We might consider more than 30–50 years due to the length of time that facilities for water resources can last, depending upon adequate repairs and replacement. The planning process is estimated to last at least 30 years, during which period humans will experience a variety of climate and social changes as shown in Fig. 2.7 (which focuses on social and environmental changes, disasters, and changes in plans for water resources).

Huge unprecedented disasters happen from time to time, and in the near future, the environment and society will no doubt affect each other. It is not just natural

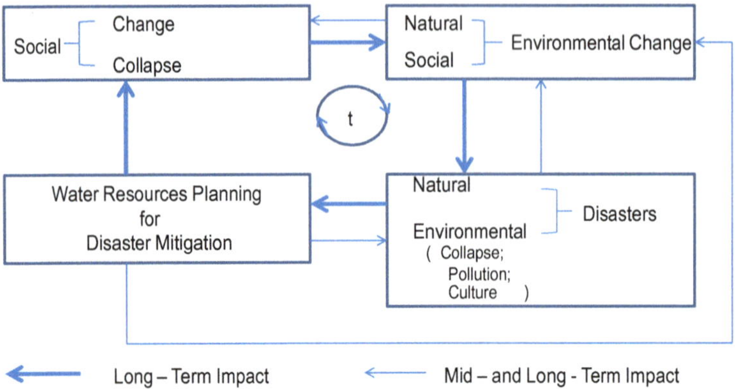

Fig. 2.7 Circulation system of mid- and long-term environmental changes and water resources planning for disaster prevention and mitigation

factors that cause disasters, but also social factors, and these two kinds of elements are in turn intertwined. Figure 2.7 shows three kinds of environmental disasters. In particular, environmental culture disasters are caused by religious, historical, and philosophical practices. Hard labours and subsequent gender issues are widely observed in India, China, and South East Asia (shown in Fig. 2.8a and b).

2.4.3 Adaptive Water Resources Planning Methodology for Social Conflict Management for Survivability

Figure 2.9 shows an adaptive water resources planning methodology for managing social conflicts (Hagihara and Sakamoto 2006).

This figure shows the following process.

(a) **Problems with identifying characteristics**: The social risks and conflicts of water resources planning shall be identified by:

- Relating them to GES environments through brain-storming,
- The KJ method, and
- The ISM (Interpretive Structuring Modelling) method.

Regional residents (including citizens and farmers) concerned with regional water resources problems will participate (Hagihara and Hagihara 2004).

(b) **Regional field GES environments research & research on conflict issues**: This step will consist of

- Making charts to locate regional characteristics of conflict,
- Creating databases for answering the 'what, why, who, where, when and how' of the problem,
- A socio-economic, statistical analysis,

Fig. 2.8 (**a**) Hard labour in Ajanta, India (Photo by Y. Hagihara in Mar. 2004). (**b**) Hard labour near Dhaka, Bangladesh (Photo by Y. Hagihara in Mar. 2007)

- Field surveys, sometimes in the form of interviews (such as those used in cultural anthropology), and
- Geophysical and ecological databases from GIS satellites.

(c) **Analysis of essential components of conflict and the regional scale of ecological risk in the GES Environment**: We determine social risk using the following kinds of analysis: multivariate analysis, time series analysis, and psychological model analysis. We examine the regional scale of ecological risk using various mathematical models, such as system dynamics, multivariate analysis, time series analysis, and mathematical ecological models. Many geophysical (including hydrological) models are also applied. These analyses should be integrated and considered comprehensively for the next stage.

(d) **Predicting and identifying targets**: Finding and formulating essential multi-objectives and multi-boundary conditions on time and space for the sake of

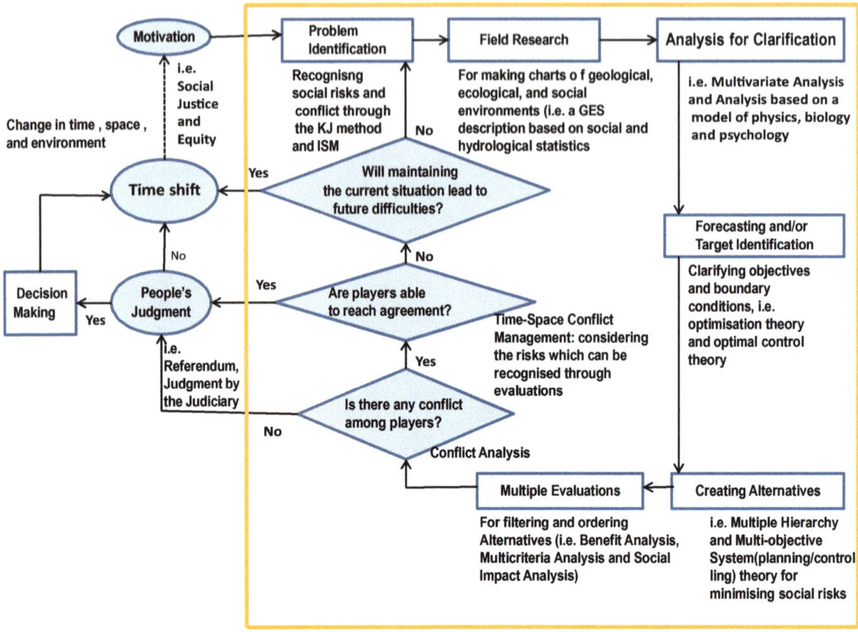

Fig. 2.9 Adaptive water resources planning methodology for managing social conflicts

conflicts mitigation. We use optimization theory and optimal control theory, and consider factors of uncertainty in order to design criteria.

(e) **Characteristic and formulation of water resources alternatives**: This aspect is based on the multiple hierarchical and multiple objects models (for planning and control) to minimise risk in GES environments and social conflicts.

(f) **Multiple evaluation of alternatives**: In order to find alternatives to conflicts (for example, between the development of water resources and the environment), a Cost-Benefit Analysis is used along with GES environments impact assessment, following an analysis for multiple criteria.

(g) **Conflict management (if conflict can be assumed)**: The blue (shadow) areas of Fig. 2.9 shall be analysed through game theory, GMCR with a mathematical psychology model, and sometimes with an evolutional game model.

2.5 Examples of Research on Conflict Management

2.5.1 *Creating a Process for Alternatives: A Hierarchical System & the Multi-objective Optimal Control Process*

2.5.1.1 The Concept of a Hierarchical System Based on the Decomposition Principle: Conflict Among Local Governments

A model is presented for sewage planning, which extends over a region controlled by a number of local governments (Hagihara and Hagihara 1981). However, the central government assigns the total grant for this region. A simple regional linear model is first set up, and using a method of decomposition, it is shown that this model itself is a decision-making process in which information is exchanged between two levels, the central and local governments. The purpose is to coordinates conflicts between local governments with the aim of controlling public water.

In this model the allocation of the grant is considered taking into account each local government's autonomous plan. That is, the grant rate is determined for each local government taking into consideration all local government grant requests. This is a decision-making process of a two-level system (Mesarovic et al. 1970) involving a framework similar to this idea for a problem of water resource allocation in the internationally shared river basin (Stone 1980).

2.5.1.2 Multi-objective Optimal Water Resources Allocation Process: Conflict Between Allocated Water and Water Conservation

A model for the multi-objective aspect of allocating water resources is presented as an optimal control process (Hagihara et al. 1981). The maximum principle, including iterative goal programming is proposed as an algorithm to solve the problem. The system equations presented are inherently general for water environmental system analysis, so they are available not only for allocating water resources, but also for sewage planning and building frameworks for regional planning. The conflicts and trade-offs between regional development projects and environmental impacts can be somewhat transformed into a decision-making process by including the residents of the river basin, which involves decision on satisfied and permitted level of objectives (Hagihara and Hagihara 1977).

As shown in Fig. 2.3, regional water resources have traditionally been allocated with the goal of minimizing costs and focusing on demand. Pollution caused by wastewater has received little consideration, and has not been included in frameworks for regional planning, as happens in contemporary developing countries. A new concept -as discussed in Sect. 2.3- is needed to allocate water resources properly, and to reduce water pollution.

2.5.2 *Multiple Evaluation Process: Decision-Making System in the Conflict Between Water Resources Development and Environmental Conservation*

2.5.2.1 Background and Goals

In recent years, water resource development projects have affected the basin environment, and it is important to solve the subsequent challenges by building consensus among stakeholders (Sato and Hagihara 2003). When making decisions on water resource development projects, the process is more important than the result. A systematic decision-making process is applied to the problem of the Yoshino River Movable Weir. A new approach to derive a value function of a stakeholder is also developed.

The following three perspectives must be considered for planning water resources development projects:

(a) The views of the people who will suffer from flood or drought, and those who are lucky to live in an environment that will not be impacted.
(b) The views of the people likely to suffer from disasters, and those who rarely suffer from it.
(c) The standpoints of people who only consider the lives of human beings, and those who consider the well-being of non-human entities, such as forests and animals.

The authors aim to develop a way to evaluate alternatives to water resource development projects based on stakeholders' perspectives, and to derive results from their feedback. The following three steps are intended to ensure the authors' goal:

(a) Firstly, the systematic decision-making process is outlined
(b) Secondly, the new approach for quantitatively evaluating each stakeholder's view is developed
(c) Finally, this process is applied to the problem of the Yoshino River Movable Weir as a case study.

2.5.2.2 Systematic Process for Decision-Making

It is necessary for stakeholders to listen to each other's opinions in order to build consensus. To achieve this goal, it is necessary to precisely describe each stakeholder's criteria for evaluation and present it to the other parties involved (Keeney and Raiffa 1993). Thus, they can discuss matters constructively.

Figure 2.10 shows the systematic process for social decision-making. In order to explain rational decision-making, some steps based on systems analysis are needed.

1. First, the various effects of water resource development projects must be clarified, and the stakeholders must be identified.

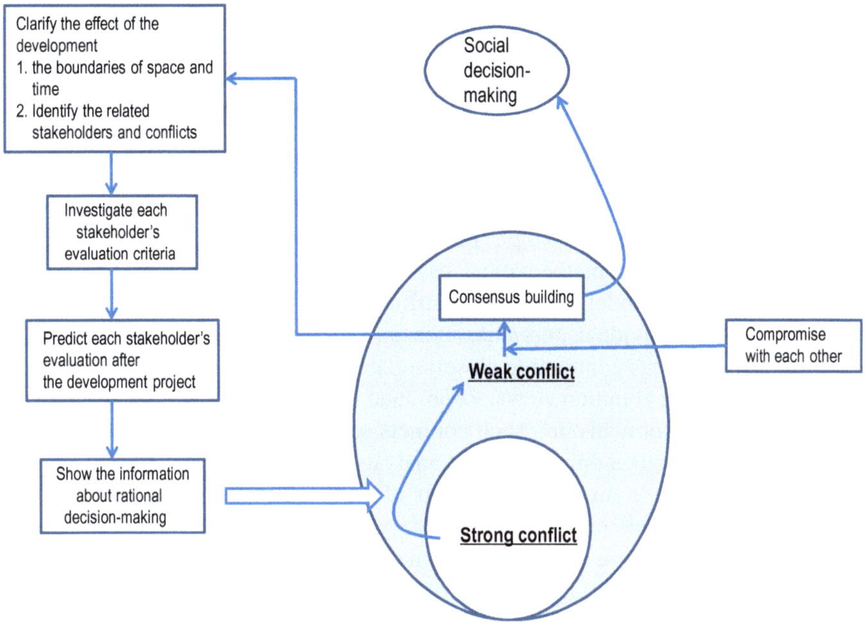

Fig. 2.10 Decision-making framework

2. Secondly, the present conditions are investigated to clarify each stakeholder's criteria.
3. Third, future estimates are analysed to predict each stakeholder's evaluation after the project is finished.

By following these steps, alternatives can be evaluated and the results shown to all stakeholders, helping to ease a tense dispute by having all parties listen to one another.

2.5.3 *The Process of Managing Conflicts: Social-Psychological Conflict Analysis and Evolutional Game Theory with GMCRII (Graph Model for Conflict Resolution II)*

2.5.3.1 Historical Nagara Observation Analysis

It is generally difficult to adjust rules and recognise the rights of all parties involved when dealing with large-scale, long-term development projects, especially in the field of water resources development (Sakamoto et al. 2004). In addition, the plan for development, which is thought to be most rational during the first stage, often strays far from residents' expectations as time passes and the environment changes.

For large-scale development projects, firstly, it shall be considered that the residents' consciousness of the disaster experience gradually decreases, and that the players, who are involved in the development project, are influenced by each other. Secondly, such a development makes a lot of players staked in, so the conflict among the players shall be cleared.

The stakeholders' changing attitudes are modelled by oblivion and mutual influence. The former describes a change in a time target of the person's oblivion, based on mathematical sociology. The latte describes the influence of all other stakeholders, based on mathematical psychology. Then, the conflict analysis is combined with the model of changing attitudes in circulation. Here, it is defined as conflict when individuals, organisations, and society have different opinions. In terms of these entities contracting closely related actors, a condition occurs where they end up sharing common views, values, and interests. Game theory is one of the effective theories for analysing such conflicts scientifically, and predicts at which point the conflict settles down. Conflict analysis (Fraser and Hipel 1984) is a form of game theory; it is an expanded form of meta-game theory on the basis of non-cooperative game.

The authors propose a dynamic mathematical system model whereby a circulation system includes a shift in time, and combines varying attitudes with an analysis of stability in conflict. This makes it possible to clarify the factors that led to a change in stakeholders' attitudes, and to identify what kind of process generated the original conflict. This model is applied to the Nagara River Mouth Weir to verify its adaptability, and to simultaneously conduct a historical analysis. Furthermore, in order to plan for the future, the authors conducted an experiment to determine how the situation might have turned out differently in the past.

The following assumptions were made in regards to the aforementioned context:

(a) As time passes, people have a tendency to forget the acuteness of their shock at the damage caused by disasters and drought.
(b) People are influenced by others' behaviour, which causes their attitude to change.
(c) Conflict occurs among a limited number of stakeholders. Thus, stability analysis is limited to the respective group of stakeholders.
(d) The outcomes are discriminated from relationships between the model for decisions (based on the behaviour decision model) and the threshold. Thus, outcomes with a probability of unrealistic equilibria are excluded from the stability analysis.

2.5.3.2 International Conflict Management on the Ganges Between India and Bangladesh

The graph model for conflict resolution is used to formally analyse the ongoing conflict between India and Bangladesh over the regulation of the Ganges River, in order to illustrate the crucial role a third party can play in resolving the disputes (Sakamoto et al. 2005; Sakamoto and Hagihara 2005, 2006). Because a third party can assist in resolving a dispute in a variety of ways, a general systems approach to conflict management with a third party is devised. Using the graph model, the strategic analysis of the India-Bangladesh conflict clearly shows that one can determine exactly how a third party can influence potential resolutions in advance.

The Ganges River is the most important source of water resources for both Bangladesh and India, and this has resulted in many hostilities between them. The distribution of water resources from the Ganges is generally benefits India, which lies upstream. However, it is difficult for Bangladesh to offer an alternative effective resolution, because it is at an economic and topographic disadvantage. It is hard to imagine that India will change its position unilaterally. The conflict between Bangladesh and India has become stagnant since they have negotiated by themselves. Thus, the participation of a third party could improve the situation.

The Graph Model for Conflict Resolution II (GMCR II) (Fang et al. 1993) is used to analyse this conflict. GMCR II is based on non-cooperative game theory, and provides algorithms to calculate equilibria among decision makers (DMs) who have different preferences. Based on the framework of GMCR II, the roles of a third party are defined and classified into three types: Arbitrator, Coordinator, and Donor, who are individually called a Third Party. The graph model is applied to the conflict between Bangladesh and India. First, the actual situation is modelled using GMCR II. Secondly, conditions necessary for improving the current dispute are analysed. Thirdly, the effectiveness of a Third Party is examined by focusing on the Coordinator's role. Chapter 17 gives a detailed analysis.

Mathematical modelling is often used to understand social systems. When it is applied to a problem of social systems, we sometimes ask: what is the relationship between mathematical stability and social stability is? There are many definitions of mathematical stability, so that we should clarify the following questions:

- What kind of social system does the mathematical model assume?
- What kind of social stability is assumed to be optimal in the respective social system?
- What characteristics does mathematical stability focus on in the study?
- What is the difference between mathematical stability and social stability?

The GMCR, evolutionary game theory, and non-linear differential equations are used to illustrate conflict among stakeholders. Firstly, a summary is provided of the relationship of these theories' mathematical stabilities, and some mathematical theories receive attention such as evolutionary game theory and differential equations. Secondly, the India-Bangladesh conflict over the Ganges River is mapped out based on the mathematical theories mentioned above. Thirdly, the potential for managing the India-Bangladesh conflict is analysed according to the relationship between social stability and mathematical stability.

2.6 Concluding Remarks

This chapter can be summarised as follows:

1. Various international problems of water resources converge at points of water resources conflict, and can affect humanity's survival. In other words, the social risk management of water resources shall result in water resources conflict management.
2. Water resources conflicts in Japan, in which there is tension between development projects and the environment, are briefly explained.
3. This chapter explained the potential of the environment to be contaminated, and paradigm-shift for water resources planning process for regional sustainability and pointed out social inequality shall be never forgotten in water resources planning.
4. Meta-methodology for managing water resources conflict was explained in three figures:
   (a) Recognising three ways of relating to the environment
   (b) A circulation system for mid- and long-term environmental changes and water resources planning for disaster prevention and mitigation
   (c) An adaptive water resources planning meta-methodology for managing social conflicts is proposed, which consists of seven processes.
5. Some examples of research were introduced with the following three processes:
   (a) A path for creating alternatives was briefly explained, along with a hierarchical system based on the decomposition principle and a multi-objective process for allocating water resources.
   (b) A multiple evaluations process, including a decision-making system to solve the conflict between water resources development and environmental conservation, with a discussion of the Yoshino-River.
   (c) A conflicts management process was briefly discussed, along with an analysis of the historical situation surrounding the Nagara-River, and conflict management of the Ganges River between India and Bangladesh.

The topics and contents of this chapter are very broad; thus, the authors strongly recommend to readers that they explore the references listed below.

References

Barrow, M., & Clarke, T. (2002). *Blue gold: The battle against the corporate theft of the world's water*. Toronto: Stoddart Publishing.

Fang, L., Hipel, K. W., & Kilgour, D. M. (1993). *Interactive decision making: The graph model for conflict resolution*. New York: Wiley.

Fraser, N. M., & Hipel, K. W. (1984). *Conflict analysis: Models and resolutions*. New York: North-Holland.

Hagihara, Y. (2008). *Adaptive system planning methodology for environment risk management.* Kyoto: Kyoto University Press (in Japanese).
Hagihara, Y. (2009). Water resources conflict management, *The Annals of the Disaster Prevention Research Institute, Kyoto University*, No. 52A, pp. 81–101.
Hagihara, Y., & Hagihara, K. (1977). Water resources allocation based on river state equations. *Studies in Regional Science, 8*, 39–51.
Hagihara, Y., & Hagihara, K. (1981). Project grant allocation process applied in sewerage planning. *Water Resources Research, 17*(3), 449–454.
Hagihara, K., & Hagihara, Y. (1991). The role of intergovernmental grants in underpopulated regions. *Regional Studies, 25*(2), 163–172.
Hagihara, K., & Hagihara, Y. (2004). The role of environmental valuation in public policymaking; The case of waterside area in Japan. *Environment and Planning C: Government and Policy, 22*, 3–13.
Hagihara, Y., & Sakamoto, M. (2006). *Conflict management: As water resources social risk.* Tokyo: Keiso (in Japanese).
Hagihara, Y., Hagihara, K., Nakagawa, Y., & Watanabe, H. (1981). A multi-objectives optimal water resources allocation process, IFAC 8th Triennial World Congress, PPCS-81-86.
Hagihara, Y., Takahashi, K., & Hagihara, K. (1995). A methodology of spatial planning for waterside area. *Studies in Regional Science, 25*(2), 19–45.
Hagihara, Y., Takahashi, K., & Hagihara, K. (1998). *Waterside planning process in urban environment by systems analysis.* Tokyo: Keiso (in Japanese).
Keeney, R. L., & Raiffa, H. (1993). *Decision with multiple objectives: Preferences and value tradeoffs.* Cambridge/New York: Cambridge University Press.
Mesarovic, M. D., Macko, D., & Takahara, Y. (1970). *Theory of hierarchical, multilevel systems.* San Diego: Academic Press.
Ohlsson, L. (1996). *Hydropolitics: Conflicts over water as a development constraint.* Dhaka: University Press of Bangladesh.
Pearce, F. (1992). *The Dammed-river, dams, and the coming world crisis.* New York: Random House.
Postel, S. (1999). *Pillar of sand: Can the irrigation miracle last?* Washington, DC: World-Watch Institutes.
Sakamoto, M., & Hagihara, Y. (2005). A study on social conflict management in a water resources development – A case of the conflict between India and Bangladesh over regulation of the Ganges river. *Journal of Japan Society of Hydrology and Water Resources, 18*(1), 11–21.
Sakamoto, M., & Hagihara, Y. (2006). Social stability and mathematical stability in conflict management. IEEE international conference on systems, man and cybernetics, pp. 5270–5274.
Sakamoto, M., Hagihara, Y., & Sugiman, T. (2004). An attempt to develop a dynamic system model of social conflict by the combination of behaviour decision model and game theory. *Asian Journal of Social Psychology, 7*, 263–284. Blackwell Publishing.
Sakamoto, M., Hagihara, Y., & Hipel, W. K. (2005). Coordination process by a third party in the conflict between Bangladesh and India over regulation of Ganges River, 2005. IEEE international conference on systems, man and cybernetics, pp. 1119–1125.
Sato, Y. (2008). Consensual alternatives evaluation under the conflict between river basin development and environment (in Japanese). *Doctoral dissertation*, Kyoto University.
Sato, Y., & Hagihara Y. (2003). Decision-making system under the conflict situation between water resources development and environmental conservation. Proceeding of the 3rd DPRI-IIASA international symposium on Integrated disaster risk management, [CD-ROM].
Stone, P. J. (1980). A systems approach to water resource allocation in international river basin development. *Water Resources Research, 16*(1), 1–13.
Watanabe, H., & Hagihara, Y. (1993). Planning and management of water resources systems. *Journal of Hydroscience and Hydraulic Engineering, SI-3*, 219–223.

Part II
Socioeconomic Vulnerability in a Regional Perspective

Chapter 3
The Depopulation Problem

Kiyoko Hagihara and Yoshimi Hagihara

Abstract The depopulation problem in Japan is considered from the viewpoint of local public goods equilibrium. If in moving from one region to another migrants do not account for the effect of their moving on the other residents, then one region may be overpopulated and the other underpopulated. In the framework of a simple model, it is suggested that the central government may be justified in using a system of intergovernmental grants to overcome these inefficiencies. In order to confirm the role of intergovernmental grants, the model is applied to a village which is designated as a depopulated area. Furthermore, the effects of the countermeasures taken for about 40 years since the first depopulation law was implemented are investigated. In the final section, it is pointed out that there is a need to take into account another situation: there are areas which are similar to depopulated areas, but are not designated as such under the law.

Keywords Depopulation problem • Intergovernmental grants • Migration • Inefficiency of population allocation

3.1 Introduction

During the periods of high economic growth in the 1960s and 1970s in Japan, outmigration from agricultural and mountain villages to urban areas increased rapidly. As a result overpopulation in the urban areas and underpopulation in rural areas have become social problems. In Japan, the problem of the underpopulated areas is called 'a depopulation problem'.

The rural areas play a very important role particularly in supplying food, conserving national land, conserving the head sources of a stream and conserving the natural environment. However, in depopulated areas it has become difficult to

K. Hagihara (✉)
School of Sociology, Bukkyo University, 96 Kita-hananobo, Murasakino, Kita-ku, Kyoto 603-8301, Japan
e-mail: hagihara@bukkyo-u.ac.jp

Y. Hagihara
Kyoto University, 264 Suemarucho, Nakagyou-ku, Kyoto 604-0901, Japan
e-mail: y-hagi@gf7.so-net.ne.jp

K. Hagihara, C. Asahi (eds.), *Coping with Regional Vulnerability*, New Frontiers in Regional Science: Asian Perspectives 4, DOI 10.1007/978-4-431-55169-0_3

manage the community and these areas have not been able to fulfil these roles. Therefore, the Japanese government has taken a number of measures to promote rural areas and to alleviate some of the problems.

Among the above roles the rural areas play a very important part in supplying water to the urban areas. Hagihara and Hagihara (1986) have suggested that people in downstream areas have a responsibility to use water resources wisely while those in upstream areas have responsibility for considering the effects of using water on the people in downstream areas (Hagihara and Hagihara 1981). From this point of view, depopulation must be regarded as a nationwide problem. So, Hagihara and Hagihara (1991) considered the depopulation problem from the viewpoint of local finance, paying particular attention to the role of intergovernmental grants.

In the next section the situation of designated 'depopulated areas' since the implementation of the depopulation laws is investigated. In the third section the inefficiency which results from free migration is considered in the framework of a simple model, using a concept of local public goods. It is suggested that the central government may be justified in using a system of intergovernmental grants to overcome these inefficiencies. In order to explore the role of intergovernmental grants, in the fourth section the model is applied to a depopulated village in the Agatsuma district of Gunma prefecture – the Agatsuma district is one of the areas which are supplying water to Tokyo. In the fifth section, the effects of the counter-measures taken for about 40 years since the first depopulation law was implemented are investigated. In the final section, it is pointed out that there is a need to take into account another situation: there are areas which are similar to depopulated areas, but are not designated as such under the law.

3.2 The Present State of Depopulated Areas and Development Policies in Japan

3.2.1 The Present State of Depopulated Areas

In the 1960s and early 1970s, net migration from rural areas to metropolitan areas such as Tokyo and Osaka occurred, creating both overpopulated and underpopulated areas throughout Japan. The rates of population increase in Tokyo and Osaka were 45 and 79 % respectively in the 1955–1975 periods. The rates of population increase in Kanagawa and Saitama, which are located next to Tokyo, were 119 and 113 % respectively in the same period. The population concentration has caused several problems, for example, average commuting hours are approximately one and a quarter hours taking overcrowded trains, roads are always congested and environmental deterioration in air and water has occurred in the metropolitan areas.

By contrast, the average rate of population decrease in rural areas was approximately 30 % during 1955–1975. As a result, those in rural areas cannot maintain the same life-styles as before 1960. Underpopulated areas are called 'depopulated' in Japan. The term 'depopulated area' is used here to describe an area where production and lifestyles have deteriorated due to outmigration. Typical examples of this deterioration include decreases in the number of fire-fighters and difficulties in financially maintaining schools. Some communities have already collapsed.

In 1970, special legislation (the Emergency Measures for the Development of Depopulated Areas Law) was introduced to stimulate the development of depopulated areas. This legislation had a time limit of 10 years and was replaced in 1980 with the Special Measures for the Development of Depopulated Areas Law. Moreover, this was replaced in 1990 with the Special Measures for the Regional Vitalization of Depopulated Areas Law, and further replaced in 2000 with the Special Measures Promoting Self-support of Depopulated Areas Law. The general outline of each law is shown in Table 3.1.

As of 2010, 775 municipalities are designated as 'depopulated areas'. They are located in 46 of Japan's 47 prefectures, the only exceptions being Kanagawa Prefecture next to Tokyo. The total population of these municipalities (2010 census) is 10.33 million. This is a mere 8.1 % of the total national population, but these areas cover 216,321 km^2, or 57.2 % of Japan's total land area (see Table 3.2).

For all these depopulated areas, the rate of population decrease was more than 10 % in both the 1960–1965 and 1965–1970 periods. Although the rate of population decrease subsequently mitigated, the population of depopulated areas decreased by 7.1 % from 2005 to 2010.

Moreover, comparing the age structure of the population in the depopulated areas with that of the nation as a whole, one finds that the proportion of the population in the younger age groups decreased, and the proportion of the population in the middle and older age groups increased in the depopulated areas, because many young people had migrated from these areas over a long period. In particular, the average number of people over-65 was 33.2 % and higher than the national average of 22.8 % in the depopulated areas as of 2010. On the other hand, the proportion of the population under-15 was 11.3 % and lower than the national average of 15.4 % in depopulated areas as of 2010 (see Table 3.3).

In most local government in the depopulated areas, own revenues are not enough for them to satisfy the expenditure requirements of their residents. Their expenditure is restricted tightly by the revenue. While local government tax revenue as a percentage of total revenue is 34.1 % nationally, local government tax revenue is only 14.0 % of total revenue in depopulated areas. On the other hand, the percentage of intergovernmental transfers including chiho-kofu zei such as rate support in Great Britain and revenue sharing in the United States of America and Canada, central and prefectural government grants was 63 %, whereas the national percentage of intergovernmental transfers is 39 % in 2011 F.Y.

The average fiscal capability index was 0.24 in depopulated areas, whereas the national average was 0.51 in 2011 F.Y.

Table 3.1 Outline of the four laws for the depopulated problems

Name of the law	The emergency measures for the development of depopulated areas law (1970)	The special measures for the regional development of depopulated areas law (1980)	The special measures for the regional vitalization of depopulated areas law (1990)	The special measures for the self-support promotion of depopulated areas law (2000, 2010)
Definition of the depopulated areas by population	Population had decreased by at least 10 % over the 5-year period between 1960 and 1965	Population had decreased by at least 20 % over the 15-year period between 1960 and 1975	Population had either	Population had either
			(1) decreased by at least 25 % between 1960 and 1985,	(1) decreased by at least 30 % between 1960 and 1995,
			(2) decreased by at least 20 % and the proportion of the population over-65 is more than 16 % by the 1985 census, or	(2) decreased by at least 25 % during 1960–1995 and the population over-65 is more than 24 % by 1995 the census,
			(3) decreased by at least 20 % and the proportion of the population over-15 and under-30 is 16 % or less by the 1985 census	(3) decreased by at least 25 % between 1960 and 1995 and the proportion of the population between over-15 and under-30 is 15 % or less, or
				(4) decreased by more than 19 % between 1970 and 1995
Definition of depopulated areas by fiscal capability index	The average fiscal capability index was less than 0.4 for three fiscal years	The average fiscal capability index averaged 0.37 or less for three fiscal years	The average fiscal capability index averaged 0.44 or less for three fiscal years	The average fiscal capability index averaged 0.42 or less for three fiscal years
Number of cities, towns and villages	1093	1157	1230	(1210) 775*

Note:
1. Fiscal capability index = standard financial revenues/standard financial needs
2. The number of cities, towns and villages has decreased because of the Japanese government's policy by uniting neighbouring cities, towns and villages, that is, the big merger of Heisei. Thus 775* is the number of cities, towns or villages after the mergers

Table 3.2 Proportion of depopulated versus non-depopulated areas in Japan in 2010

Area	Depopulated areas	Non-depopulated areas
Items		
Number of cities, towns and villages (percentage)	775 (45.1 %)	945 (54.9 %)
Population per 10,000 (percentage)	1.033 (8.1 %)	11.773 (91.9 %)
Area in km^2 (percentage)	216,321 km^2 (57.2 %)	161,629 km^2 (42.8 %)

Table 3.3 Average number of population over 65 and under 15, 1970–2010

		1970	1975	1980	1985	1990	1995	2000	2005	2010
Population over −65	The whole nation	7.1	7.9	9.1	10.3	12.0	14.5	17.3	20.1	22.8
	Depopulated areas	10.1	11.9	13.8	15.8	19.1	23.4	27.4	30.6	33.2
Population under-15	The whole nation	27.7	24.8	21.5	20.7	21.7	21.7	20.2	17.4	15.4
	Depopulated areas	20.5	19.8	18.0	15.8	14.6	14.4	14.3	12.9	11.3

Note:
1. Figures based on national census statistics
2. Depopulated areas correspond to official depopulated areas as of 1 April 2013

3.2.2 Countermeasures

Up to now, four laws on the depopulated areas have been implemented (see Table 3.1).

Measures introduced under four Laws for the depopulated areas are being implemented on a planned and comprehensive basis at the national, prefectural and municipal levels. The specific procedures involved are as follows.

Guidelines for the vitalization of depopulated areas are designed at the prefectural level and are used as the basis for local development planning by individual municipalities in the depopulated areas. These municipal activities are supported by development plans formulated by prefectural governments on the basis of the same guidelines. The planning horizon of all these guidelines and plans is 5 years.

Typical measures which are implemented under depopulated area revitalization guidelines and plans are as follows. Firstly, road improvement is pushed forward in the depopulated areas. This is because the road quality in these areas is worse than that in other areas. A special system has been set up whereby prefectural governments are responsible for the construction of trunk roads in local municipalities.

Secondly, an amalgamation of primary and junior high schools is sometimes carried out. This measure is taken in cases where it has become difficult to maintain schools due to the decrease in the number of school-aged children. In some areas, however, the amalgamation is impractical because of geographical or other considerations. Therefore, there are some schools which must be kept to provide good educational services. In these cases, measures are taken to improve the quality of education.

Thirdly, the maintenance of fire-fighting capabilities is considered. In order to counter the possible deterioration in fire-fighting capability due to the decrease in the number of fire-fighters, efforts are made to improve mechanical efficiency through the provision of pump vehicles and other equipment.

The maintenance of adequate medical services is also a very serious problem because the level of medical services is very low in many communities in the depopulated areas. To deal with this problem, a number of measures are being implemented including the construction of medical treatment facilities, the introduction of regular medical visits and the stationing of district nurses.

The measures described here represent only a few examples of the project that are now currently being implemented. Other governmental activities related to the improvement of depopulated areas range from the construction of public facilities to the promotion of industry. The central government gives fiscal and financial support for these measures including grants-in-aid, loans and preferential tax treatment.

3.3 A Mechanism for Creation of Underpopulated Areas and Intergovernmental Grants

3.3.1 A Simple Model for the Allocation of Population Between Regions

Tiebout (1956) developed a formal model of population distribution based on the concept of all individuals locating in accord with their preferences for local public goods and services. He suggested that, if there were enough communities, individuals would reveal their true preferences for public goods by their choice of community in which to live. Where there is a wide range of choice, all those deciding to live in the same community would have essentially the same tastes, and there would be no problem of reconciling conflicting preferences. Moreover, it is often asserted that such a local public goods equilibrium would be Pareto-efficient.

However, a number of authors have questioned the validity of this 'metropolitan' version of the Tiebout model of efficient local public service provision, and have constructed the 'regional' model of local service provision (see Mieszkowski and Zodraw 1989). Many researchers (e.g. Buchanan and Wagner 1970; Buchanan and Goetz 1972; Flatters et al. 1974; Boadway and Flatters 1982; Stiglitz 1977, 1983a, b,) have attempted to clarify the welfare implications of the Tiebout hypothesis. They have suggested that non-optimality may occur because, in moving from one region to another, migrants do not account for the effect of their moving on the other residents in the regions they leave or enter. Therefore, the Tiebout type of free market equilibria may not be Pareto-efficient. If this externality is not internalized by migrants' decision-making, then one region may be overpopulated and the other underpopulated.

In this section some of the problems which are associated with the regional distribution of population will be explored by using a very simple model based on Boadway and Flatters (1982).

Suppose a nation consists of two regions, A and B, and there is a total population N^T who are costlessly mobile between regions. Region A is richly endowed with land on natural resources, L, while B is the poorer region, i.e. $L_A > L_B$. Each region produces a single homogeneous commodity Y, which can be allocated for use in the private sector or in the public sector within the region. Thus, the marginal rate of transformation between private goods and public sector goods is unity.

Each region has a concave production function:

$$Y_i = f^i(N_i, L_i) \quad i = \mathrm{A, B} \tag{3.1}$$

$$\text{where}: \quad f_N^{'} > 0 \text{ and } f_{NN}^{'} < 0.$$

It is assumed that individuals have identical preferences, represented by the utility function:

$$U(X_i, G_i) \quad i = \mathrm{A, B} \tag{3.2}$$

where: $X_i =$ the per capita consumption of private goods

$G_i =$ the quantity of public sector goods provided in region i.

Assuming that the value of output of a region accrues entirely to its own residents, per capita private goods consumption X_i is given by:

$$\{f^i(N_i, L_i) - G_i\}/N_i$$

If each regional government aims at maximizing the welfare of its residents for a given level of N, it will solve the following problem:

$$\max_G U\left[\frac{f(N, L) - G}{N}, G\right] \tag{3.3}$$

where subscripts which represent region have been dropped.

The first-order condition yields:

$$N\frac{U_G}{U_X} = 1$$

where: $U_G = \partial U/\partial G$ and $U_X = \partial U/\partial X$.

This is the familiar result that the sum of the marginal rates of substitution equals the marginal rate of transformation.

Next, the maximum per capita utility, $V(N)$, which is attained under a given level of N in each region is defined as follows:

$$V(N) = \max_{G} U\left[\frac{f(N,L) - G}{N}, G\right] \tag{3.4}$$

Differentiating (3.4) with respect to N with the aid of the envelope theorem yields:

$$V''(N) = \frac{U_X}{N^2}[G - (f - Nf_N)] \tag{3.5}$$

If it is assumed that $U_X > 0$, at the optimum per capita consumption of private goods equals the marginal product of labour, i.e.:

$$\frac{f(N,L) - G}{N} = f_N$$

or:

$$G = f - Nf_N$$

Thus, if the level of public expenditure is fixed, but the population is variable, the population that maximizes consumption per capita is such that rents equal public goods expenditure. This has been dubbed the 'Golden Rule' result (Flatters et al. 1974) and the 'Henry George' theorem (Stiglitz 1977).

The local government determines its public expenditure levels to maximize the welfare of its residents and can finance these services with either a head tax or a property tax. The Henry George theorem states that decentralized provision of local public services will be efficient if taxes on land rents are used to finance local public expenditures. However, Stiglitz 1977, showed that the simple decentralized version of the Henry George theorem does not hold when sites vary in quality or natural resources and regions are not of the same size.

The function $V(N)$ may take many forms (see Atkinson and Stiglitz 1980). Firstly, it is assumed that the graphs of $V(N)$ are single-peaked as in Fig. 3.1. In this situation free migration ensures that people allocate themselves among regions until $V(A)$ equals $V(B)$ and there is a unique stable equilibrium, N^e. On the other hand, in Fig. 3.2 there are three possible equilibria: $N^{e'}$ is a stable equilibrium; the other internal equilibrium $N^{e''}$ is unstable; the corner one is stable but region B will be completely depopulated,

i.e.:

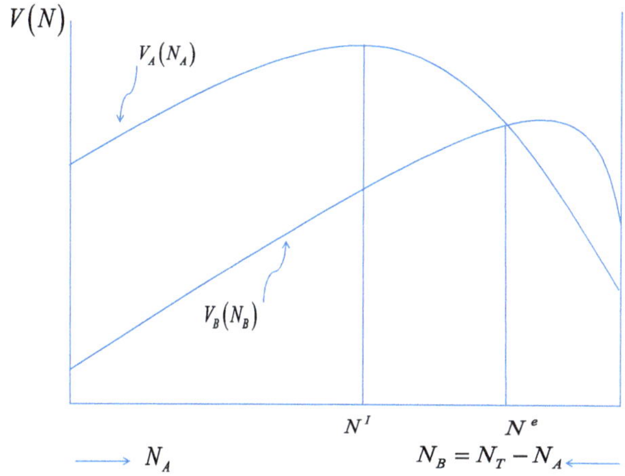

Fig. 3.1 Stable equilibrium

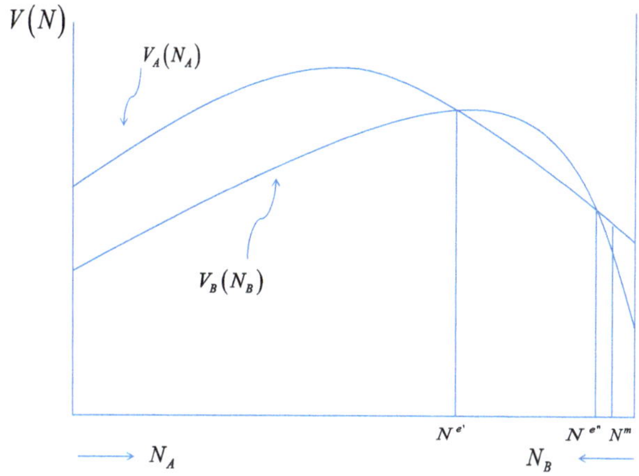

Fig. 3.2 Unstable equilibrium

$$V_A(N^T) > V_B(0)$$

Even in the stable case, the above local public good equilibria N^e and $N^{e'}$ may not be Pareto-efficient. As mentioned earlier this arises because migrants who move from one region to another do not take into account the effect of their moving on the other residents in the region they leave or enter. Now let us calculate the benefits of the existing residents when one additional migrant enters the region. From Eq. (3.5) we get:

$$NV'(N) = U_X\left(\frac{G-R}{N}\right) \tag{3.6}$$

where $R = f - Nf_N$

Therefore, in the equilibrium when the maximum utility of a region is attained, the following equation holds:

$$\frac{G_A - R_A}{N_A} = \frac{G_B - R_B}{N_B} \tag{3.7}$$

$G_i/N_i\,(i = \mathrm{A, B})$ in Eq. (3.7) is the per capita tax payment in region i. Each additional resident contributes to the financing of G_i thus reducing the tax bill of existing residents. On the other hand, $R_i/N_i\,(i = \mathrm{A, B})$ is the per capita share of the rent generated in region i. The more residents there are, the smaller the share of the rent going to each person. From the 'Golden Rule', a marginal residents reduces the public goods expenditure of existing residents by R_i/N_i

3.3.2 The Role of Intergovernmental Grants

There have been several commonly suggested rationales for intergovernmental grants. In general, they fall into two categories—economic justification and political or institutional justification. The achievement of a minimum level of each service throughout a nation is a goal which is based upon the political or institutional justification. One economic justification for grants relates to inter-jurisdictional responsibilities or spillovers. A second economic justification for grants regards fiscal imbalance. And the local public equilibrium which is attained by free migration may be Pareto-inefficient with respect to the level and choice of public goods. There is no reason that condition (3.7) is attained in the free migration equilibrium N^*. Assuming that the marginal migrant is indifferent between residing in A and B, the net benefits to the nation of moving an individual from region A to region B is:

$$\{G_A/N_A - G_B/N_B\} - \{R_A/N_A - R_B/N_B\} \tag{3.8}$$

The first bracketed term is the fiscal externality effect, while R_A/N_A and R_B/N_B are per capita land rents in the two regions. From Eq. (3.8), the fiscal externality effect will be zero when head taxes are equalized across regions. However, the equalization of head taxes across regions does not guarantee efficiency, as differences in per capita land rents will influence the distribution of the population and cause inefficiencies. In this model, region A with relatively high resource rents will tend to be overpopulated because of 'rent-seeking' behaviour on the part of migrants (Mieszkowski and Zodraw 1989). Resource-rich regions can impose taxes on the rents accruing to their natural resources and thus achieve relatively high levels of services or low levels of taxes on mobile factors such as labour and capital. In this situation, capitalization occurs and reflects the change in land rents that occur as a result of capital migration induced by tax differentials relative to the tax in region B. Hence, free migration will generally lead to an inefficient allocation of population over the nation. The appropriate grants would ensure Eq. (3.8) equals zero at the optimum (see Boadway and Flatters 1982; Hartwick 1980). Therefore, although it is difficult to determine the timing of intervention and the amount of grants, a system of intergovernmental grants from central government is introduced to eliminate inefficiency.

Figure 3.3 illustrates the operation of a system of intergovernmental grants in the case of Fig. 3.1. N^{e*} in Fig. 3.3 shows the equilibrium after the optimal transfer from region A to region B has been implemented.

It is possible that in the case of Fig. 3.2 the complete depopulation of a region can occur. However, this is rarely observed, presumably because there are migration costs which consist of not only monetary costs but also psychological costs through a person's attachment to their home region (see Boadway and Flatters 1982). In Fig. 3.2, N^m is the after-migration equilibrium in the case. Figure 3.4

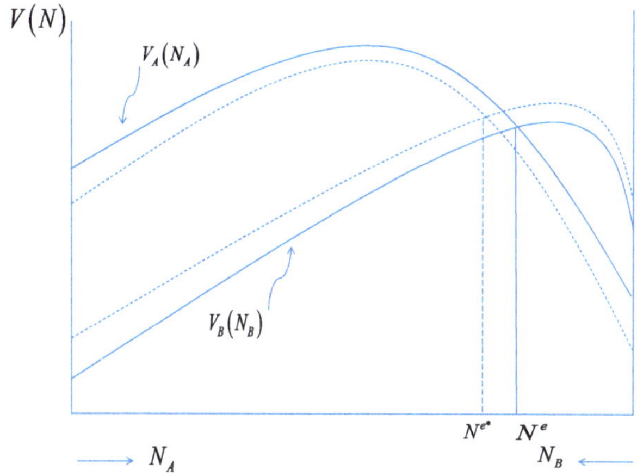

Fig. 3.3 The operation of a system of intergovernmental grants in the case of Fig. 3.1

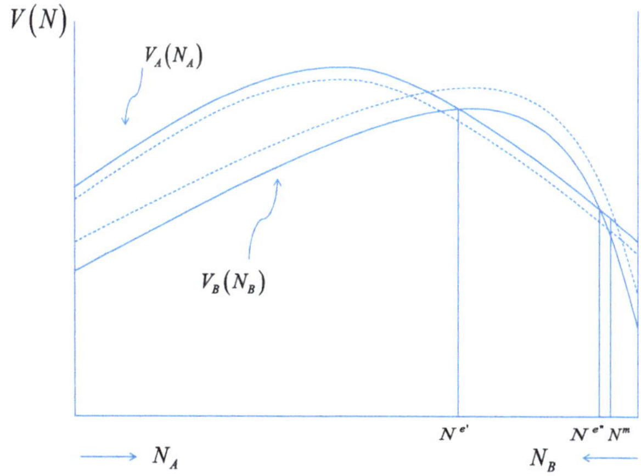

Fig. 3.4 The operation of a system of intergovernmental grants in the case of Fig. 3.2

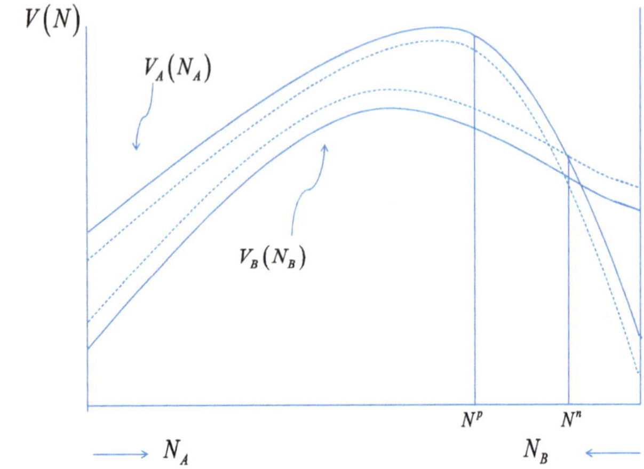

Fig. 3.5 The stable case

illustrates the operation of a system of intergovernmental grants in the case of Fig. 3.2. The dashed line shows the situation after allocation of intergovernmental grants. If the initial allocation of population is N^m, then there is a possibility that people move from B to A, but no possibility that people move from A to B.

Figure 3.5 illustrates the other stable case. From the above consideration, the condition for stability of the internal equilibrium can be written:

$$\frac{dV_A(N_A)}{dN_A} + \frac{dV_B(N_B)}{dN_B} < 0 \tag{3.9}$$

This condition occurs in Figs. 3.1 and 3.3, but not in Figs. 3.2 and 3.4. Although $dV_B(N_B)/dN_B > 0$ in the Fig. 3.5, since $|dV_A(N_A)/dN_A| > |dV_B(N_B)/dN_B|$, condition (3.9) occurs in this case. In other words, the utility level in the poorly endowed region increases slowly with population increases while it decreases rapidly with the increase of population in the richly endowed region.

3.4 A Case Study

3.4.1 Structure of Revenues and Expenditures

In order to confirm the role of intergovernmental grants, the model is applied to a village named Kuni-mura in the Agatsuma district of Gunma prefecture. The village is designated as a depopulated area by every depopulation law. In 1965, the population of the village was 3091 and the population density was 15 persons/km^2. In 1982, the population was 2354.

As previously noted in most local governments, own revenues are not enough for them to satisfy the expenditure requirements of their residents. The expenditure is

Table 3.4 Per capita revenue in 1982: comparison between Kuni-mura and other cities, towns and villages

	Local tax	Chihokofuzei	Grants-in-aid from central government	Grants-in-aid from prefecture	Local debt	Total revenue (yen)
Kuni-mura	49,181 (7.0)	288,361 (41.1)	147,144 (21.0)	41,086 (5.9)	102,039 (9.5)	700,744 (100.0)
Average value in other towns	82,500 (37.1)	32,900 (14.8)	33,100 (14.9)	11,200 (5.0)	20,700 (9.3)	222,700 (100.0)

Note 1: Values in parentheses represent percentage proportion to total revenue

restricted tightly by the revenue. Firstly, settled accounts of revenue and expenditure from 1965 to 1984 F.Y. are investigated (see Hagihara 1985a; Hagihara and Hagihara 1991). From the investigation, the proportion of local revenue decreased greatly during 1965–1970, but since 1970 it has remained on the same, though low, level. The percentage of intergovernmental transfers including chiho-kofu zei such as rate support in Great Britain and revenue sharing in the United States of America and Canada, central and prefectural governments' grants, and local debt permitted administratively by the central government, has become very large since the laws of the depopulated areas were enforced in 1970.

Table 3.4 shows a comparison between Kuni-mura and other cities, towns and villages in each component of per capita revenue. Central government stipulates a standard level of each public service throughout a nation. Intergovernmental transfers are needed to achieve the standard level of public services. Even though there is a small population in Kuni-mura, the village must provide several public services such as police and educational services, one teacher is needed to teach two pupils as well as to teach forty pupils. Consequently a large per capita intergovernmental transfer is allocated to Kuni-mura to achieve the standard level of each public service. Per capita revenue in Kuni-mura is ranked very high, for most of revenue in the village consists of intergovernmental transfers (see Table 3.4).

From percentage annual changes in expenditure, education, construction and promotion for agriculture account for a very large percentage of total expenditure. Moreover, total and per capita cumulative expenditure on construction has increased since the laws of the depopulated areas have been enforced (see Hagihara 1985a).

3.4.2 Utility Level

In 1965, the population of Kuni-mura was 3091. Over twenty years from 1965 approximately 700 people migrated from the village. Most of these people consist of the younger generation. So the index for percentage of old age, i.e. over-65

Table 3.5 The level of utility for 1982

	(a)	(b)	(c)	(d)	(e)	(f)
Kuni-mura	1.4	74.0	8.8	8.0	43.3	47.8
Average value in Gunma prefecture	1.7	96.5	30.3	23.1	78.5	86.5
Average value in Japan	1.9	92.5	46.8	31.4	90.0	88.0

Notes: (a), per capita income, million yen; (b), the ratio of the population who are served by the municipal water services to total population; (c), the ratio of the distances of paved roads to total roads; (d), the ratio of the distances of improved roads to total roads; (e), the ratio of the population who are served by municipal waste disposal services to total population; and (f), the ratio of floor area which is built of reinforced concrete to total floor area of primary school buildings

population/under-14 population × 100, was 70.37 in 1982, while average values in Gunma prefecture and Japan were 52.09 and 38.71, respectively. The number of school-aged children per primary and junior high school in the village was 61 in 1982, while the average value in Gunma prefecture was 472. Indices of gross product in industry and agriculture in 1965 and 1970 were 161 and 214 respectively, while average indices in Gunma prefecture were approximately 200 and 345 respectively. Per capita annual income has changed from about 150,000 yen in 1965 to about 1.4 million yen in 1982, while average per capita annual income in Gunma prefecture changed from about 170,000 yen to about 1.7 million yen.

The typical components of residents' utility are: (1) per capita income; (2) the ratio of the population who are served by the municipal water services to total population; (3) the ratio of the distance of paved roads to total roads; (4) the ratio of the distances of improved roads to total roads; (5) the ratio of the population who are served by municipal waste disposal services to total population, and (6) the ratio of floor area which is built of reinforced concrete to total floor area of primary school buildings. Among the above indicators, all except (1) correspond to effects of counter measures such as road improvement, improvement of educational services, and construction of public facilities. The value of these components are shown in Table 3.5.

3.4.3 Utility Level and Population

The relation between per capita expenditure and population for 70 cities and towns in Gunma prefecture is investigated (see Hagihara and Hagihara 1991). The result of the investigation shows that the per capita expenditure of towns with small populations are large and the larger the population becomes, the smaller the per capita expenditure becomes. This implies that the supply cost of public services is high if the population is small. However, if the population is very large, congestion in public good use occurs. As a result of congestion, the supply cost of public services becomes high. Therefore, it is reasonable to assume that the graph of $V_i(N_i)$, $(i = \mathrm{A}, \mathrm{B})$ is single-peaked as in Figs. 3.1, 3.2, 3.3, 3.4, and 3.5. From Table 3.5, the utility level in the villages is lower than the average level of other

Table 3.6 Change in per capita expenditure and change in population (000 s yen)

	N town		T town		Kuni-mura	
	Population	Per capita expenditure	Population	Per capita expenditure	Population	Per capita expenditure
1960	22,853	–	15,214	–	3530	–
1965	21,591	24,950	13,775	36,320	3091	43,368
1970	21,155	42,268	12,314	61,278	2610	118,984
1975	20,608	93,137	11,227	176,211	2450	258,048
1980	20,456	137,929	11,065	263,220	2386	378,229
1982	20,319	189,051	11,114	281,975	2354	502,608

regions. Hence, the village in question can be regarded as region B. Since the population in region B has decreased it seems that the present state is not N^e, $N^{e'}$, and $N^{e''}$, but, for example, N^l, or N^m, N^n. The relation between the change in per capita expenditure and the change in population is shown in Table 3.6.

Both N and T towns are located in the same district as Kuni-mura. As mentioned above, the standard level of public services is guided by the central government. Three towns in Table 3.6 try to provide approximately the same per capita level of public services. It is implied from Table 3.6 that the supply cost of providing public services has increased as the population of the village has decreased. It appears, therefore, that the present state is not N^l but N^m or N^n. If the present state is N^m, intergovernmental grants will not play a role to make V_A equal to V_B, although the difference between the two regions becomes smaller. On the other hand, if the present state is N^n, intergovernmental grants will work efficiently to make V_A equal to V_B.

Since the population in region B is still decreasing, it is hard to say which is closer to the present situation. Suppose, in Fig. 3.5, N^P is the initial allocation of population before the law of depopulated areas was implemented. While the population of region B is decreasing now, the effects of intergovernmental grants will appear soon and then it is expected that the situation will move towards a new equilibrium. On the other hand, in the case of Fig. 3.4, if N^m is the initial allocation of population before the law, region B itself must try to raise utility levels with its own efforts as well as being allocated appropriate intergovernmental grants. For example, town T was designated as a depopulated area by the laws of depopulated areas in 1970. Since town T is located at the foot of Mount Asama, a famous volcano in Japan, the town has developed tourism for ten years under the old laws in 1970. As a result, town T was not designated as depopulated area by the new laws in 1980. Therefore, Kuni-mura itself should have promoted new industries which have advantages compared to other areas (Hagihara 1985b). In this situation, $V_B(N_B)$ may shift upwards without intergovernmental grants. However, Town T was designated as depopulated area again by the laws in 2010.

Table 3.7 Amounts of investment in projects addressing depopulation problems (Unit: a hundred million; %)

		(b)									
	(a)		(b1)	(c)	(d)	(e)	(f)	(g)	(h)	(i)	Total
(1)	17,524 (22.2)	39,197 (49.6)	156 (0.2)	8945 (11.3)		953 (1.2)	9470 (12.0)		190 (0.2)	2739 (3.5)	79,018 (100.0)
(2)	48,257 (27.8)	85,942 (49.5)	769 (0.4)	17,983 (10.4)		2457 (1.4)	17,085 (9.8)		412 (0.2)	1534 (0.9)	173,669 (100.0)
(3)	106,604 (29.3)	142,673 (39.3)	2477 (0.7)	64,057 (17.6)	11,308 (3.1)	6211 (1.7)	24,864 (6.8)		1186 (0.3)	6384 (1.8)	363,286 (100.0)
(4)	69,629 (28.4)	91,919 (37.5)	5046 (2.3)	49,657 (20.3)	9521 (3.9)	5330 (2.2)	13,298 (5.4)	2142 (0.9)	1003 (0.4)	2629 (1.1)	245,128 (100.0)
(5)	19,584 (29.8)	20,352 (31.0)	1956 (3.0)	10,350 (15.7)	4736 (7.2)	3116 (4.7)	5791 (8.8)	688 (1.0)	414 (0.6)	709 (1.1)	65,737 (100.0)
Total	261,598 (28.2)	380,083 (41.0)	10,403 (1.1)	176,557 (19.0)		18,067 (1.9)	73,338 (7.9)		3205 (0.3)	13,995 (1.5)	926,839 (100.0)

Notes:
1 Source: Ministry of Internal Affairs and Communications (2014)
2 Values in parentheses indicate percentage
(1) The emergency measures for the development of depopulated areas law (1970–1979);
(2) The special measures for the regional development of depopulated areas law (1980–1989);
(3) The special measures for the regional vitalization of depopulated areas law (1990–1999); (4) The special measures for the self-support promotion of depopulated areas law (2000–2009); (5) The special measures for the self-support promotion of depopulated areas law (2010, 2011, 2013)
(a) Industrial development; (b) Improvement of communication and transportation system; (b1) Improvement of information and communication; (c) Provision of living environment; (d) Improvement of welfare for aged people; (e) Provision of medical services; (f) Improvement of educational services; (g) Promotion of local culture; (h) Development of towns and villages; (i) Others

3.5 Effects of the Countermeasures

According to the above-mentioned grounds for intergovernmental grants, the total amount of investment in depopulated areas under four laws since 1970 was approximately 93 trillion yen. Among the various areas of investment from providing better living environments to promoting of industry, investments in traffic-communication system account for approximately 50 % of total investments until 1989. The amounts of investments in communication and information projects, improving living conditions, and medical services have increased since 1990. Further, the amounts of investment in improving the living conditions and welfare for the elderly has increased under the current law (See Table 3.7).

With the aid of these investments in local roads, the ratio of the distances of improved roads to the total distance of local roads in depopulated areas was 9.0 % in 1970, increased to 22.7 % in 1980, and further increased to 51.1 % in 2005. Likewise, the ratio of the distances of paved roads to the total distance of local roads in depopulated areas was 2.7 % in 1970, increased to 30.6 % in 1980, increased further to 68.3 % in 2005. Since the ratio of the distances of improved roads and paved roads to the total distance of local roads in Japan are 55.0 % and 75.9 %, respectively, for 2005, the roads in depopulated areas have been considerably improved.

As for the level of medical services, the number of depopulated areas with no local doctors has decreased from 1168 in 1978 to 565 in 2009, whereas the number of non-depopulated areas with no local doctors has decreased from 582 to 140 respectively. Thus, there is still a difference between depopulated areas and non-depopulated areas.

3.6 Concluding Remarks

Total population in Japan has already begun to decrease. Both urban and rural areas share this problem, particularly with an increasing population over 65 years of age and a decreasing population under 15 years of age. Since the Lehman Crisis, Japan's fiscal balance has been in a significant deficit. With an aging population, the difference between social security expenses and revenues from social insurance premium begins to grow. Actual pension reserves have been below forecasts since FY 2010.

The greatest factor pushing up the outstanding debt of central and local governments is the primary fiscal balance. Between 2007 and 2011, Japan's outstanding debt increased almost at the same pace as the debt of the European periphery countries. Japan did not change its debt burden, despite increasing social security expenses, and Japan must not change it.

On the other hand, it is necessary to build a less imbalanced and stable local tax system in order to allow local governments to local needs by providing the services they are responsible for at their own discretion. Comparing local tax revenue amounts, Tokyo, the highest, was approximately 2.5 times larger than the smallest amount of local tax revenue, which is collected by Okinawa prefecture. The central government collects revenue resources that would essentially be attributable to local tax revenue and reallocates them as local allocation tax to local governments that have weaker financial capabilities.

The central government's tax revenue has declined extremely since the collapse of the so-called 'economic bubble'. Nevertheless, the central government must take some measures to promote revitalization of rural areas, as well as some urban areas. Further, the central government has promoted the merger of cities, towns and villages since 1999 as a part of decentralization, and the designated depopulated towns or villages have been included in non-depopulated cities.

The depopulated areas and non-depopulated areas were clearly distinguished before 'the big merger of Heisei', but the formerly designated depopulated areas become now one part of a merged city or town after the big merger. For example, Kuni-mura, town mentioned in Sect. 3.4, was included in Nakanojo-machi located next to Kuni-mura because of 'the big merger of Heisei'. Nakanojo-machi is designated as depopulated area and some measures have been executed by both central government and Gunma prefectural government. The case of Kuni-mura is better than other towns such as Keihoku-cho, which has been included in the city of Kyoto due to the merger of Heisei. Since Kyoto city is not designated as depopulated and does not receive government grants for depopulation projects, Keihoku-cho is not given adequate measures for revitalization by the central government. There are many towns in similar situations to Keihoku-cho.

In addition to this situation, there are some areas of non-depopulated cities that meet the legal definition of depopulated areas by population, like Kumogahata area of Kyoto city. In this case, the same ideas presented in this chapter are needed to keep the area sustainable and survivable. We will discuss this case in Chap. 16.

Acknowledgements We gratefully acknowledge the helpful comments on earlier drafts by S. Mahay, W. Jarvie, M. M. Hufschmidt, and F. H. Bollman. A part of this chapter has originally published in *Regional Studies*, 25(2), pp. 163–172. We would like to thank the publishers for their permission to use the material here.

References

Atkinson, A. B., & Stiglitz, J. E. (1980). *Lectures on public economics*. Maidenhead, Berks: McGraw-Hill.

Boadway, R., & Flatters, F. (1982). Efficiency and equalization in a federal system of government: A synthesis and extension of recent results. *Canadian Journal of Economics, 15*, 613–33.

Buchanan, J. M., & Goetz, C. J. (1972). Efficiency limits of fiscal mobility: An assessment of the Tiebout model. *Journal of Public Economics, 1*, 25–43.

Buchanan, J. M., & Wagner, R. W. (1970). Efficiency basis for federal fiscal equalization. In J. M. Margolis (Ed.), *The analysis of public output* (pp. 139–162). New York: Columbia University Press.

Flatters, F. R., Henderson, J. V., & Mieszkowski, P. M. (1974). Public goods, efficiency and regional fiscal equalization. *Journal of Public Economics, 3*, 99–112.
Hagihara, K. (1985a). Kaso mondai no kezaigakuteki kousatu (An economic study of the depopulation problem). *Studies in Regional Science, 15*, 185–211.
Hagihara, K. (1985b). The role of intergovernmental grants for environmental problems. *Environment and Planning C: Government and Policy, 3*, 439–50.
Hagihara, Y., & Hagihara, K. (1981). Project grant allocation process applied in sewerage planning. *Water Resources Research, 17*, 449–54.
Hagihara, K., & Hagihara, Y. (1986). Pricing policies for conservation of water resources and environment. *Environment and Planning C: Government and Policy, 4*, 19–29.
Hagihara, K., & Hagihara, Y. (1991). The role of intergovernmental grants in underpopulated regions. *Regional Studies, 25*(2), 163–172.
Hartwick, J. M. (1980). The Henry George rule, optimal population and interregional equity. *Journal of Public Economics, 13*, 695–700.
Mieszkowski, P., & Zodraw, G. R. (1989). Taxation and the Tiebout model: The differential effects of head taxes, taxes on land rents and property taxes. *Journal of Economic Literature, 27*, 1,098–146.
Ministry of Internal Affairs and Communication. (2014). Kaso taisaku no genkyou (The present state of measures for depopulation problem). http://www.soumu.go.jp/maincontent/000276127.pdf. Accessed 7 July 2014.
Stiglitz, J. E. (1977). The theory of local public goods. In M. S. Feldstein & R. P. Inman (Eds.), *The economics of public services* (pp. 274–333). London/New York: Macmillan/Wiley and Halsted Press.
Stiglitz, J. E. (1983a). Public goods in open economies with heterogeneous individuals. In J.-F. Thisse & H. G. Zoller (Eds.), *Locational analysis of public facilities* (pp. 55–78). Amsterdam: North-Holland.
Stiglitz, J. E. (1983b). The theory of local public goods twenty-five years after Tiebout: A perspective. In G. R. Zodraw (Ed.), *Local provision of public services: The Tiebout model after twenty-five years* (pp. 17–53). New York: Academic.
Tiebout, C. M. (1956). A pure theory of local expenditure. *Journal of Political Economy, 64*, 416–424.

Chapter 4
Regional Vulnerability of the Tokyo Metropolitan Area to Flood and Earthquake Disasters

Sotaro Tsuboi and Chisato Asahi

Abstract Until the 1970s, the majority of flood disasters occurred in the lowlands to the east of Tokyo. In recent years, however, with the increasing frequency of localized torrential downpours – referred to as 'guerrilla rainstorms' – locally concentrated, devastating damages have been suffered. With respect to measures against such damages, underground reservoirs and rivers are being constructed. As a soft measure, flooding hazard maps are being made public. Because of its densely concentrated urban structure, the Tokyo Metropolitan Area is also at high risk of fires and building collapses during earthquakes. Furthermore, a large-scale earthquake is expected to occur there in the near future. As mitigation measures, earthquake-proof reinforcement is being subsidized and disaster prevention activities and education are being conducted by community organizations. Although a variety of disaster prevention measures against floods and earthquakes are being conducted in the Tokyo Metropolitan Area, lowland areas could still suffer serious damage due to their densely concentrated urban structure. Since disasters affect society as a whole, vulnerability can be defined not only by such urban structure and mitigation measures, but also by the pre-disaster welfare state of individual households, capacity of administrative bodies and cooperation within them. Thus, we should clarify the respective roles of affected individuals, areas, and administrative bodies so that they complement each other. However, even administrative responses to disasters (rescue and assistance) have limits. Therefore, individuals should be prepared to protect their own lives and assets (self-help) and respond cooperatively in collaboration with their neighbours.

Keywords Tokyo metropolitan area • Earthquake • Flood • Self-help • Resilience • Mitigation • Community • Hazard map

S. Tsuboi (✉)
Disaster Reduction and Human Renovation Institution, 1-5-2 Kaigan-dori, Wakinohama, Chuo-ku, Kobe 651-0073, Japan
e-mail: tsubois@dri.ne.jp

C. Asahi
Division of Urban Policy, Faculty of Urban Liberal Arts, Tokyo Metropolitan University, 1-1 Minami-osawa, Hachioji, Tokyo 192-0397, Japan
e-mail: asahi@tmu.ac.jp

K. Hagihara, C. Asahi (eds.), *Coping with Regional Vulnerability*, New Frontiers in Regional Science: Asian Perspectives 4, DOI 10.1007/978-4-431-55169-0_4

4.1 Introduction

Suffering from natural disasters refers to phenomena in which earthquakes and heavy rains occur in vulnerable areas and directly or indirectly impact buildings and people, thus making daily socioeconomic activities difficult. With respect to 'natural causes' of vulnerability to disasters, the topography of Tokyo metropolitan area (Fig. 4.1) comprised alluvial lowlands in the east and a hilly plateau in the west, make the eastern area particularly vulnerable to flooddisasters due to river inundations (Figs. 4.2 and 4.3). In addition, this area has complex underground infrastructure and is the hypocentre of large-scale earthquake disasters. In contrast, with respect to 'social causes', overconcentration of Tokyo's population accelerated when the Tokyo Olympic Games were held in 1964; urban areas that were developed without systematic urban planning still exist. Characterized by densely concentrated, old residential

Fig. 4.1 The study area (Tokyo metropolitan area)

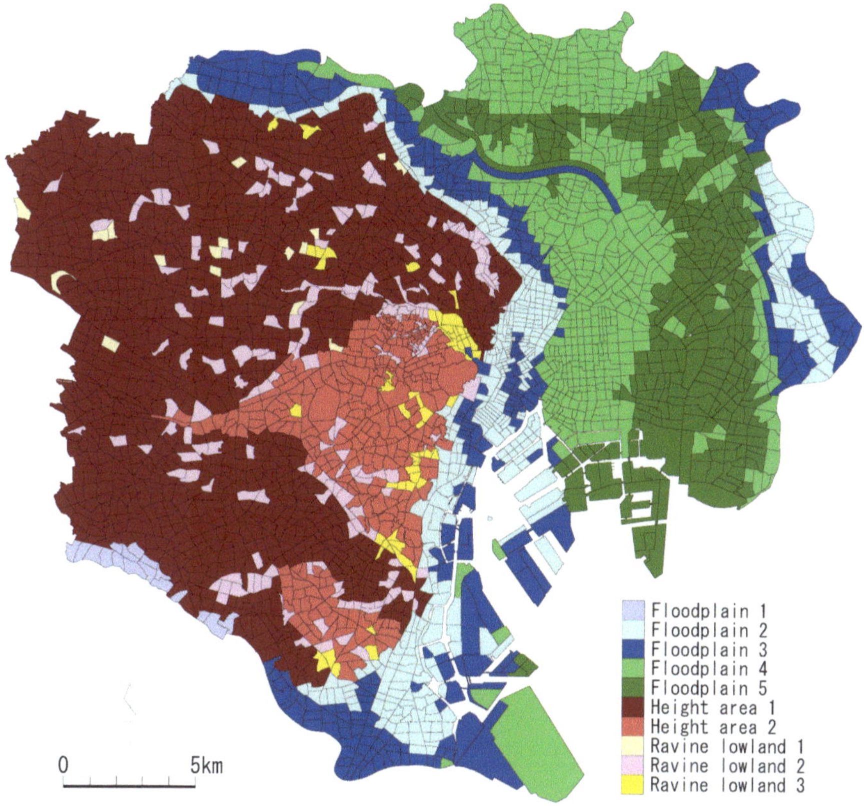

Fig. 4.2 The soil structure (Tokyo metropolitan area)

buildings, these areas have high risk of building collapse during fires and earthquakes. Since many cities in the world are located in river and coastal areas, they commonly have high risk of floods and earthquakes; prevention measures against these disasters are cornerstones of these cities' urban planning policies.

4.2 Risk of Flood Disasters in Tokyo

Although there has been an increase in flood control measures for rivers and waterfront areas in cities, catchments are also inherently at risk of being damaged by urban flooding that occurs annually in several regions nationwide. 'On-site' infrastructures, such as levees and underground reservoirs, have been continuously constructed by government agencies in order to prevent urban flooding. However, in recent years, such efforts have also been made at 'off-site' infrastructures; for example, making flood hazard maps available to the public.

Fig. 4.3 The rivers (Tokyo metropolitan area)

Generally, it is easy to underestimate the risk of flooddisaster caused by a typhoon or concentrated heavy rain because of the relatively large amount of real-time data provided by weather stations and satellites before the occurrence of a disaster, and because flood disasters differ from earthquake disasters in terms of predictability, spontaneity and regional impacts. However, the management of flood-related risks is an important challenge, given Japan's terrain and annual exposure to passing typhoons.

4.2.1 Characteristics of Flood Disasters in Tokyo

In Tokyo, the flood damage caused by Typhoon Ida in 1958 was characterized by the inland flooding of small to medium rivers in western Yamanote, where housing and land development was taking place. This differed from the disasters that occurred before, where water damage due to flooding was common in the eastern lowlands. This damage, initially called the 'Yamanote flood damage', gradually expanded in scale to the western suburbs.

The flood damage that occurred in Tokyo after torrential rain in 1963 and Typhoon Number 4 in 1966 can be classified into the following four categories based on their locations and characteristics: (1) flood damage in areas of Tokyo below sea level, (2) Yamanote flood damage, (3) flood damage exacerbated due to large scale development such as new towns and (4) flood damage from large river flooding and/or levee failure. However, beginning in the 1970s, flooding in Tokyo presented new challenges characterized by flood damage in small to medium-sized rivers in metropolitan areas. The years around the 2000s were further characterized by increasingly diversified instances of localized flood damage from inland waters, caused by increased occurrences of sudden heavy rain.

Tables 4.1 and 4.2 show the trends of wet surface areas and the total number of flooded houses in Tokyo metropolitan area from 1963, when statistical data collection of flood damage began, to 2000.

Although the tables show an overall decrease in the wetted surface area and the total number of flooded houses in recent years, the observations for each ward show that both indexes peaked in the mid-1980s for Adachi ward, indicating that the region is prone to flooding. The total flood area within Adachi, Katsushika and Edogawa wards, located on the Eastern low-lying left bank of the Ara River, exceeds 60–68.0 % from 1966 to 1970, 60.0 % from 1971 to 1975 and 65.0 % from 1981 to 1985. These records indicate the serious potential for disastrous flooding and related damage in these areas.

The Kanda River, the Yata River, and a portion of the Shibuya River in central Tokyo and its western suburbs have traditionally been subject to regular flooding. However, since the 1970s, flooding has expanded in small to medium rivers in the central wards, creating new flooding challenges for Tokyo.

The Shinagawa and Ota wards are located in the low-lying basins of the middle and lower reaches of the Meguro, Tachiai, and Nomi Rivers. Since the 1980s, flood damage in these areas has expanded in the Shinagawa and Ohta wards, indicating that flooding and related damages have been moving further downstream. Increasing urbanization and the continued maintenance of drainage facilities in the middle reaches of these rivers has resulted in the increased flow incurred downstream and the inundation of levies due to reverse flow of sewage and poor drainage. By classifying flood damage in Tokyo, which occurred after the latter half of the 1970s by the degree of regional development and the amount of inundation, it was possible to characterize flood damage into the following four categories: (1) damage with a large amount of inundation in regions undergoing urban development, (2) damage with a small amount of inundation in regions undergoing urban development, (3) damage with a large amount of inundation in currently developed areas and (4) damage with small amounts of inundation in already developed areas.

The first category primarily includes areas such as the new residential development in low-lying areas of the small to medium rivers in the western suburbs of Tokyo and in inland basin areas on the plateau. The second category relates

Table 4.1 The trends of wet surface areas in Tokyo metropolitan area (ha)

	1963–1965	1966–1970	1971–1975	1976–1980	1981–1985	1986–1990	1991–1995	1996–2000	Total (ha)
Chiyoda	0.0	0.0	0.0	1.3	74.2	18.6	34.9	2.1	131.1
Chuo	0.0	5.9	0.0	0.0	0.0	0.0	0.0	0.4	6.3
Minato	0.0	1.0	1.0	0.2	7.0	0.5	25.1	19.0	53.7
Shinjyuku	0.0	81.2	38.4	45.7	293.0	35.0	32.9	7.8	534.0
Bunkyo	4.0	18.8	72.7	48.2	362.6	17.9	31.4	0.5	556.1
Taito	2.0	1.0	4.0	0.8	143.2	16.4	55.6	0.8	223.8
Sumida	0.0	52.8	0.0	0.0	33.2	0.0	0.0	0.0	86.0
Kouto	632.0	8.0	1.2	0.1	28.2	0.2	1.3	9.8	680.8
Shinagawa	6.0	24.4	73.2	0.8	93.6	36.9	50.8	112.2	397.9
Meguro	0.0	2.6	11.3	11.8	147.3	16.2	2.4	6.9	198.5
Ohta	1851.0	437.6	311.3	16.1	193.9	32.3	2.5	16.6	2861.2
Setagaya	409.0	136.4	121.8	38.0	82.5	7.9	3.0	3.2	801.8
Shibuya	0.0	6.0	0.0	0.1	18.8	31.5	0.0	1.0	57.4
Nanano	72.0	55.0	1.9	39.6	133.6	20.2	49.2	9.5	381.0
Suginami	1030.0	186.2	73.2	77.6	97.5	16.9	16.2	4.6	1502.2
Toshima	92.0	0.0	20.0	35.4	57.3	40.7	3.5	3.1	251.9
Kita	326.0	120.0	5.5	18.7	5.5	2.1	8.3	4.1	490.2
Arakawa	136.0	29.0	299.1	0.0	102.1	1.3	9.3	4.3	581.0
Itabashi	213.0	107.9	263.0	100.9	116.1	2.9	1.6	2.5	808.0
Nerima	386.0	274.0	79.4	66.7	35.5	10.1	14.8	0.7	867.3
Adachi	611.0	1790.3	1485.3	289.3	2798.1	30.3	234.2	0.5	7239.0
Katsushika	669.0	289.9	144.2	6.0	554.6	3.4	17.1	0.2	1684.4
Edogawa	1635.0	1207.4	437.5	1.2	412.3	5.6	0.0	0.0	3699.0
Total	8074.0	4835.4	3444.0	798.5	5790.1	346.8	593.8	209.9	24,092.4

Table 4.2 The total number of flooded houses in Tokyo metropolitan area (houses)

	1963–1965	1966–1970	1971–1975	1976–1980	1981–1985	1986–1990	1991–1995	1996–2000	Total
Chiyoda	0	0	0	35	659	224	251	34	1203
Chuo	0	453	0	0	0	0	0	15	468
Minato	0	54	35	25	71	25	155	819	1184
Shinjyuku	0	2003	1453	2073	5701	721	1137	99	13,187
Bunkyo	707	1619	1113	426	10,630	1054	2017	64	17,630
Taito	181	88	144	0	1450	54	1080	114	3111
Sumida	0	2088	0	0	897	0	0	0	2985
Kouto	26,504	210	28	1	1298	24	132	174	28,371
Shinagawa	305	693	1482	41	1295	1323	161	2919	8219
Meguro	363	415	106	278	3819	402	22	345	5750
Ohta	24,582	10,536	4544	672	6490	2525	175	1131	50,655
Setagaya	2132	2815	1504	611	2783	619	177	122	10,763
Shibuya	0	160	0	0	1021	2166	0	103	3450
Nanano	3559	765	120	733	2457	1590	2626	345	12,195
Suginami	21,262	1690	2010	698	1351	653	533	230	28,427
Toshima	3744	0	320	1250	1664	1562	162	442	9144
Kita	16,694	1272	833	358	186	148	343	121	19,955
Arakawa	3125	1158	284	0	4331	100	65	109	9172
Itabashi	5971	3959	4971	3915	1624	157	169	198	20,964
Nerima	3397	4149	1967	1098	1341	816	205	456	13,429
Adachi	27,635	45,653	17,679	6056	27,363	5154	3280	14	132,834
Katsushika	8725	11,895	5715	199	17,979	330	404	13	45,260
Edogawa	15,946	3684	5194	45	12,557	876	0	2	38,304
Total	164,832	95,359	49,502	18,514	106,967	20,523	13,094	7869	476,660

primarily to flood damage in low-lying regions to the east of Tokyo in the Ayase, Ara, Naka, and Edo river basins and the damage from relatively shallow but broad-reaching flooding in areas of flat terrain. The third category relates to flood damage with the highest frequency of incidence, mainly in the western suburbs during the latter half of the 1970s. This is a typical form of the so-called 'urban-type water damage'. In other words, it follows a trend in which there is very little time to warn of any impending flooding and in which the same areas experience repeated damage from flooding. Together with inland flooding, this indicates a high ratio of flooding above floor level for houses subject to flooding. The fourth category relates to inland flood damage in the lower reaches of the Nomi and Meguro rivers of the Jonan district, which are below sea level. Although pumping has been used to drain these areas, the urbanization of the watershed and the accompanying expansion of impervious areas has rapidly increased the runoff coefficient and resulted in cases where it exceeds the capacity of drainage facilities.

The second and fourth categories include the Adachi, Katsushika and Edogawa wards in the eastern lowland belt as well as the Ota ward, positioned in the lower

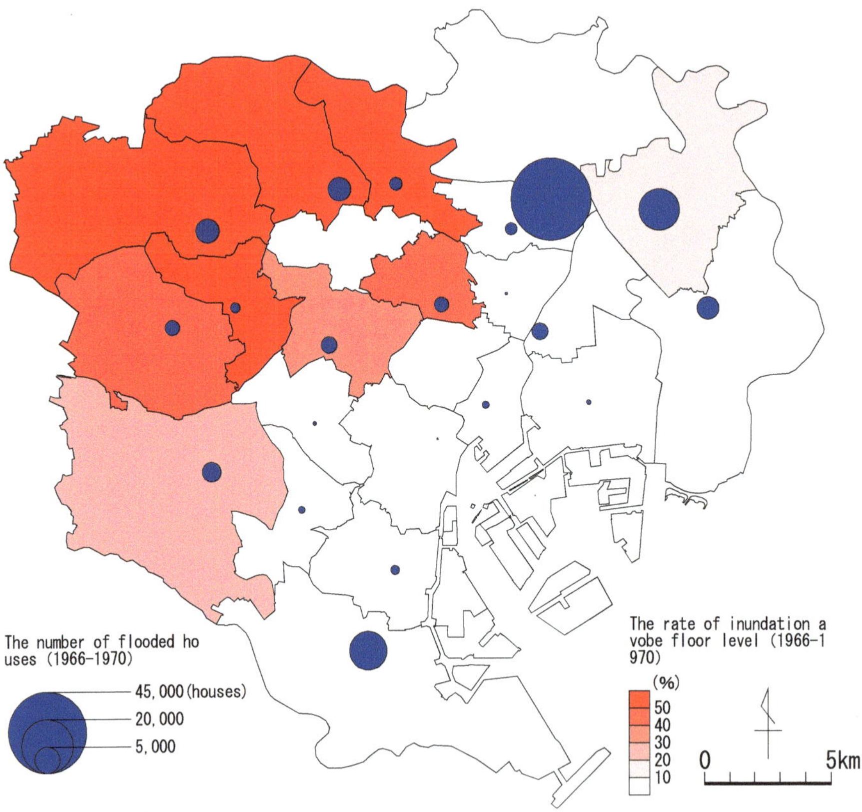

Fig. 4.4 Total number of flooded houses and the rates of inundation above floor level (1966–1970)

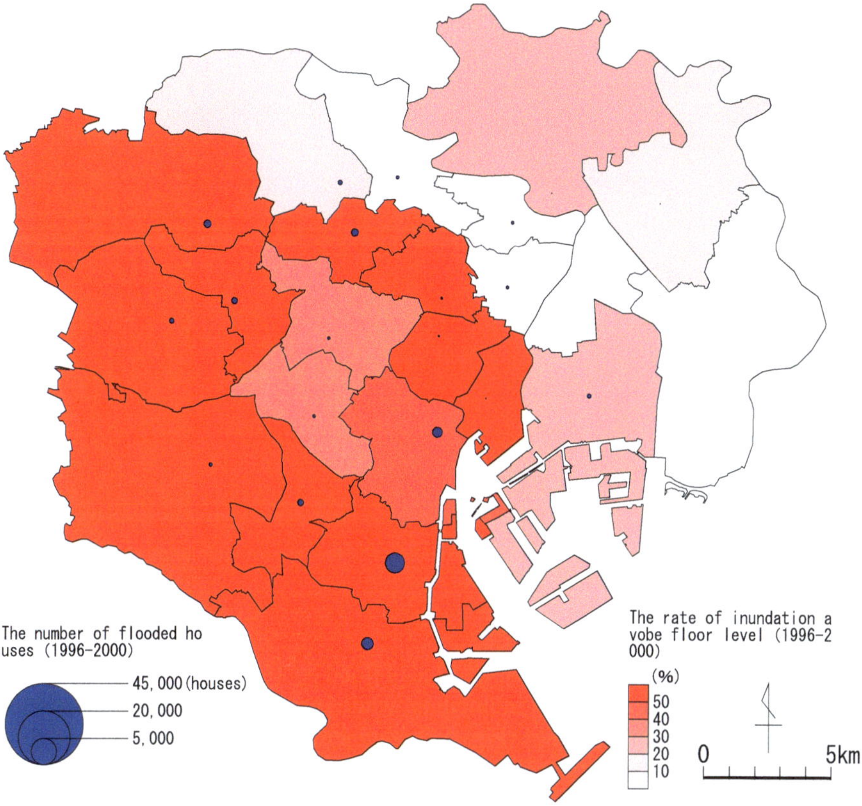

Fig. 4.5 Total numbers of flooded houses and the rates of inundation above floor level (1996–2000)

reaches of the Nomi River. These areas are characterized by extensive flooding with the ‘pooling’ of floodwaters, rarely above the first floor of buildings. With regard to the third category, the rapid influx of rainwater into rivers, which result from increased impervious areas, has led to the overflowing of the Kanda and Old Tanida rivers, resulting in damage in the central wards such as Shinjuku and Bunkyo. In addition, as time passes, these areas are showing increasing rates of inundation above floor level (Figs. 4.4 and 4.5).

On the other hand, in recent years, there has also been a typical form of urban flooding damage characterized by small affected areas but a high incidence of inundation above floor level. This applies to the western wards of Nakano, Shinagawa, Toshima, Meguro, Suginami, and Itabashi. In addition, while there are few case studies of the first category of flood damage in these wards, urban flooding damage can also be seen in the western cities of Hino, Kokubunji, and Machida. These wards are characterized by small areas of flooding and some instances of flooding above the first floor of buildings. Although it can be said that flooding in Tokyo’s western suburbs since the 1970s had these characteristics,

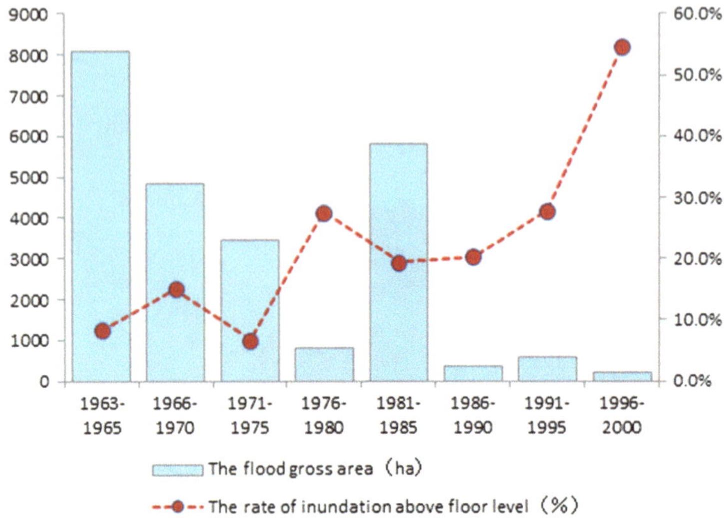

Fig. 4.6 Change in the total number of flooded house and the rates of inundation above floor level for the Tokyo metropolitan area (1963–2000)

'on-site' measures such as the maintenance of flood control facilities in recent years have decreased the area subject to flooding, while increasing the rate of inundation above the first floor of buildings (Fig. 4.6). An increase in this trend can be seen in the expanded flood damage from Jonan to areas of central Tokyo. In other words, there is a visible trend towards locally devastating damage due to flooding, which indicates heightened danger for underground areas and subways.

4.2.2 Countermeasures for Flood Disaster in Tokyo

Flood control measures in Tokyo have been making progress in conjunction with the Rivers Division of the Department of Construction and the City Planning Bureau. Both 'on-site' and 'off-site' measures have been utilized in a way that is consistent with the concept of comprehensive flood control measures. On-site measures are comprised of two components: (1) maintenance of flood control facilities and (2) maintenance of rainwater runoff control facilities. The first component primarily includes rehabilitation projects for rivers, along with projects to reduce sewer discharge during flooding events. The second includes storage and infiltration. Storage refers to the facilities for storage of floodwaters, such as disaster prevention ponds, storage dikes, storage tanks and storage at individual houses. Infiltration refers to permeating basins, permeable trenches, permeating basins of road, permeable pavement and U-shaped permeable grooves. Furthermore, current plans propose to gradually and systematically improve flood control standards. Storm water storage facilities installed for basin management in the provisional plan (50 mm plan) will become permanent storm water storage facilities when the default plan (50 mm plan) is achieved. Meanwhile, storm water facilities installed as basin management measures in the long-term plan (75 mm plan) will be used as permanent rainfall storage facilities. In the basic plan (100 mm), which is the eventual goal, 10 mm portions of the 100 mm that can be withstood are planned for the watershed countermeasures (e.g., storm water-runoff control facilities).

In terms of off-site countermeasures, integrated flood control measures began to be implemented in 1977. Basins for which the installation of storm water runoff control facilities was proposed were divided into three classifications based on the type of measure: (1) water retention areas, (2) retarding areas and (3) low-lying areas. In water retention areas, water retention functions were preserved and strengthened by installing runoff control facilities and through the preservation of urban growth control areas and natural land. For retarding areas, storm water retardation is preserved by the maintenance of urban growth control areas and the improvement of embankment control and agricultural environments. For low-lying areas, water resistant buildings are promoted along with the maintenance of inland water removal facilities, the installation of various types of storage facilities, and soft story buildings. Moreover, when compared with large rivers, the small to medium sized rivers have short flood arrival times because of the decreased rainwater flooding area, due to the original structures and residential developments. There is also little time for flood warnings; once a flood occurs, the damage tends to become widespread in a short period of time. Therefore, it is critical that the transmission of information be improved and residents take voluntary action. However, flood awareness and knowledge based on experience is difficult to pass on given the typical patterns of urban residence and lifestyle and the combination of new and old residents. An underdrainage system stretches around much of the urban centre, and the residents, especially new ones, have no awareness of the drainage and the risk of flooding. As the drainage system has been highly effective for flood protection and environmentalhealth, urban residents have come to pay less attention to water after use or non-use like rainwater. In actuality, the underdrainage system is almost exceeding its limit in capacity, as mentioned above in Sect. 4.2. That is one of the urban-specific vulnerabilities even with a highly constructed drainage system.

In a situation like this, comprehensive flood control measures often take the form of gathering information on rainfall and water levels and establishing provisional and predictive information systems and a warning/flood prevention framework by providing figures showing previous flooding and releasing flood prediction figures and hazard maps.

As a flood damage countermeasure, Tokyo inaugurated the Tokyo Urban Flood Damage Countermeasures Committee in response to flood damage sustained as a result of torrential rain in September 2000. In addition to re-investigating flood prevention facilities, flood hazard maps have been prepared and are currently available in all wards.

4.3 Risks of Earthquake Disaster in Tokyo

Japan is known as one of the world's most earthquake-prone countries. It is predicted that, within the next 30 years, there is a 70 % chance of a magnitude 7 earthquake hitting the Southern Kanto region, which includes Tokyo. Since the 1960s, rapid urban development has proceeded apace in Tokyo, and there now exist several densely built-up areas. Thus, there are concerns about the danger that shaking from an earthquake could collapse buildings, trigger fires, and cause enormous damage.

4.3.1 Degree of Regional Risk in Tokyo

Adoption of the Tokyo metropolitan earthquake disaster prevention bylaws has led to the prioritization of earthquake disaster countermeasures and earthquake disaster education for residents. Education has been conducted every 5 years since 1975 with the goal of creating an index for disaster resistant cities. Risk evaluations made available to the general public identify one of five risk levels (1: lowest danger level to 5: highest danger level) for all of the districts of approximately 5000 towns in Tokyo. For each risk level, up to four risk indices may be assigned: building collapse, fire danger, evacuation risk, and comprehensive risk reflecting the sum of all three risk indices.

Among these, the building collapse index reflects building damage due to earthquakes and the extent of slope danger. This is determined by the ground strength, presence of active faults, building type (e.g., wooden, reinforced concrete), and year of construction. There is a trend towards high danger in dense residential areas with old wooden buildings. Similarly, the fire danger index reflects the degree of fire danger, as estimated using measurements of the ease with which fires may start or spread due to earthquake. There is a tendency for regions with densely packed wooden structures to have a high risk of fire, while regions with non-combustible building materials such as concrete, wide roads and areas with parks are at lower risk, due to the lower demand on firefighting operations.

The evacuation risk index reflects the evaluation of the time required to reach a nearby evacuation site (broad evacuation areas) and the number of residents to be evacuated. This risk tends to increase when shelters are relatively far away and where there exist factors along roads that obstruct the evacuation of residents. Regional assessments are not damage estimates for a particular earthquake, but evaluations of the relative danger and vulnerability of that region.

Figures 4.7, 4.8, and 4.9 indicate the degree of risk for the Tokyo metropolitan area. Based on these figures, the Ara river basin is currently at high risk for both flood and earthquake damage.

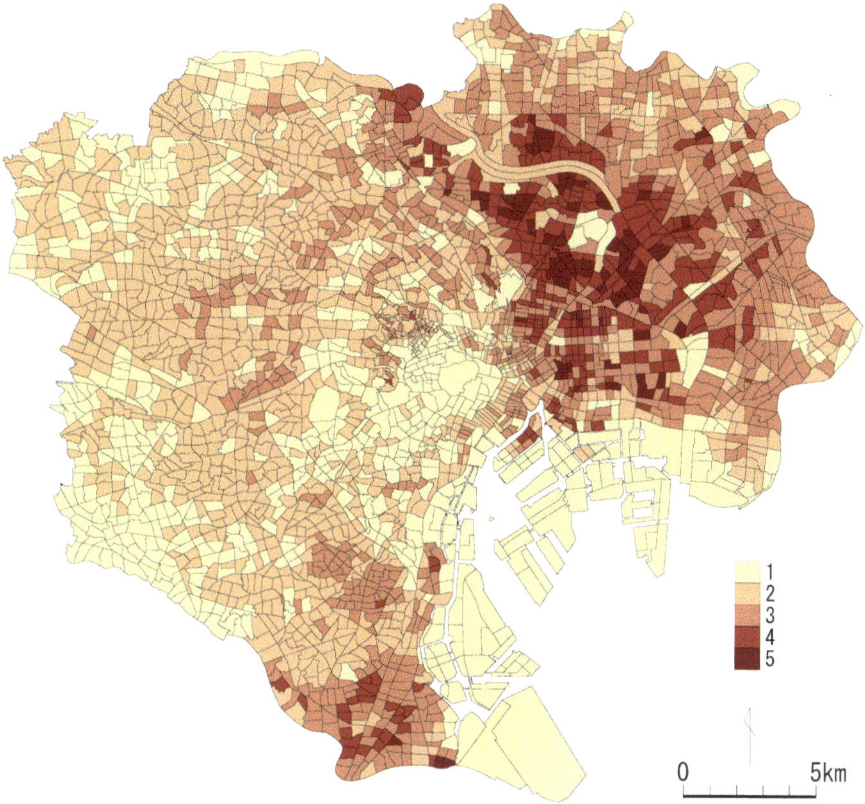

Fig. 4.7 The risk of building collapse

4.3.2 Countermeasures for Earthquake Disaster in Tokyo

As an on-site countermeasure for earthquake disasters in Tokyo, a plan to promote disaster-resistant cities was formulated in 1995 and based on the lessons of the great Hanshin–Awaji earthquake. The plan spurred efforts to improve the resilience of urban areas against disasters. Plans are being revised based on the indicators for an impending earthquake that could strike directly below the capital and the latest results of regional risk measurement surveys with a focus on 'fireproofing urban areas' and 'creating cities that are fireproof and indestructible'.

Fireproofing urban areas includes active and multi-layered implementation of projects with high maintenance effects, regulations, and guidance plans, along with maintenance that aims for a 65 % rate of both non-combustible areas and the interception of spreading fire. Plans to create fireproof and indestructible cities include the promotion of policies for earthquake resistance based on revised seismic retrofitting promotion plans as well as earthquake resistance for buildings

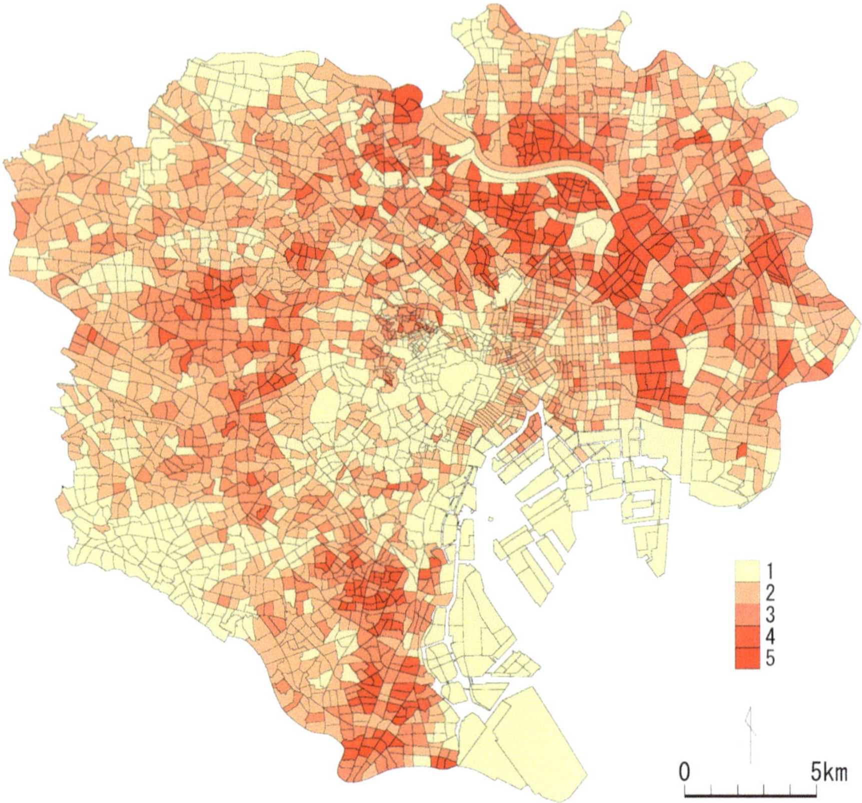

Fig. 4.8 The risk of fire

on emergency transportation routes and wooden residences in order to preventing roadblocks before they occur.

Additionally, raising consciousness regarding the ideas of 'one protects one's own life' and 'one's village is protected by its residents' is an off-site countermeasure for earthquake-related disasters that will ensure that households, regions, and workplaces play a role in their own self-defence against disasters. For that purpose, pamphlets are being handed out to Tokyo residents, videos are being created, lectures and disaster prevention fairs are being held in Tokyo, and disaster prevention organizations are disseminating information via radio, TV, and the Internet. Moreover, Tokyo tourist information centres distribute disaster prevention information to tourists in several languages, both in person and online.

When an earthquake occurs, the cooperation of regional residents and the active pursuit of disaster prevention, such as initial fire extinguishing and rescue and relief, can mitigate earthquake damage. Unfortunately, achieving cooperation among residents is challenging in a metropolitan area, because many local communities have declined as a result of the process of rapid urbanization during the period of high economic growth as well as the declining birth rate and aging population today. On

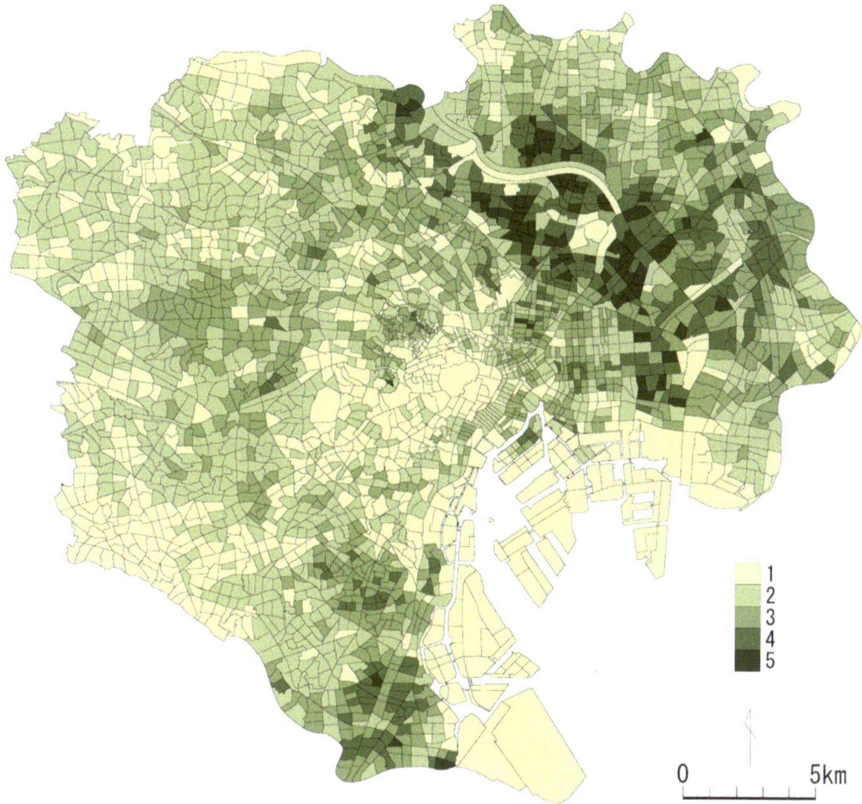

Fig. 4.9 The integrated risk

the other hand, civic organizations are important centres for regional disaster prevention activities. These organizations are required to implement 'regional cooperative recovery activities' during times of recovery, and their efforts enormously affect the vulnerability of residents in the metropolitan area. However, civic organizations face several problems, such as the aging of their members, the shortage of leaders, undeveloped facilities, and low rates of participation in training exercises. Currently, Tokyo is aiming to address these issues by activating the local voluntary disaster management organizations and making an effort to promote their growth.

Table 4.3 Estimated damage from flooding of the TONE and ARA rivers

The river	Embankment spot	The flooding area (km^2)	The flooding number of households	The number of assumption dead (person)	The number of isolate persons (10 thousand)
TONE river	Saitama-Kazo	530	86	2600	110
ARA-river	Tokyo-Kita	110	51	2000	86
ARA-river	Tokyo-Sumida	90	42.6	3500	73

4.4 Countermeasures Against Disaster in Metropolitan Area

4.4.1 Damage Estimation of Flood and Earthquake Disaster in Metropolitan Tokyo

With respect to expected damage from large-scale flooding, river development is underway, considering the probability of a once-in-200-years occurrence. However, serious damage would still be expected near the Tone and Ara Rivers that flow through the eastern part of Tokyo. As this area is below sea level, large-scale flood damages could persist for more than 2 weeks. As many as 1.1 million people could be in isolated conditions (Table 4.3). Moreover, the damage estimates naturally include subway flooding. Although measures such as installing sealing strips at subway entrances have been implemented in recent years, there remain concerns about the downtown subway being submerged. This could happen only a few hours after a flood occurs due to breaches in riverbanks, among other causes.

With respect to earthquake damage, an earthquake in the northern part of the Tokyo Bay (hypocentre: 30 km underground; magnitude: 7.3) would likely cause serious damage. Although recent earthquake- and fire-proof measures are expected to reduce the damage caused by collapsing buildings and fires, there remains a concern that a high percentage of buildings could be completely destroyed in the eastern area, which is characterized by weak alluvial ground (Table 4.4).

4.4.2 Shift from 'Rescue and Assistance' to 'Cooperation and Self-Help'

Tokyo still faces a high risk of damage from floods and earthquakes; thus, both soft and hard countermeasures are being pursued. However, risk management in recent years is shifting from 'prevention', which aims to achieve zero risk, to fail-safe 'mitigation'. Although government-led preservation of human life during a disaster is the most significant mission, at the same time, demands are also increasing for

Table 4.4 Damage to building due to the Kanto earthquake 1923 and The Tokyo bay northern earthquake (M7.3)

		The Grate kanto earthquake of 1923	The Tokyo bay northern earthquake (M 7.3)					
	Alluviums (m)	Total collapse (%)	Total collapse houses	Total collapse rate (%)	The burning rate (%)	The wooden houses	The non-wooden houses	The number of houses
Chiyoda	5–10	5.2	828	7.7	0.0	2079	8741	10,820
Chuo	5–15	0.4	1942	12.3	0.5	4970	10,748	15,718
Minato	5–25	2.1	2043	8.1	0.8	8996	16,344	25,340
Shinjyuku	0–5	0.7	3605	7.3	3.9	27,174	22,238	49,412
Bunkyo	5–10	0.6	3548	10.3	6.1	19,668	14,642	34,310
Taito	5–30	4.8	6680	18.0	6.6	17,466	19,624	37,090
Sumida	30–40	13.1	9902	21.3	19.7	27,072	19,525	46,597
Kouto	20–70	10.2	8010	14.3	5.9	26,491	29,559	56,050
Shinagawa	5–20	0.6	5223	8.6	31.9	40,408	20,596	61,004
Meguro	5–10	0.1	2516	4.8	20.8	33,680	19,271	52,951
Ohta	5–25	4.6	11,043	7.9	24.0	91,716	48,191	139,907
Setagaya	0–5	0.4	6027	3.4	13.4	119,895	59,795	179,690
Shibuya	5–10	0.6	2578	7.0	8.7	18,732	18,323	37,055
Nanano	0–5	1.2	2218	3.5	10.7	44,826	19,253	64,079
Suginami	0–5	3.2	3690	3.0	19.8	88,815	34,315	123,130
Toshima	0–5	0.2	1675	3.3	2.5	33,108	18,091	51,199
Kita	5–30	6.4	2687	4.1	0.9	44,901	20,373	65,274
Arakawa	5–40	16.6	7212	17.9	13.7	27,093	13,122	40,215
Itabashi	5–30	0.5	1612	1.7	0.7	61,303	32,806	94,109
Nerima	0–5	0.1	1942	1.3	2.2	110,805	35,146	145,951
Adachi	10–55	8.2	10,082	7.3	10.2	98,033	39,722	137,755
Katsushika	10–65	1.0	7446	7.4	10.2	73,023	28,196	101,219
Edogawa	5–65	1.5	8744	7.0	11.4	89,681	34,993	124,674
Average	–	4.4	111,254	6.6	11.1	1,109,935	583,616	1,693,551

information that residents can use themselves to become abreast of the potential for disasters and not remain overly dependent on the government.

Strengthening regional resilience, a concept that has attracted attention in recent years, means not only post-disaster resilience, but also adaptive capacities that prevent pre-disaster hazards from developing into calamities. The adaptive capacity should reflect and help to mitigate the locality-specific vulnerability. One of the greatest vulnerabilities in a highly urbanized area like Tokyo is the fragmentation of local communities due to high population fluidity. This characteristic makes it difficult for urban areas to develop social capital, which has come to be known as an essential component of voluntary and cooperative disaster prevention activities.

The role of administrative bodies as an alternative means to promote local communities is important in developing social capital and mitigating vulnerability. Specifically, strengthening administrative disaster prevention support for pre-disaster cooperation and self-help is an urgent issue. This involves such measures as individual stockpiling in preparation for disasters and public support of 'small communities', such as neighbourhood and self-government associations.

References

Bureau of Urban Development Tokyo metropolitan government. (2010). *Summary of propulsion plan for creation of disaster-proof Districts (Bousai Toshi Zukuri Suishin Keikaku).* (in Japanese).

Bureau of Urban Development Tokyo metropolitan government. (2013). *Survey report of degree of regional risk 7th (Jishin Ni Kansuru Chiiki Kikendo Sokutei Chousa).* (in Japanese).

Expert Examination Committee on Countermeasure for Great Flood, Central Disaster Prevention Council. (2010). *Special committee report: Inundation of metropolis, countermeasures for damage* mitigation (*Daikibo Suigai Taisaku Ni Kansuru Senmon Chousakai Hokoku Shutoken Suibotsu- Higai Keigen No Tameni Torubeki Taisaku Towa*). (in Japanese).

Expert Examination Committee on Countermeasure for Near Field Earthquake of Metropolis, Central Disaster Prevention Council. (2005). *On the results of damage estimation* (*Higai Soutei Kekka Ni Tsuite*). (in Japanese).

Miyamura, T. (1985). *The flood disaster* (*Suigai*), Tokyo: Chuko Shinsho 786. Chuo Kouronsha. (in Japanese).

Miyata, T. (1969). Variation and runoff and flood damages caused by urbanization in the Shakujii River Basin Tokyo. *Geographical Review of Japan (Chirigaku Hyouron), 42–10*, 667–680 (in Japanese).

Moroi, T., & Takemura, M. (2002). Re-evaluation on the damage statistics of wooden houses for the 1923 kanto earthquake and its seismic intensity distribution in and around Southern Kanto District. *Journal of Japan Association of Earthquake Engineering, 2*(3), 35–71.

Nakabayashi, I. (2005). *Planning of disaster proof city: Community development for disaster prevention and reconstruction in Science for City (Toshi No Kagaku)* (pp. 149–176). Center for Urban studies, Tokyo metropolitan University Press. (in Japanese).

Nakabayashi, I. (2010). National wide regional improvement for mitigation from huge earthquake disaster, basic issue for mitigation from the next Tokyo earthquake. *Journal of Economics, 242*, 24–40 (in Japanese).

Nakano, T. (1963). *Land areas below sea level in Japan* (*Nihon No Zerometoru Chitai*). Tokyo: University of Tokyo Press. (in Japanese).

Okuma, T. (1981). *Historical change of flood control of the TONE river and flood disaster (Tonegagawa Chisui No Hensen to Suigai)*. Tokyo: University of Tokyo Press. (in Japanese).

Takahashi, H. (1971). *Metamorphosis of National Land and Flood disaster (Kokudo No Henbo To Suigai)*, Tokyo: Iwanami Shinsho 793. Iwanami shoten. (in Japanese).

Chapter 5
Regional Vulnerability in Okinawa Prefecture

Daisuke Kamiya

Abstract The remote island in Okinawa prefecture is roughly typified by the island which is the tourist resort and is not the tourist resort. Both islands have some severe problems. The former islands have problems; drought water, waste disposal and succession of tradition. On the other hand, decrease in population and aging are progressing in the latter islands.

The influence that tourism promotion gives for water shortage is analysed in Sect. 5.2. This result expresses the carrying capacity in the small island of water resource. The vulnerability for tsunami disaster in depopulated and aging remote island was analysed in Sect. 5.3. It make clear that problem structure of the social network which paid its attention to the support person required at the time of disaster using investigation about residents daily life activities and association with the neighbours.

Keywords Remote island • Tourism • Disaster mitigation • Tsunami

5.1 Introduction

Okinawa prefecture is located in the far south of Japan (See Fig. 5.1). It composes hundreds of the Ryukyu Islands in a chain over 1000 km long. As of 2013, total population of Okinawa prefecture is about 1.42 million and population density is 622/km^2. Since the end of World War II in 1945, Okinawa has been under the reign of U.S.A. for 27 years. After U.S. retroceded Okinawa to Japan in 1972, prefectural income has been the worst or the second worst in Japan though Japanese government has taken a number of measures to promote the economy of Okinawa prefecture.

In this chapter, water shortage problem is described first from the viewpoint of tourism which is promoted to activate the Okinawa's economy, and second, disaster risk mitigation especially in remote islands is considered.

D. Kamiya (✉)
University of the Ryukyus, Okinawa 903-0213, Japan
e-mail: d-kamiya@tec.u-ryukyu.ac.jp

K. Hagihara, C. Asahi (eds.), *Coping with Regional Vulnerability*, New Frontiers in Regional Science: Asian Perspectives 4, DOI 10.1007/978-4-431-55169-0_5

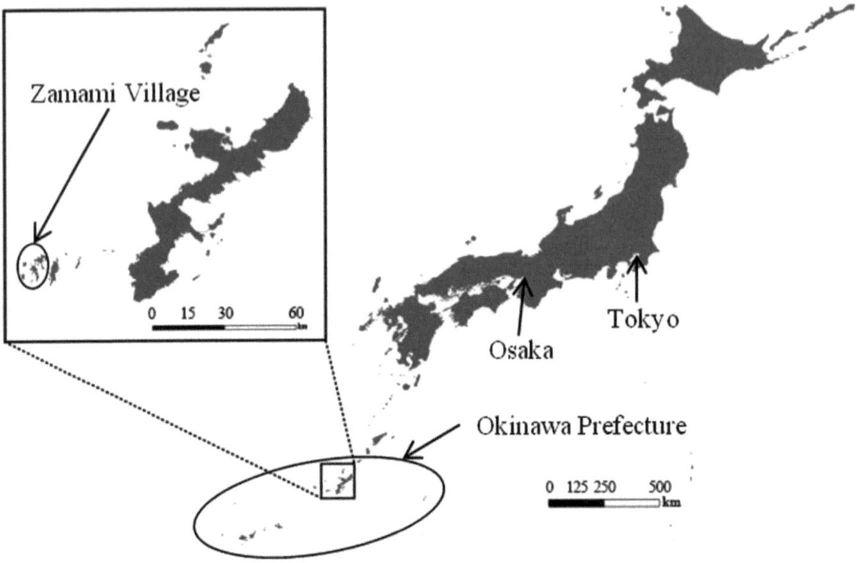

Fig. 5.1 Location of Okinawa Prefecture in Japan

5.2 Analysis of Water Shortage Problem Considering Influence of Tourism in Okinawa Archipelagos

Okinawa is composed of many small islands in a subtropical oceanic climate, and since ancient times has been troubled by water shortages. Many water re-source developments (such as dams and ocean water desalinization plants) were constructed there after World War II, and because of this there have been no water supply restrictions on Okinawa's main island since 1994. Water supply restrictions mean that water supply stops 8 or 12 h a day. Figure 5.1 shows the location of the prefecture.

Water demand increases with changing lifestyles and increasing numbers of tourists. Tourism is one of Okinawa's main industries; more than 6.4 million tourists visited in 2013. Almost tourists are domestic. A host of measures are being taken with the goal of attracting ten million tourists. Promotion of tourism is a particularly important measure for the remote islands. This is because, as income from sugar cane production decreases, younger people leave the islands in search of work, and the island populations subsequently decrease. Tourism is an effective industry for helping slow this trend. However, tourism promotion in Japan causes problems such as increased risk of water shortage, increased waste and destruction of scenery. This chapter specifically discusses issues surrounding water shortage on tourism-oriented islands.

5.2.1 *Problems of Water Supply Works and Water Shortage*

5.2.1.1 Problems Concerning Water Supply Works

Table 5.1 shows a summary of the water supply works on remote islands in Okinawa. Desalination facilities give a stable supply of water, but their costs are high, which subsequently increases water prices. The water rate for home is 1000–1500 yen/10 m^3 on Okinawa's main island. However, the water rates in municipalities using desalination are more than double those on the main island. Kita-daito Village has the highest water rate, at approximately six times that of the cheapest in Higashi Village on Okinawa's main island. Public waterworks of remote islands rely on public funding for water supply operation. Works using desalination facilities face issues of increased financial burden as water use rises. Water rates have to be increased to resolve the issue. Okinawa, however, has the lowest per-capita income in Japan, so these increases pose difficulties. Understanding of

Table 5.1 Summary of water supply system in remote islands

Municipality name	Start year	Water source	Purified water system	Water rate (Yen/10 m^3)	Works	Population (2010)
Iheya Vil.	1970	Dam, underground water	Electrodialysis	2100	1	1363
Izena Vil.	1971	Dam, underground water	Slow filtration	2300	1	1537
Kita-Daito Vil.	1985	Sea water	Desalination	3535	1	592
Minami-Daito Vil.	1974	Sea water	Desalination	3354	1	1397
Kumejima Town	1975	Dam, underground water	Slow filtration	1449	1	8488
Aguni Vil.	1974	Sea water	Desalination	3250	1	849
Tonaki Vil.	1987	Sea water	Desalination	2620	1	475
Tokashiki Vil.	1976	Dam	Rapid sand filtration	1687	1	785
Zamami Vil.	1974	Dam, underground water	Rapid sand filtration	1917	2	939
Miyakojima City	1965	Underground dam	Slow filtration	1757	1	52,203
Tarama Vil.	1974	Underground water	Desalination	2780	1	1249
Ishigaki City	1962	Dam, underground water	Slow filtration	1338	2	46,933
Taketomi Town	1957	River water	Slow filtration	1401	7	4021
Yonaguni Town	1959	River water	Slow filtration	1150	1	1673

Source: Okinawa prefecture (2010): summary of Okinawan water supply residents unconnected to the tourism industry is needed regarding conserving use of desalination systems to offset tourism increases. Some islands are composed of eight municipalities, some of which have water purification plants. However, owing to the low population figures, there is typically just one person in charge of the water service.

Water purification methods must be changed because of water shortages and deterioration of the intake water quality, but this is a difficult change financially, technically and with regard to human resources in a small municipality. This consequently places a greater burden on residents.

5.2.1.2 Structure of Water Shortage Problem

We conducted an interview-based survey with personnel in charge of water services. Influence on regional society of the water shortage risk and water resource development was structured using the KJ method (Fig. 5.2). Natural environments serve as a tourist resource, and tourism promotion will contribute to an island's economy. However, water demand will increase along with an increase in tourists. Water supply restrictions thus become highly likely because the quantity of water resources is limited. Residents endure a burden in that water supply restrictions will occur if they do not respond to this issue. For a municipality with remote islands and/or waterworks, the residents of one island may be able to use water normally while residents of another island have to live with limited water supply. This difference leads to feelings of unfairness. A typical example is Zamami Village. This village supplies water for two islands and residents pay the same water rate. However, the frequency and periods of water supply restrictions differ. Elderly residents of this island said there were no water supply restrictions when there were few tourists.

Dam construction and setting of desalination facilities both affect residents and the environment. Dam construction destroys forests and the digging sediment is used for landfill on the coastline because there is sparse flatland on small islands. The natural environment is disrupted to supply the water that tourists use. In other words, projects for promoting tourism reduce the value of tourist resources.

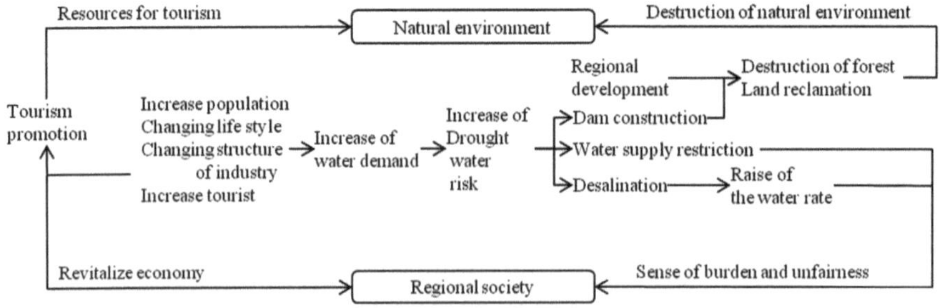

Fig. 5.2 Social problems considering water supply on small island

Revitalizing the economy and the issue of water shortage are closely related for small islands, and tourism promotion is an actual social problem. Therefore, it is not appropriate to consider the problem of water shortage as one related to dam and water supply management. This problem must be handled as a part of regional planning. In particular, there is a strong need to project reasonable numbers of tourists in view of water resources. This is the carrying capacity for tourism-oriented islands.

5.2.2 Analysis of Water Consumption of Accommodations

Table 5.2 shows quantities of water use according to the scale of the accommodation. Figure 5.3 shows the relation of water use quantity to the number of guests in large-scale hotels. The operational rate is higher in hotels above and to the left of the regression line. Those in the lower region have a low rate. The number in parentheses in the index of symbols indicates the quantity of water use per night per person.

The quantity of average daily water consumption in residences is 150–250 l per person on remote islands. Guests at large-scale resort hotels on small islands will use 2–10 times that rate. Considering the standards of small water supply systems in Japan, the operational rate of accommodations must be up to 50 %. However, the operational rate of accommodations in Okinawa is approximately 75 % (2013), which is one of the factors leading to the possibility of water supply restrictions.

5.2.3 Actual Water Use Amid Water Supply Restrictions in Zamami Island

The total area of Zamami Island is approximately 6.7 km^2, with a population of 578. Approximately 50,000 tourists come to this small island annually, mainly for scuba diving and swimming in the ocean. There is a coral reef near the island. Water supply restrictions were periodically implemented every year during 2001–2009, with the 325-day restriction in 2003–2004 being of special note. Water

Table 5.2 Quantity of water use in accommodations

Staying capacity scale	Mean (litter/person/stay)	95 % confidence interval	The number of sample
Large (300 people more)	744	696–792	148
Middle (100–299 people)	551	507–596	109
Small (less than 100)	253	235–271	178

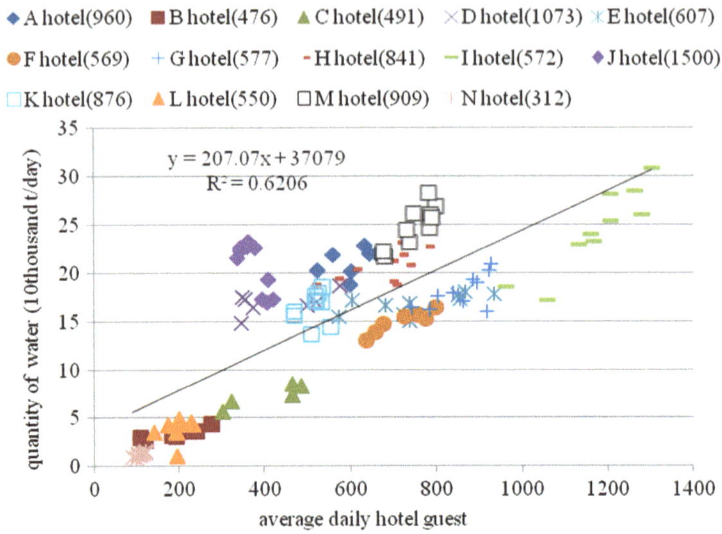

Fig. 5.3 Relation of quantity of water use and average numbers of hotel guests

consumption according to user category was analyzed from April 2004 to November 2007. This was divided into tourist facilities (accommodation and diving shops) and others, and summed (Tables 5.3 and 5.4). Daily water consumption per tourist is the quantity representing total tourist consumption divided by the number of tourists. Residents of Zamami consume approximately 175 l/day; low in comparison with residents of other islands. Some months are subject to stronger restrictions than normal. Water consumption increased in months with higher numbers of tourists than in an average year. Even accommodations with large tanks for storage have little advantage if there are water supply restrictions. Many families on the island also use wells, but seawater mixes with well water when there is little rain. In such cases, when the well water is not potable, domestic water supply consumption increases.

5.2.4 Scenario Analysis Considering Tourism Promotion

5.2.4.1 Estimation of Water Supply Restriction Risk Considering Tourist Increase and Saving-Water by Residents

Under dam management rules, water supply restrictions are implemented when the rate of water storage of Zamami Dam drops below 50 %. In the surveyed period the following restriction periods were implemented: July 18, 2003 to June 10, 2004 (325 days); October 17, 2005 to March 13, 2006 (142 days); November 27, 2006 to January 22, 2007 (57 days); and April 2, 2007 to July 6, 2007 (57 days). Zamami

Table 5.3 Tourists' water consumption

	Total consumption (m^3)		Daily consumption par tourist (m^3)	
Month	Normal	Restriction	Normal	Restriction
April	1844	1823	0.573	0.541
May	2155	2202	0.574	0.584
Jun	1785	1874	0.404	0.434
July	3168	3843	0.427	0.418
August	3782	3613	0.394	0.325
September	2896	2601	0.421	0.408
October	2528	2270	0.592	0.495
November	2294	1974	0.797	0.712
December	1339	1326	0.608	0.617
January	1635	1563	0.824	0.926
February	1516	1285	0.576	0.573
March	1736	1444	0.518	0.408
Average	2223	2151	0.559	0.537

Table 5.4 Water consumption of others (e.g., house, company, school)

	Total consumption (m^3)		Daily consumption par residents (m^3)	
Month	Normal	Restriction	Normal	Restriction
April	3105	3148	0.168	0.167
May	3350	3232	0.175	0.166
Jun	2919	3029	0.158	0.161
July	3797	4408	0.196	0.230
August	4190	3778	0.217	0.198
September	3748	3351	0.200	0.181
October	3226	3091	0.166	0.162
November	3182	2950	0.169	0.159
December	2698	2313	0.141	0.121
January	3020	2822	0.158	0.147
February	3168	2923	0.181	0.167
March	3166	2707	0.165	0.142
Average	3297	3146	0.175	0.167

has been aiming to increase its number of tourists and reduce water supply restrictions. We conducted analyses in two scenarios—when tourists increase by 10 % and when residents save water—and the number of tourists is the actual, current average value. Analysis considered that the social situation changed and precipitation did not change. Figure 5.4 shows scenario analysis results. When the number of tourists increased by 10 %, new water supply restrictions were implemented from January 18 to February 24, 2005 (38 days). The water supply restrictions of 2005 were extended for 87 days, and it is 229 days.

Here we discuss the results of residents' water-saving efforts. We examined what quantity of water needed to be saved to avert water supply restrictions in 2005,

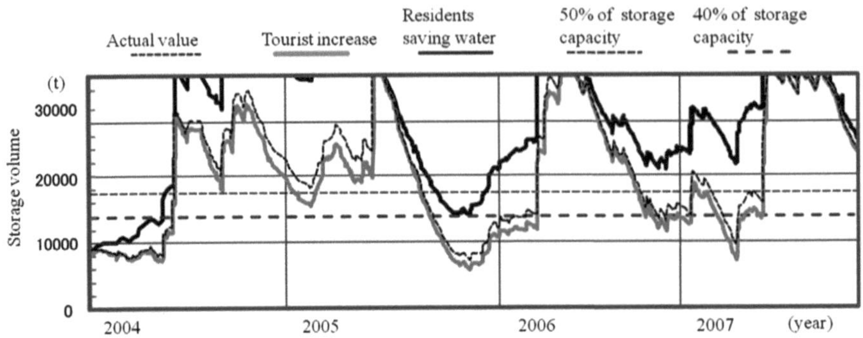

Fig. 5.4 Result of scenario analysis

the year that had the lowest storage volume of those we surveyed. This analysis was done under the condition that the storage rate is no less than 40 %, which was the actual amount under dam control. It was found that residents must reduce consumption by approximately 16.2 l/day. This results in residents of Zamami using less water than residents of other islands. Table 5.4 shows that even if water supply restrictions are implemented, a residential water-saving rate of only around 10 % can be achieved. If a higher-than-average numbers of tourists visit the island, it is thought that water supply restrictions cannot be avoided. Moreover, if water resource development, including construction of desalination facilities, does not take place, tourists will be required to conserve water.

5.2.4.2 Acceptable Numbers of the Tourists

The acceptable number of tourists was calculated as per the carrying capacity. Normal daily consumption per tourist, which is shown in Table 5.3, was used for this calculation. One tourist uses an average of 559 l/day. Water supply restrictions can be avoided when there is a decrease of approximately 12,500 tourists (about 25 %). In other words, a suitable number of tourists for this island is estimated at approximately 40,000.

If the aim is a self-sufficient economy by means of the tourism industry, it will be difficult to implement a policy for reducing numbers of tourists. However, understanding may be needed that the tourism industry brings about water restrictions for residents. Construction of a water pipe on the bottom of the ocean connecting Okinawa's main island and Zamami Island was considered as a means of averting water supply restrictions, but the plan was abandoned because of high construction and maintenance costs. Introduction of desalination facilities has also been discussed. Additionally, collection of an island entry tax was considered for addressing the problem of raising the water rate, and raising water rates for business and tourism use. A new vision of water supply works for the whole of Okinawa is being discussed for lowering the water rate on remote islands. Whatever path is taken, consensus-building will be indispensable.

5.3 Regional Analysis for Disaster Risk Mitigation in a Depopulated and Aged Island Society

5.3.1 Paradigm Shift from the Conventional Disaster Reduction Plan to the Integrated Disaster Risk Management

The experiences from the Great East Japan Earthquake (2011) and Great Hanshin Earthquake (1995) alerted us to the need for preparation for and mitigation of damage resulting from earthquake- and tsunami-related disasters. These disasters also showed the limits of the capacity of disaster prevention facilities and the importance of disaster mitigation, such as refuge action. The tsunami triggered by the Great East Japan Earthquake, which far exceeded common projections, devastated many structures. The Japanese government, in response, changed its way of projecting tsunami. Past tsunami estimates assumed a scale of tsunami as had been generated repeatedly in the past (Level 1 tsunami). Future estimates will assume tsunami of the highest rank in consideration of every possibility based on scientific knowledge (Level 2 tsunami). For a Level 2 tsunami disaster, the fundamental step in protecting human life is evacuating to higher ground without hesitation, swiftly and autonomously, as soon as strong or extended shaking from the earthquake is felt (Central Disaster Management Council 2011).

Okada et al. (2005) showed the way of thinking of the integrated disaster risk management in the twenty-first century as a paradigm shift from the conventional disaster reduction plan. Table 5.5 points out that advance preparedness and adaptive and community-based management systems become more important. The risk management cycle is also important. It is a check-action-plan-do cycle. First a regional diagnosis must be made because it is critical to know the regional weaknesses and vulnerability. Pre-assessment is also part of the first step.

More detailed countermeasures for disasters on the community scale are also important. Self-initiated and co-operative measures for tsunami are more important in peripheral areas, such as remote island regions, than in urban areas, because islands have a more difficult time receiving external support.

This section covers evaluation of regional vulnerability for disaster risk management on a small island within Okinawa Prefecture, Japan. Special emphasis is placed on co-operative evacuation for vulnerable people, such as elderly and disabled from tsunami.

Table 5.5 Conventional disaster plan and twenty-first century integrated disaster risk management

Twentieth century	Twenty-first century
Reactive	More proactive
Emergency and crisis management	More risk mitigation + preparedness approach
Countermeasure manual approach	More anticipatory/precautionary approach
Predetermined planning	More comprehensive policy-bundle approach
Sectoral countermeasure approach	More adaptive management approach
Top-down approach	More bottom-up approach

5.3.2 Disaster Risk and Vulnerability on a Remote Island

Depopulation and aging is progressing in most of Japan's remote islands and mountainous areas, and there are concerns about the full extinction of communities in various parts of the country. Moving of communities has been discussed for some mountainous areas. However, moving an entire community from a remote island leaves the island uninhabited, while its history and culture are lost. Moreover, there is the problem of territorial waters. Some islands are promoting tourism to counteract the problems associated with depopulation and aging. However, population is decreasing in line with decrease public works projects on many islands, and many younger people are leaving because there is little work available.

On the other hand, unique history and culture are often handed down because they exist on a remote island. Representative examples of this are districts preserving traditional architectures. Taketomi and Tonaki Islands are certificated preservation districts in Okinawa Prefecture. These islands house a great deal of traditional Okinawan architecture. Many tourists go to Taketomi, because of the high number of resorts there or on nearby islands. Yet there are few tourists visiting Tonaki because it take 2 h by boat to reach it from the city of Naha (Okinawa's prefectural capital) (Fig. 5.5).

The study area was the village of Tonaki, which presents some dire conditions that portend disaster in that it is a remote, aged and depopulated island. The village is constructed on one island. The official population of Tonaki is 452, and the aging rate (the population rate of 65 years or over) is 33 % (average Japanese aging rate: 23 %) (Fig. 5.6). A staff member of the village government office reported that the actual population on the island was at most 350.

The north and south side of the island have low mountains, and the village is formed on the tombolo where sand deposited between mountains (Fig. 5.7). The elevation of the village is therefore less than about 10 m.

Its traits of being an aged, depopulated, remote island with traditional architecture all pose weaknesses and vulnerabilities to natural disasters. Okinawa's traditional architecture is fragile and not resilient against earthquakes. Disaster risks for this village are earthquakes, tsunami and typhoons. The entire village is within the assumed area of tidal inundation caused by a typhoon or tsunami (Fig. 5.8). Predictions also state that a tsunami would arrive at this island approximately 16 min

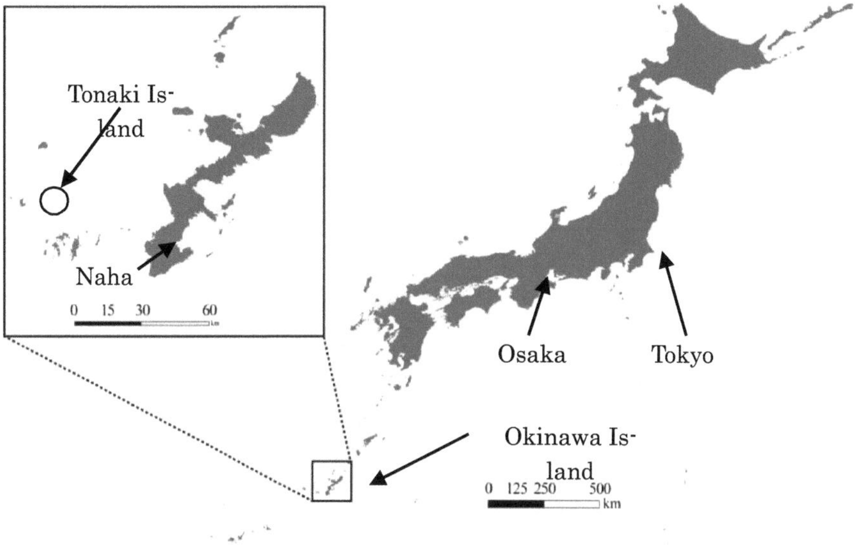

Fig. 5.5 The location of Tonaki Island

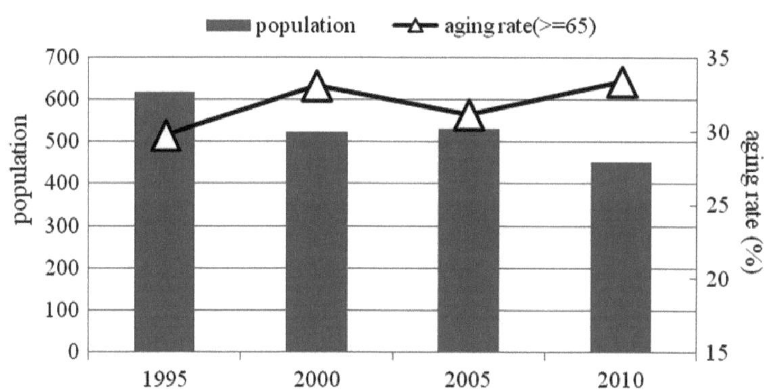

Fig. 5.6 Population and aging rate of Tonaki

after an earthquake (Okinawa Prefecture 2012). An earthquake measuring 6 or stronger on the Japanese Meteorological Agency seismic intensity scale is predicted to strike Tonaki.

Tsunami evacuation building standards in Japan mandate that a building be over five-stories and built with steel-reinforced concrete. The island, however, has no five-story buildings. A Cultural Properties Protection Act enforcement order forbids construction of high-rise buildings which are three-stories or over in the village. In this situation we presented a discussion on co-operation and human power in times of disaster. Therefore, we surveyed weakness and vulnerability factors in the village, and considered how to reduce risk.

Fig. 5.7 Tonaki village

We conducted field and interview surveys in 2008 and 2009. Survey items were as follows.

- Wooden building or concrete building: possibility of building collapse during an earthquake
- Width of roads and distribution of cement-block wall: possibility of road confinement when buildings and walls collapse
- Family structure: household only containing elderly, who are vulnerable when disasters occur

There are only 195 households in the village. Of these, 59 (30 %) house only elderly people, and 23 (12 %) of those have only one resident. A potential road traffic evaluation at the time of an earthquake was carried out using distribution of walls made of concrete blocks and road width (Fig. 5.9). The means of evaluation was based on occurrences in the Great Hanshin Earthquake. There are two evacuation zones located on hills. A school and a welfare centre are earthquake-resistant buildings. The evaluation results showed that road blockage would occur near houses of elderly people, which means that village personnel and others would not be able to get to these houses by car to provide help.

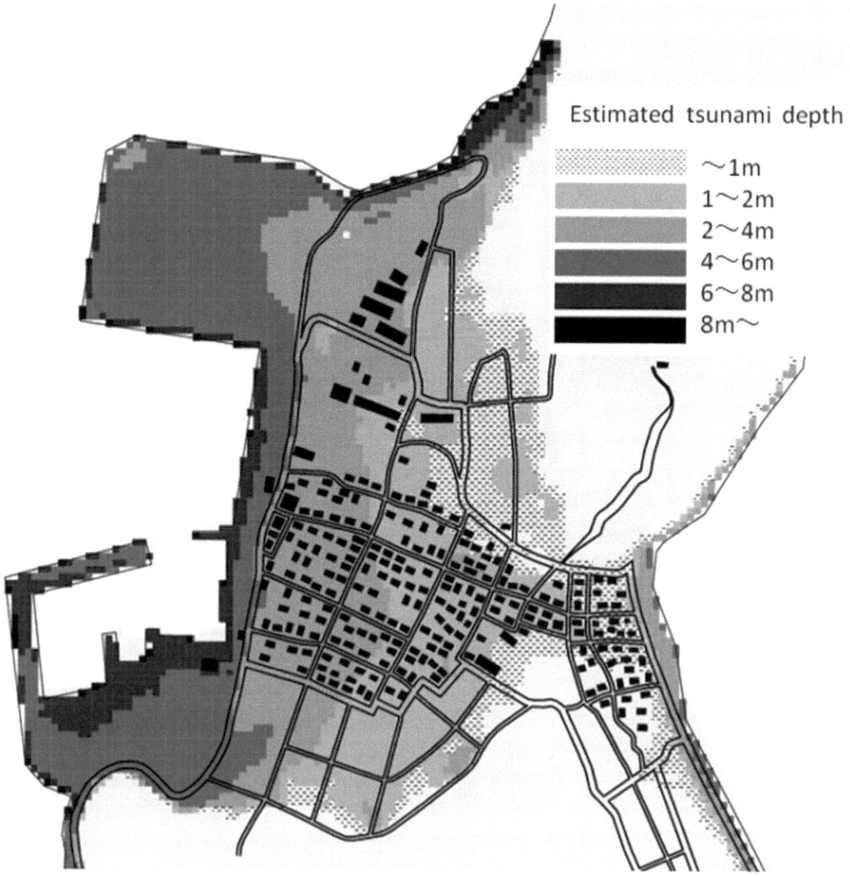

Fig. 5.8 Tsunami disaster risk identification map

5.3.3 Analysis of Evacuation from Tsunami

An evacuation simulation assuming a tsunami when an earthquake was generated at or near the island was carried out. We calculated the evacuation time of all residents and evaluated the possibility of safety evacuation in consideration of the tsunami's expected time of arrival (16 min). Some findings from the simulation were as follows.

- Residents would begin evacuating as soon as the earthquake strikes.
- Residents would evacuate via the shortest route to an evacuation zone.
- A questionnaire revealed people who cannot evacuate under their own power.
- There are all residents in each home.
- Table 5.6 shows the evacuation speed that was found.

Fig. 5.9 A potential road traffic evaluation and residence house

Table 5.6 Evacuation speed

	Flat road (m/s)	Step and up stare (m/s)
Youth	1.30	0.65
Age person	0.80	0.40

Four case simulations, shown in Table 5.7, were carried out. Table 5.8 shows results of the simulations.

Table 5.8 shows results in the best-case scenario. Specifically, these results are based on a condition in which a supporter is matched with a vulnerable person. Supporter is a person helping vulnerable person. Matching is decided based on minimizing total evacuation time. Daily interaction is needed between the supporter and vulnerable person in order for this desirable situation to occur.

5.3.4 *Risk Reduction with Residential Network*

Matching of a supporter and a vulnerable person is important for supporting the evacuation of the vulnerable person at the time of a disaster. To shorten the evacuation time, matches between nearby residents are desirable. A neighbourly relationship between supporter and vulnerable person is important. The study included a questionnaire on neighbourly relations and daily activity. We collected

Table 5.7 Simulation cases

Case	Evacuation space	Evacuation
1	2 hills	All residents evacuate by each person
2	2 hills, school, welfare centre	All residents evacuate by each person
3	2 hills, school(3 F), welfare centre(2 F)	The supporter give a vulnerable person on one's back and evacuate. (Evacuation speed is the same as the elderly person)
4	2 hills, school(3 F), welfare centre(2 F)	Two supporters support a vulnerable person with a wheelchair. (Evacuation speed is the same as the youth)

Table 5.8 Results of evacuation simulation

Case	Result
1	54 persons which are vulnerable person and elderly are not completed of evacuation
2	All vulnerable people are not completed of evacuation before tsunami arrived
3	All residents are able to evacuate one minute before a tsunami arrived
4	All residents are able to evacuate 3 min 48 s before a tsunami arrived

the questionnaire from 125 of the 158 households on the island. Figure 5.10 shows the totals recognized.

There is a major difference between those barely recognized and those recognized by many people. Participation in residents' festivals and events is shown to clarify the reason for this difference. It is thought that participation in events strengthens bonding social capital. Village cleaning is carried out three times a week on the island, and an athletic meet on the beach and boat race festival are held annually. About 80 % of those who are recognized by many people participate in the events, while 40 % of those who are hardly recognized participate. Of those who are hardly recognized, 70 % are elderly, and all members have low awareness of elderly people staying alone.

Table 5.9 shows the typical daily life of a person recognized by many people, and Table 5.10 shows that of a person who is hardly recognized. Table 5.9 reflects typical days of more than 100 people seen as recognized by many. Table 5.10 shows a day of those recognized by few people. Those recognized by many go to a workplace on weekdays and leave the island on vacation days. Some of those who are hardly recognized spend the entire day indoors.

Table 5.11 shows the professions of supporters (those who wrote they can support a vulnerable person, are younger than 44 and do not have a vulnerable person in their own family).

Most supporters are teachers at the elementary and junior high school. Small islands have many temporary teachers and most move their teaching locations within 1 year. Even regular teachers move to a different school within 3 years. Many teachers are unmarried. On vacation days teachers also often go to an island with a bigger city. Therefore, there are few connections between supporters and vulnerable people. Teachers know only one in 15 vulnerable people.

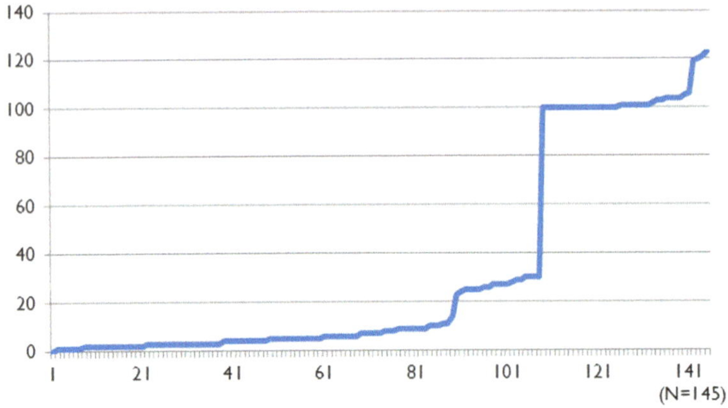

Fig. 5.10 Number of people known to other residents

Table 5.9 Daily life of person recognized by many people

	4(44)		5(50)		63(31)	
ID(age)	Week day	Holiday	Week day	Holiday	Week day	Holiday
Time						
1 a.m.	House	Another	House	House	House	Another
2 a.m.	House	Another	House	House	House	Another
3 a.m.	House	Another	House	House	House	Another
4 a.m.	House	Another	House	House	House	Another
5 a.m.	House	Another	House	House	House	Another
6 a.m.	House	Another	House	House	House	Another
7 a.m.	House	Another	Work	House	House	Another
8 a.m.	Work	Ship	Work	House	Work	Another
9 a.m.	Work	Ship	Work	Ship	Work	Another
10 a.m.	Work	Ship	Work	Ship	Work	Another
11 a.m.	Work	House	Work	Ship	Work	Another
12 a.m.	House	House	House	Ship	Work	Another
1 p.m.	Work	House	Work	Another	Work	Another
2 p.m.	Work	House	Work	Another	Work	Another
3 p.m.	Work	House	Work	Another	Work	Another
4 p.m.	Work	House	Work	Another	Work	Another
5 p.m.	Work	House	Work	Another	Work	Another
6 p.m.	Work	House	Work	Another	Work	Another
7 p.m.	House	House	House	Another	Work	Another
8 p.m.	House	House	House	Another	Work	Another
9 p.m.	House	House	House	Another	House	Another
10 p.m.	House	House	House	Another	House	Another
11 p.m.	House	House	House	Another	House	Another
12 p.m.	House	House	House	Another	House	Another

Another: On another island (e.g., in Naha)

Table 5.10 The daily life of the person recognized by small number of people

	4(44)		5(50)		63(31)	
ID (age)	Week day	Holiday	Week day	Holiday	Week day	Holiday
Time						
1 a.m.	House	House	House	House	House	House
2 a.m.	House	House	House	House	House	House
3 a.m.	House	House	House	House	House	House
4 a.m.	House	House	House	House	House	House
5 a.m.	House	House	House	House	House	House
6 a.m.	House	House	House	House	House	House
7 a.m.	House	House	House	House	House	House
8 a.m.	House	House	House	House	House	House
9 a.m.	Farm	Farm	House	House	House	House
10 a.m.	Farm	Farm	House	House	Daughter's house	Daughter's house
11 a.m.	Farm	Farm	House	House	Daughter's house	Daughter's house
12 a.m.	House	House	House	House	House	House
1 p.m.	House	House	House	House	House	House
2 p.m.	House	House	House	House	House	House
3 p.m.	House	House	House	House	House	House
4 p.m.	Farm	Farm	House	House	House	House
5 p.m.	Farm	Farm	House	House	House	House
6 p.m.	Farm	Farm	House	House	House	House
7 p.m.	House	House	House	House	House	House
8 p.m.	House	House	House	House	House	House
9 p.m.	House	House	House	House	House	House
10 p.m.	House	House	House	House	House	House
11 p.m.	House	House	House	House	House	House
12 p.m.	House	House	House	House	House	House

Table 5.11 Supporter's professions

Job	Number of person
Public office	1
Teacher	12
Office worker	1
Unknown	1

It is thought that countryside areas, particularly small communities such as remote islands, have stronger bonds between people than in cities. However, it was shown that there are few of the connections between supporters and vulnerable people that are necessary for reducing disaster risk. Teachers are younger people, who are precious commodities in a depopulated area. Therefore, the social network of teachers and vulnerable people is important for reducing disaster risk.

References

Central Disaster Management Council. (2011). *Report of the committee for technical investigation on countermeasures for earthquakes and tsunamis based on the lessons learned from The 2011 Off the Pacific Coast of Tohoku Earthquake.*

Okada, N., Tatano, H., Stuzuki, Y., Hagihara, Y., Hayashi, Y., & Hatayama, M. (2005). *Integrated disaster risk management: Research on methodology of urban diagnosis* (Vol. 48C, pp. 1–5). Kyoto: Annuals of Disaster Prevention Research Institute, Kyoto University.

Okinawa prefectural enterprise bureau. (2010). *Summary of Okinawan water supply.*

Okinawa prefectural government. (2012). *Tsunami assumption.*

Chapter 6
Health and Environmental Risks Related to Water Supply and Sanitation in the Socio-environment of Rural Bangladesh

Akira Sakai, Kunio Takahashi, Maiko Sakamoto, Yoshimi Hagihara, and Kiyoko Hagihara

Abstract Ensuring safe water supply and proper sanitation is a basic requirement for maintaining public health and safety, as well as the foundation of sustainable development. Although arsenic contamination mitigation from tube wells is an urgent issue related to public health in Bangladesh, a significant portion of the population still subsists on arsenic contaminated water. One of the primary reasons for this is the difficulty involved in accessing safe water, even in areas where arsenic mitigation facilities exist. As for sanitation, although coverage of sanitary facilities has been increasing in the country, existing toilets have various disadvantages; one of which is that improper human excreta management is causing surface water pollution, hindering the use of this water as an alternative drinking water source. To maintain soil quality and food productivity in the future, it is expected that human excreta will be used to fertilize agricultural land. In this chapter, socio-environmental problems and the risks related to water supply and sanitation will be identified and their interrelationships will be discussed. Additionally, the factors considered during the selection of technical options aimed at reducing risks will be discussed based on the authors' sanitation improvement-related field activities.

A. Sakai (✉)
University of Marketing and Distribution Sciences, Kobe 651-2188, Japan
e-mail: akira_sakai@red.umds.ac.jp

K. Takahashi
Japan Association of Drainage and Environment, Shinjuku, Tokyo 162-0067, Japan
e-mail: nqg48559@nifty.com

M. Sakamoto
University of Tokyo, Kashiwa, Chiba 277-8653, Japan
e-mail: m-sakamoto@k.u-tokyo.ac.jp

Y. Hagihara
Kyoto University, 264 Suemarucho Nakagyou, Kyoto 604-0901, Japan
e-mail: y-hagi@gf7.so-net.ne.jp

K. Hagihara
School of Sociology, Bukkyo University, 96 Kita-hananobo, Murasakino, Kita-ku, Kyoto 603-8301, Japan
e-mail: hagihara@bukkyo-u.ac.jp

K. Hagihara, C. Asahi (eds.), *Coping with Regional Vulnerability*, New Frontiers in Regional Science: Asian Perspectives 4, DOI 10.1007/978-4-431-55169-0_6

Keywords Arsenic contamination • Safe water supply • Sanitation • Socio-environment

6.1 Introduction

Safe water supply and sanitation are fundamental human needs and provide the foundation of sustainable development. In the rural areas in Bangladesh, they are also among the key issues related to improvements in the living environment. However, in addition to their quantitative macro goals, such as millennium development goals (MDGs), local communities need to set realistic and achievable goals to improve their living conditions in ways that complement their local environments.

A particular issue related to drinking water in Bangladesh is the arsenic contamination found in water extracted from the tube wells that have been constructed throughout the country. The number of such wells is currently approximately 10 million, and about 30 % produce water that exceeds the national guidelines for acceptable arsenic content (50 μg/L). The affected areas cover 70 % of the entire country and affect in excess of 30 million people. As a considerable segment of the affected population has poor access to safe drinking water, the risks of contracting arsenic-related ailments, including cancer, are considered to be high.

As for access to sanitation facilities, although the national coverage ratio has increased in recent years, primarily through the installation of pit latrines, the pit latrine concept has not been effectively disseminated in Bangladesh, and poor management practices can be observed throughout the country.

In this chapter, the authors will identify the socio-environmental problems that hindered improvements in living conditions, along with risks related to water supply and sanitation based on conditions in villages where the authors have been undertaking field studies. It is considered that the reduction of health and environmental risks related to water supply and sanitation will improve the social welfare of local residents. The authors will also discuss locally appropriate technologies that can be employed to reduce risks related to water supply and sanitation.

6.2 Relationship Between Water Supply and Sanitation

Fundamental improvements in human welfare can only be accomplished by satisfying basic human needs. These include safety, nutritious food, clean water, clothing, shelter, and healthcare. Although, during the undeveloped phase of human existence, people depended primarily on the natural environment to obtain the resources they needed for survival, appropriate social infrastructures are indispensible to securing such basic requirements at present. Water supply and sanitation refers to the social infrastructure necessary to secure clean water, health,

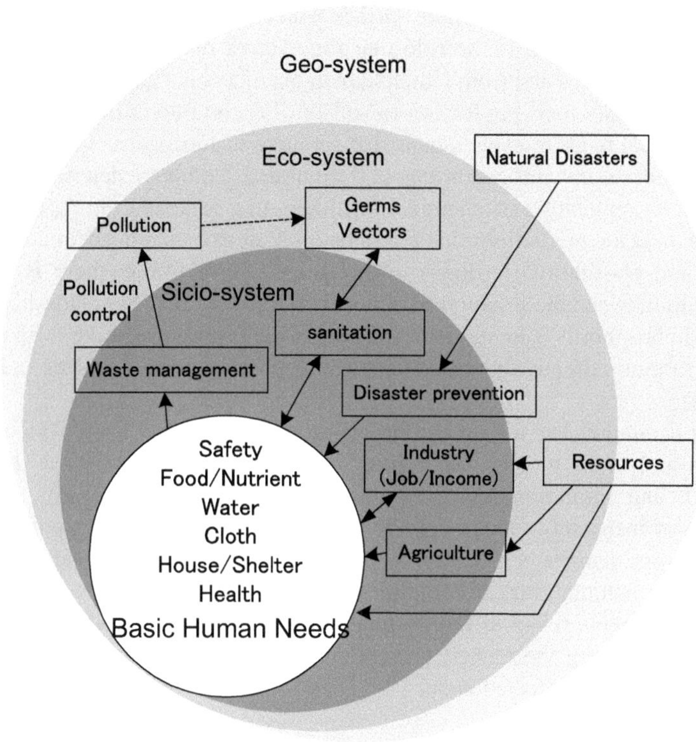

Fig. 6.1 Basic human needs and social infrastructure

and safety. Figure 6.1 shows an overview of the natural systems, geo-system and eco-system, and socio-system infrastructures – including water supply and sanitation – that are necessary to secure basic human needs.

In rural areas in Bangladesh, water supply and sanitation relate directly to social development issues such as poverty alleviation, advances in the status of women, health risk reduction, and environmental resource conservation. In situations where water supplies and sanitation are unsafe, it is easy to contract infectious diseases. Additionally, if it takes excessive amounts of time to collect safe water, lost income and economical hardship can result, which may further hinder efforts to alleviate poverty. Furthermore, if social mores and customs require women to collect and transport household water, it is difficult to advance gender equality. Gender inequality also imposes significant social costs, including high infant mortality, making other forms of social progress difficult. Therefore, in order to alleviate poverty it is necessary to recognize that water supply safety and sanitation are prerequisites for social infrastructure development.

Herein, the authors examined the importance of human excreta management, which is a key aspect of sanitation. There are two paths by which excreta relates to what humans intake orally. One path is the pollution of drinking water sources due to poor management of human excreta, especially in areas where tube-well water is contaminated by arsenic and where surface water is an alternative drinking water resource. The other path is the role that the excreta of all living things play in natural nutrient cycles and from which human excreta is not excluded. This natural nutrient cycle links food production to soil fertility, and diverting human excreta from the natural nutrient cycle means that a source of fertilizer is wasted.

In Bangladesh, the soil conditions of agricultural land have deteriorated due to the excessive application of chemical fertilizers that followed the "green revolution", and there are predictions that it will be difficult to maintain adequate levels of nitrogen and phosphorus fertilizer in the future. Furthermore, there is a strong relationship between the production of nitrogen fertilizer and Bangladesh's energy crisis, and phosphorus is being depleted worldwide. Therefore, there is an increasing need to more effectively utilize human excreta as organic fertilizer and/or soil conditioner.

Proper sanitation has a strong relationship with water and food, which people intake through their mouths. Therefore, water supply and sanitation should be integrated, and sanitation improvements should not be seen merely from the viewpoint of increasing toilet coverage, it is also necessary to ensure appropriate human excreta management aimed at water quality preservation and stable food production by maintaining soil conditions. Figure 6.2 shows how living environment improvements relate to family finances and how poverty alleviation can be pursued by resolving the related vicious cycles in the rural areas of Bangladesh, especially in arsenic-affected areas.

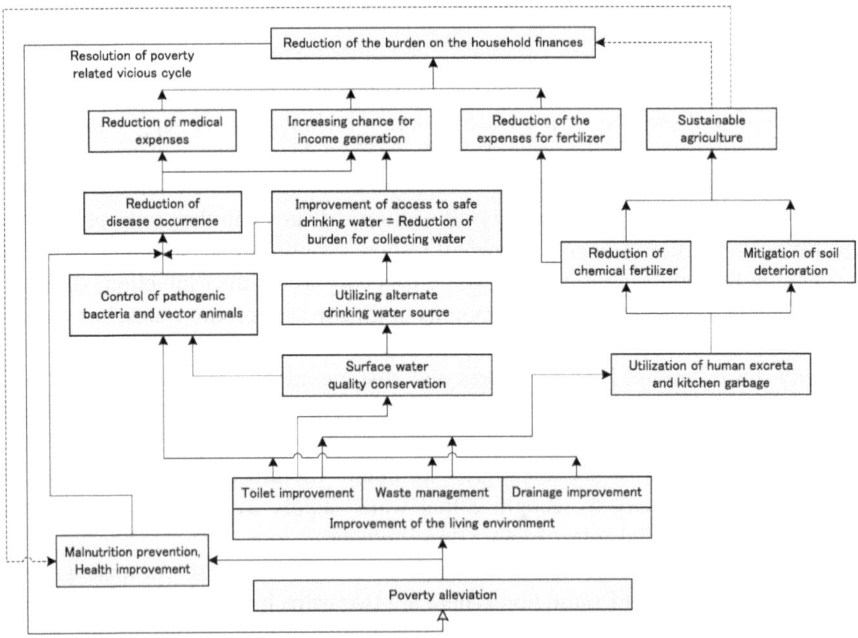

Fig. 6.2 Relationship between living environments and household finances

6.3 Socio-environmental Problems Related to Water Supply and Sanitation

6.3.1 State of Water Supply and Sanitation

This section focuses on a rural village named Basailbogh in Srinagar Upazila, Munshiganj District, which is located approximately 30 km south of the national capital of Bangladesh. This district, which is near the Ganges River, is a flood-prone area. There are slightly more than 100 households in the village, most of which are located on raised ground called "bari" in order to avoid flood damage, as are most other housing sites in Srinagar. The main occupations of the villagers are farming and small business, primarily retailing. The arsenic contamination mitigation status and sanitation level of this village will be described in the following subsections.

6.3.1.1 Water Supply

Prior to the detection of arsenic contamination in the region, shallow tube wells (STWs) were the primary source of water for all villagers. However, once it was determined that such wells were contaminated with arsenic, the government installed three deep tube-wells (DTWs) to supply water in this village. Taking into consideration the physical and psychological stresses related to collecting and transporting water, it was proven that those wells were unable to provide safe drinking water for the entire village. To compensate for the lack of wells, some families installed their own DTWs. However, most such well were on private property, and therefore could not be used by other villagers.

A foreign research organization previously installed two DTWs as compensation for land use during its research, but one of the DTWs had to be abandoned due to the bad taste of the water produced. As of 2006, there were nine active DTWs in the village. Additionally, while it has been reported that some family-sized arsenic filtration devices were provided by a foreign non-government organization (NGO) just after the arsenic contamination was detected, those devices can no longer be found, and DTWs were the only arsenic mitigation option available in this village until recently, when a local NGO installed a arsenic iron removal plant (AIRP) for household use. Yet, even today, some families continue to subsist on arsenic-contaminated STW water.

Numerous families need to collect safe drinking water from DTWs located far away from their houses. In most cases, the task of collecting and transporting water is regarded as "women's work", and the efforts involved can be the cause of physical and psychological stress among the women engaged in such labour. Additionally, when collecting water, the routes taken by women from their homes to the water source will change depending on the season. For example, during the dry season they must traverse steep slopes between the main road in the village and the "bari" to collect water. In contrast, most of the routes between the main road and

Table 6.1 Drinking water sources and water collection in Basailbogh

Drinking water source	Owner or time for collecting	Total	Collector		
			Female	Male	Both
Deep tube well	Tube well owner	4	–	–	–
	Within 10 min	36	32	4	–
	More than 10 min	26	20	3	3
Shallow tube well	Tube well owner	12	–	–	–
	Within 10 min	5	5	–	–
	More than 10 min	2	1	–	1
Total		85	58	7	4

"baris" are submerged during the rainy season, and villagers must build and traverse bamboo bridges. Both seasons impose hard physical burdens on persons carrying heavy water vessels. Note that the weight of the water they carry sometimes exceeds 20 kg, and it is usually necessary to collect and transport water twice daily.

Another burden is the psychological stress that Muslim women feel when they walk through places where numerous men can stare at them. These heavy stress levels might be among the reasons why some families still consume arsenic-contaminated water, and it is clear that releasing women from the burden of collecting water would improve their home lives and allow them to have more creative time. Table 6.1 shows the source, collection time and gender of the drinking water collectors recorded during a survey conducted in December 2006. These results show that 19 of the 85 families that responded the survey still consume STW water. The survey results also showed that the maximum time required to collect safe water exceeds 30 min.

During the survey, a number of villagers expressed doubts regarding the safety of DTWs and it was necessary to test the quality of water treated by AIRP. The water quality tests performed using an arsenic field test kit showed that all of the village DTWs had levels between 1 and 5 μg/L, and therefore complied with the Bangladeshi drinking water standard of 50 μg/L, but since DTWs are not always safe, periodic monitoring of arsenic concentrations remains essential. However, despite the need for monitoring, it is seldom undertaken.

It should be also noted that the AIRP did not remove arsenic sufficiently to satisfy the water quality standard. The following reasons were considered likely: the existence of phosphorus (which prevents arsenic removal), the low frequency of maintenance, and device overload caused by the fact that several families were using the AIRP, which was designed for single family use. The NGO that installed the device and its users should recognize the limitations and applicability of that technology.

6.3.1.2 Sanitation

As for sanitation, most of the families use sealed pit latrines and some families are now using flush toilets linked to septic tanks. Additionally, a Japanese NGO collaborated with a local NGO in the construction of five new toilets in the village, the design of which is based on the "ecological sanitation" concept. Hereinafter, these will be referred to as "EcoSan" toilets. EcoSan technology aims at increased agricultural use of human excreta as fertilizer and soil conditioner on farms in order to improve the organic content of the soil.

Actually, local government organizations in Bangladesh are encouraging people to construct pit latrines and are providing materials for pits in order to increase sanitation coverage. As a result, the number of the people who defecate in the open or use hanging latrines is declining rapidly. For example, in this village, 62 (73 %) families are now using sealed pit latrines, while 12 families use flush toilets linked to septic tanks, and five families use EcoSan toilets. This means that only six (7 %) families use improvised toilets, which include open defecation, hanging latrines, and broken or damaged latrines.

However, the existing pit latrines installed in this area are prone to a number of defects and disadvantages. For example, the pit is usually filled to its capacity within months, or within a couple of years, at most. To allow continuous use, residents must remove the contents and transport them to another location for disposal. However, as local villages do not have any systems capable of transporting excreta, or places to sanitarily dispose it in ways that do not result in adverse environmental impacts, the most common solution is simply to build a new pit latrine, leaving the old pit and its contents. Since most of the housing sites in this village have not sufficient space to allow pit latrines to be constructed one after another, it is not uncommon to see abandoned pit latrines in housing areas.

Leaving night soil in abandoned pits results in the waste of valuable biomass resources that could be utilized as fertilizer, soil conditioner or a fuel source. Another problem is that some people in this village create discharge holes at the bottom of the pits when they construct latrines the edge of their bari so that, during the flood season when the pits are submerged, and the owners can discharge the pit contents into the water. This way of managing human excreta is intentionally incorrect and extremely unsanitary. In this area, when flood levels rise above the raised ground level of the bari, pit latrines are very unsafe because residents are unable to use the latrine and it becomes highly likely that night soil in the pits will overflow. This happens more frequently at households on bari that are not sufficiently high, which directly relates to the affluence of the householder. Furthermore, the lack of any treatment at the pit bottoms means that the untreated human excreta is free to seep into the ground, thereby contaminating shallow groundwater and surrounding ponds, some of which are water sources for domestic use, such as cooking, washing, and bathing.

Increasing pit latrine coverage provides people who do not habitually defecate at fixed locations the opportunity to do so in a sanitary manner and a private space to

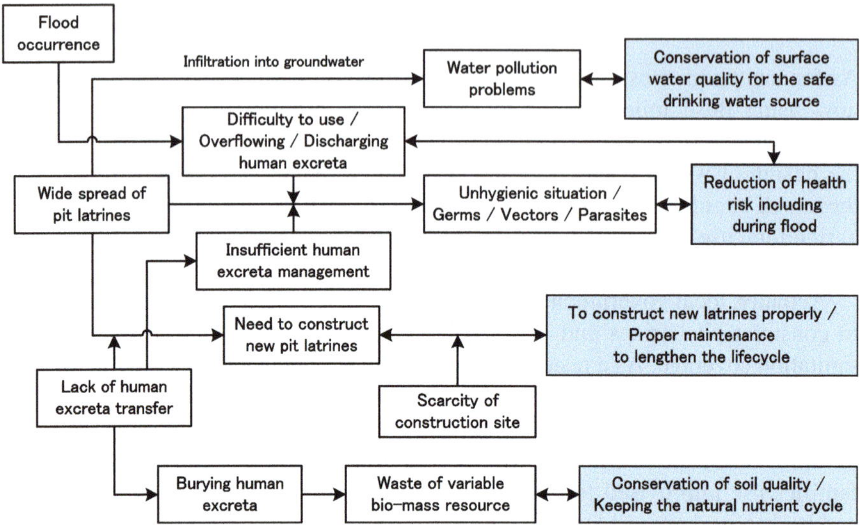

Fig. 6.3 Problems related to current sanitation situation (Revised from Sakai et al. 2007a)

defecate. Nevertheless, when improperly managed, they can result in insanitary living environments, excess environmental loads, and resource wastage (both the human excreta itself and resources needed to construct another pit latrine), as shown in the Fig. 6.3 (Sakai et al. 2007a). To ensure that pit latrines function effectively as sanitary facilities that enhance the living environment, careful consideration must be given to appropriate site selection, adequate capacity, and ensuring proper maintenance conducted in line the design of the latrine.

6.3.2 Socio-environmental Problem Identification

As mentioned in the previous section, various problems related to drinking water safety and sanitation remains unresolved. Some families still consume arsenic contaminated shallow tube well water and most housewives continue to bear the burden of long trips to collect and transport safe water from the limited number of safe water sources. Furthermore, as previously discussed, sanitation has direct interrelationships with water quality preservation, agricultural land productivity, and the natural nutrient cycle. In fact, the current sanitation situation might actually hinder the creation of safe, hygienic, and productive living environments. In the following subsections, factors that relate to failure situations will be examined based on problems encountered at safe water supply projects and current sanitary situations.

6.3.2.1 Low Awareness Levels

Communication failures are often related to low awareness levels on the part of local residents regarding the issues. Since arsenic contamination awareness-raising programs were not implemented in an organized way, the villagers were forced to piece together fragmentary information about arsenic contamination and its related health risk from mass media sources. As a result, their knowledge is not systematic and the level of awareness remains low. Based on the survey conducted in December 2006, the relationship between the drinking water sources and villager anxiety regarding the state of their drinking water is shown in Table 6.2. Although more than 60 % of the respondents expressed anxiety, 14 of the 19 respondents reported that they were unconcerned about the fact that they were still consuming arsenic contaminated STW water. This can be attributed to a low awareness level.

As for sanitation, more than half of the respondents chose "3" as the level of anxiety they felt, which indicates that their interest and awareness of the importance of sanitation is low compared to their anxiety of drinking water issues.

6.3.2.2 Inadequate Technology Transfer

When installed facilities have been abandoned, the reason can often be attributed to inadequate technology transfer together with a lack of clear ownership. An arsenic removal filter that had been purchased and installed by a foreign NGO is a typical example. Inadequate technology transfers can also lead to improper management. The pit latrines problems shown in Fig. 6.3 are examples of typical cases where the facilities were not properly constructed and/or maintained. Pit latrines should have sufficient capacity to hold human excreta for up to 2 years, and twin pits are preferable so they can be used in turn. These design concepts were not disseminated properly into this country. The lack of technical option variety can also cause problems related to sanitation. Before the installation of EcoSan toilets, other

Table 6.2 Drinking water sources and the degree of related anxiety

		Degree of worry over drinking water					
Drinking water source	Time for water collecting	Anxious (5)	4	3	2	Anxious (1)	Total
Private deep well		2			1	1	4
Public deep well or private deep well owned by other	Less than 10 min	18	2	4	4	8	36
	10 min or longer	24	1		1		26
Private shallow well		3			5	4	12
Private shallow well owned by others		2			3	2	7
Total		49	3	4	14	15	85

Anxiety was evaluated in five grades. (5) means "very anxious" while (1) means "unconcerned" technical options were seldom considered in preference to pit latrines, even though pits have obvious disadvantages in flood prone districts.

6.3.2.3 Insufficient Communication and Limited Community Participation

In light of the circumstances described above, it is assumed that communication deficiencies between villagers and facility suppliers resulted in the problems. Insufficient community participation in the decision-making process might have also led to a lack of a sense of ownership, while insufficient community participation might have prevented the creation of an appropriate plan. This is especially relevant in site selection for arsenic mitigation facilities as a badly situated site will increase the stress experienced by the women who must collect and transport water.

6.3.2.4 Lack of Responsible Management Organization

Although it might have been said that DTWs require minimal maintenance work, periodical arsenic testing is necessary and the results must be reported to the users. Thus, a management organization should have been arranged within the local community. In the village, available information related to DTW safety has been confusing because no responsible organization existed to provide accurate information. Therefore, despite the fact that most of the people depend on the public DTW, users were unable to obtain accurate information related to their safety.

As this case shows, the lack of a monitoring and information system exacerbates a variety of problems. Furthermore, a monitoring system is also necessary to ensure actual sanitary conditions are known, prevent improper construction (such as pits connected with water sources) and to ensure proper maintenance (such as fixing broken toilets); all of which would have the beneficial affect of expanding the influence of the system, even if the number of the households failing to accept proper management is small. Various identified problems and their causes are shown in Fig. 6.4.

6.4 Water Supply and Sanitation Related Risks

Related to the above-mentioned socio-environmental problems, a number of health and environmental risks have been imposed on the people residing in rural areas. Arsenic drinking water contamination is one of the more serious health risks as it brings with it the possibility of increased numbers of arsenicosis and cancer victims. This health risk is exacerbated by the limited availability of alternative safe drinking water due to surface water pollution caused by improper human excreta management. Furthermore, improper sanitation facilitates the spread of

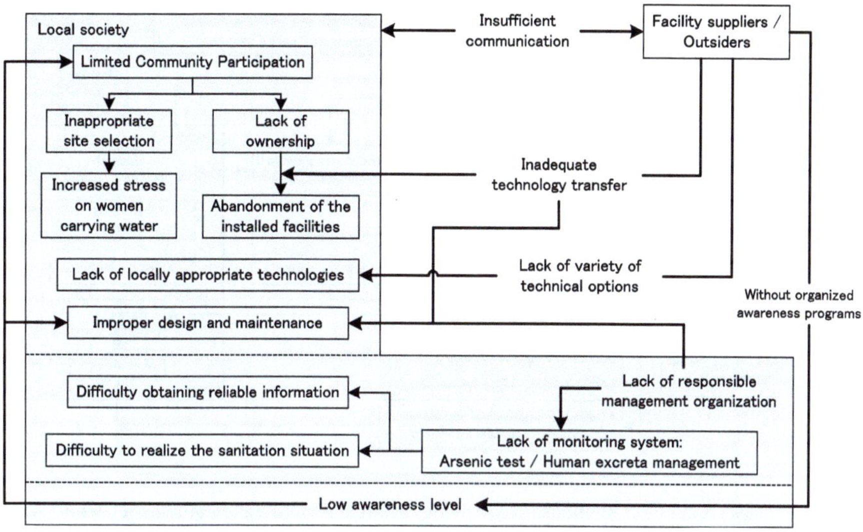

Fig. 6.4 Identified problems and their related causes

parasites and vectors carrying germs. Parasites cause malnutrition, which increases individual vulnerability to disease; so a person suffering from malnutrition is more likely to be affected by arsenic poisoning.

In addition to the scarcity of alternative drinking water sources, surface water pollution restricts domestic water utilization for cooking, cleaning, and bathing purposes.

As the authors mentioned previously, if sanitation facilities forcibly divert human excreta from the natural nutrient cycle, farmers are obliged to apply chemical fertilizers, which will cause the organic content of the soil to decrease and, in turn, degrade soil productivity in areas where alternative organic sources are unavailable. Such soil deterioration will accelerate malnutrition.

The cause and effect relationship involving health and environmental risks related to water supply and sanitation is shown in Fig. 6.5. As can be seen in the figure, health risk directly relate to economic loss, poverty, and high infant mortality rates, while environmental risks also exacerbate poverty by causing production shortfalls. Thus, water supply-related risks are integral parts of the poverty-related vicious cycle.

Since Basailbogh is located in one of Bangladesh's many flood prone areas, it is highly likely the regular floods impact sanitation conditions in the village. At such times, villagers are often unable to use the installed facilities, and the human excreta in the pits can be intentionally or accidentally discharged. Such flooding also makes it difficult to maintain a safe drinking water supply, which in turn exacerbates health risks related to water supply and sanitation.

From this risk identification process, it can be recognized that the development of an integrated scheme aimed at risk reduction through local living environment

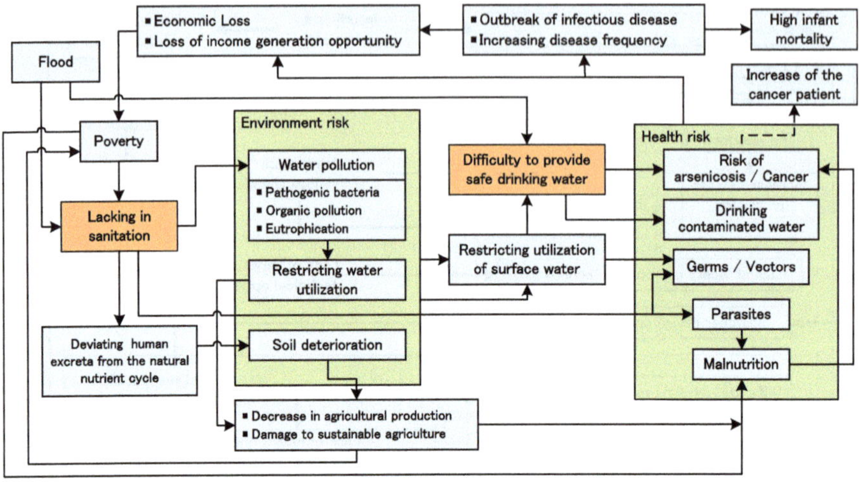

Fig. 6.5 Cause and effect relationship involving health and environmental risks related to water supply and sanitation

improvements is necessary. For example, when providing arsenic mitigation facilities and/or sanitation facilities, it is necessary to put full consideration on vulnerability factors such as poverty, resource and finance limitations, and geographical conditions such as flood recurrences and groundwater salinity.

6.5 Locally Appropriate Technologies

In this section, the authors will discuss locally appropriate water supply and sanitation technologies that can be used to cope with related health and environmental risks, thereby clarifying the design requirements for multi-purpose and multi-stage alternatives.

Technical options applied to local communities must match the characteristics and requirements of the local society, and must take into consideration the limitations of available materials, capital, and human resources. In other words, locally appropriate technology refers to technology that satisfies the abovementioned requirements and which would improve the health and welfare for people in the local society, if used.

6.5.1 Water Supply

Drinking water supply facilities that mitigate arsenic contamination are categorized into the arsenic removal from contaminated tube-well water and the changeover to

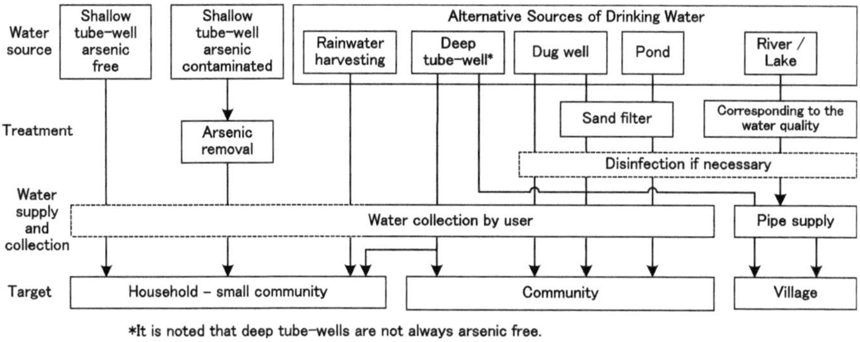

Fig. 6.6 Technical options to mitigate arsenic contamination

alternative drinking water sources. Such alternative drinking water sources are shown in Fig. 6.6. In this figure, treatment, supply method, manpower or pipe supply, and target size, is also shown from household to village level. Since the stress that accompanies water collection labour depends on the facility location, site selection is one of key subjects for consideration when contemplating facility introduction, especially if the target is community sized or larger. When selecting a water source, information regarding safety, water quality, iron or salinity content, taste, quantitative stability, especially during dry season, must be collected and thoroughly analyzed.

6.5.2 Sanitation

For the present, household-sized sanitary facilities should be given priority in rural areas because most toilets constructed for common use are eventually abandoned due to the lack of ownership and proper maintenance. However simply installing a toilet, which just provides the fixed place to defecate, does not result in proper sanitation – even though it can have a major impact on the hygienic behaviour of rural villagers. Appropriate human excreta management is mandatory for complete sanitation to be assured. Therefore, any technical options considered must be capable of fulfilling this requirement.

In locations where agriculture is a primary occupation, preference should be given to sanitary human excreta management methods that utilize resources derived from human excreta as organic fertilizer or soil conditioners. The EcoSan toilet is designed to process human excreta into a useable resource as part of efforts to control pathogenic infection. In addition to its resource utilization, the EcoSan toilets installed in Basailbogh have been accepted because they do not produce bad odour or attract flies. Therefore, safety and comfort is also assured (Sakai et al. 2007b).

6.5.3 Common Factors to be Considered

When selecting technical options in Bangladesh, the following common factors should be considered: energy and resource consumption, environmental impact, operation and maintenance, useful service life, and flood resistance. The authors considered these factors, together with locally appropriate technology requirements and ultimately selected the EcoSan toilet type and design for their project areas, which includes Basailbogh, because it requires almost no energy and has minimal environmental impact.

Since one local post-defecation custom was incompatible with the normal EcoSan toilet method of rendering excreta harmless, a design adjustment was necessary, and it was also necessary for users to endure an inconvenient change their usual behaviour. However, after gaining an understanding, through actual practice, that the behaviour change was necessary in order to obtain good quality resources from human excreta, the users accepted the inconvenience.

The socio-cultural aspects of the local society, which include human relations and/or original decision-making processes, will have an impact on community-based maintenance and affordability options. As such, they will also affect the selection of technical options and management systems. Concurrently, it must also be noted that the installed options will have social impacts on local practices and society.

6.5.4 Design of Multi-purpose and Multi-stage Alternatives

As has been discussed above, when it is necessary to develop alternatives that will improve the living environment, reduce environmental impacts, and minimize health risks, the most beneficial alternatives will fulfill multiple purposes, and hence have multiple stages. These stages relate to the priorities in which the issues are to be addressed and their sub areas. Therefore, it is important to clarify the priority of "what" kind of facility to construct and "where" to construct it. The evaluation method used is one of the subjects. The priority and constraints regarding capital will determine "when" to construct the facilities. In addition to the construction process, the establishment of a management system and a responsible local organization must be addressed.

When designing alternatives, it is also important to address the interrelationships between issues. For example, when selecting surface water (such as a pond) as an alternative drinking water source, it is necessary to preserve water quality by installing appropriate sanitation facilities. Another example of the interrelationship is the way in which an installed option causes a separate issue. For example, if installing a facility that allows users to use increased quantities of water, such as a water pipe system, it is important to realize that without an appropriate drainage system, that option can result in an insanitary environment caused by the accumulation of used water. Therefore, it is required to visualize and take countermeasures for every potential impact for newly installed technical options.

6.6 Conclusion

In this chapter, the authors have identified the following problems and risks through their field survey experiences and activities:

- Problems related to the current water supply and sanitation situation: Fig. 6.3 for sanitation.
- Socio-environment problems that have caused various failures: Fig. 6.4.
- Environmental and health risks: Fig. 6.5.

Figures 6.3, 6.4, and 6.5 show that the current water supply and sanitation situation has wide-ranging interrelationships with living environment requirements, environmental risks, and health risks. For example, water supply and sanitation have a naturally close relationship due to the need to ensure water quality preservation via appropriate sanitation, especially when surface water is to be utilized as an alternative drinking water source. In addition to this relationship, the interrelationships between water supply and sanitation extend to environmental and health factors, as shown in Fig. 6.5. Therefore, it is necessary to carefully consider water supply and sanitation within the context of improving local living environments and reducing relevant risks.

The requirements for locally appropriate technologies and the necessity of multipurpose and multi-stage alternatives were discussed, and it was stressed that consideration must be given to the relationships between social aspects and installed options. More specifically, it is inevitable that some social aspects will affect the options selected, and that the installed options will have inevitable impacts on the local society.

Acknowledgments The authors would like to express deep gratitude to Mr. Tofayel Ahmed, a local member of the Japan Association of Drainage and Environment (JADE), a Japanese NGO, for his meaningful contributions during authors' field survey. Basailbogh Village is one of the JADE project sites that are being developed in collaboration with the Society for People's Action in Change and Equity (SPACE), a local NGO. The authors would also like to thank Mr. Azahar Ali Plamanik, the executive director of SPACE, his staff, and the residents of Basailbogh for their cooperation.

References

Sakai, A., Takahashi, K., Ahmed, T., Sakamoto, M., & Hagihara, Y. (2007a). Issues on safe water supply and sanitation and local people's welfare in a rural Bangladesh. In *International conference on water & flood management, 12–14 March 2007* (Vol. 2, pp. 537–544). Dhaka, Bangladesh.

Sakai, A. et al. (2007b). Challenges to overcome the problems related with sanitation in rural area of Bangladesh. *Paper presented to the first annual conference, Bangladesh Regional Science Association, 16–17 March 2007*. Dhaka.

Chapter 7
Safe Drinking Water and Sanitation Awareness in Rural Bangladesh

Akira Sakai, Maiko Sakamoto, and Kunio Takahashi

Abstract Safe drinking water and sanitation are closely related to the environmental and health risks faced by villagers in the rural areas of Bangladesh where tube-well arsenic contamination is an issue of particular concern. Since a significant percentage of the people affected by this problem have poor access to safe drinking water, the risk of increasing numbers of people suffering from arsenicosis is considered to be high. Regarding access to sanitation facilities, while the coverage ratio of improved facilities is increasing, improper management of existing facilities is often observed in rural areas. In this chapter, the authors conducted questionnaire surveys related to safe drinking water and sanitation awareness in two rural Bangladeshi villages with differing socio-environments. Based on the results of those surveys, the current status and local awareness levels of issues concerning drinking water and sanitation in their villages will be compared while taking socio-environmental differences into consideration. The authors will also discuss how the questionnaire survey results were used to help the villagers discuss alternatives that will improve their living environments and reduce health risks through the mitigation of arsenic contamination and sanitation improvements.

Keywords Arsenic contamination • Health risk • Awareness on safety of drinking water

A. Sakai (✉)
University of Marketing and Distribution Sciences, Kobe 651-2188, Japan
e-mail: akira_sakai@red.umds.ac.jp

M. Sakamoto
University of Tokyo, Kashiwa, Chiba 277-8653, Japan
e-mail: m-sakamoto@k.u-tokyo.ac.jp

K. Takahashi
Japan Association of Drainage and Environment, Shinjuku, Tokyo 162-0067, Japan
e-mail: nqg48559@nifty.com

K. Hagihara, C. Asahi (eds.), *Coping with Regional Vulnerability*, New Frontiers in Regional Science: Asian Perspectives 4, DOI 10.1007/978-4-431-55169-0_7

7.1 Introduction

In rural areas of Bangladesh, safe water and sanitation are among the more urgent issues related to human living environments, and it is necessary for local societies to improve their own living conditions by facing such problems and understanding their characteristics.

One distinct drinking water-related issue in Bangladesh is arsenic contamination in the tube wells that have been constructed throughout the country since the 1980s. Currently, the number of such wells is approximately ten million; of which 30 % produce water with arsenic levels that exceed the nation's safe concentration guidelines (50 μg/L). The affected areas cover 70 % of the entire country. A considerable percentage of people in the affected regions has difficulty accessing safe drinking water, and are therefore forced to utilize arsenic contaminated water. Therefore, the risk of contracting arsenicosis and cancer is considered high.

As for access to sanitation facilities, while pit latrines are the most common type of sanitary facility in use, the proper maintenance and management of such latrines has not been disseminated properly in rural Bangladesh, and improper management is commonly observed in rural areas.

The authors have been discussing the socio-environmental problems observed in villages where they have been engaged in field activities and the risks they have identified related to water supply safety and sanitation (Sakai et al. 2008). As part of this effort, the authors conducted questionnaire surveys related to safe drinking water access and sanitation in two rural villages that have significantly different socio-environments and safe drinking water options, and where the ratio of families that utilize an acceptable level of sanitation is also significantly different. In the following section, the current status and local awareness states related to drinking water and sanitation in the two villages mentioned above will be examined in relation to their socio-environment differences. Then, based on the questionnaire survey results, the authors will discuss potential ways to develop integrated schemes that will result in improved living environments through arsenic contamination mitigation and sanitation improvements.

7.2 Study Area and Questionnaire Survey

The authors conducted a questionnaire survey related to safe water supply and sanitation for entire families in Basailbogh Village in Srinagar Upazila, Munshiganj District, and Bansbaria Village in Keshobpur Upazila, Jessore District of Bangladesh. Basailbogh is located 30 km to the south of Dhaka, the capital city of Bangladesh, while Bansbaria is 160 km to the southwest of Dhaka. Both villages are located in flood-prone and arsenic-affected areas.

As for their socio-environments, primary occupations are shown in Fig. 7.1. As can be seen in the figure, most families in Bansbaria subsist on agriculture, as

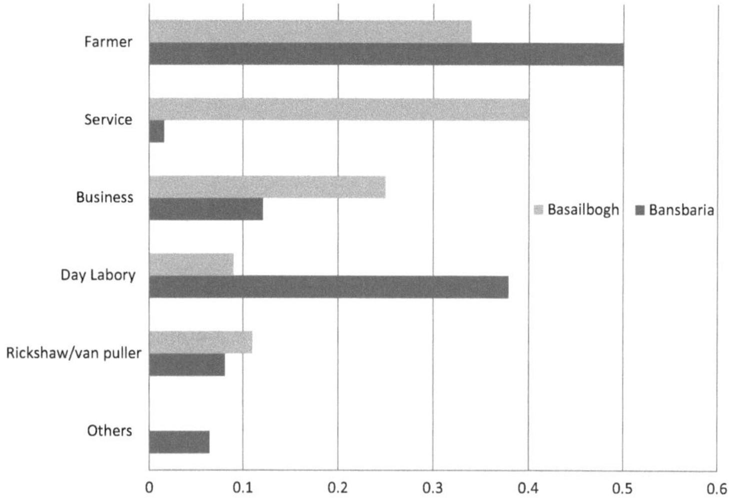

Fig. 7.1 Primary householder occupations (Multiple answers were permitted)

almost 50 % of the respondents list farmer and 40 % of them list farm-related day labour work as their primary occupation. In contrast, in Basailbogh, where most of the villager's homes are surrounded by farmland, the percentage of persons engaged in small business and service industry work exceeds 60 %. However, since the village is located near Dhaka city, a sizable percentage of the population works in the city, and population movements occur at a high frequency.

Literacy rates are also different between the two villages, with 60 % literacy observed in Basailbogh compared to 38 % in Bansbaria. To simplify, while the occupations of Basailbogh villagers are becoming more urbanized, Bansbaria remains a typical poor rural village.

The surveys were conducted in August and September 2007 (Basailbogh) and in March 2009 (Bansbaria). The questionnaire items consisted of personal data and questions related to drinking water, sanitation, and daily life. Survey responses were received from 118 out of 123 families in Basailbogh and 124 out of 125 families in Bansbaria, respectively.

7.3 Current Drinking Water and Sanitation Conditions

In Basailbogh, there are a total of seven deep tube wells (DTWs) capable of providing safe drinking water. However, 30 families continue to obtain water from arsenic-contaminated shallow tube wells (STWs). This is primarily because the task of collecting and transporting safe water from distant DTWs is burdensome. As a result, 25 % of the families in the village do not have access to a safe source of drinking water (Shibata et al. 2009), even though two arsenic iron

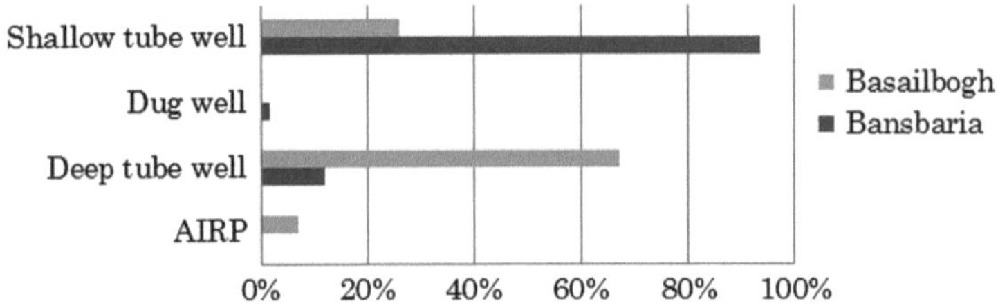

Fig. 7.2 Comparison of primary drinking water sources

removal plants (AIRPs) were installed for families by a non-government organization (NGO) in 2006 to remove arsenic from STW water.

In contrast, as shown in Fig. 7.2, more than 90 % of the families in Bansbaria utilize arsenic contaminated STW water. There is no source of arsenic-free drinking water in Bansbaria, which means that most villagers drink arsenic-contaminated STW water.

It should also be noted that, even though there are a few DTWs in neighboring villages, due to the aforementioned difficulties involved in collecting and transporting DTW water, Bansbaria villagers seldom avail themselves of those sources. Furthermore, women villagers responding to a participatory rural appraisal (PRA, see Fig. 7.3) conducted by the authors in August 2008 said that psychological and social barriers prevented them from obtaining this safe drinking water. Such barriers included a lack of understanding and cooperation on the part of their family members (including their husbands) that persisted even if the women desired to acquire safe water for their children.

As shown in Fig. 7.4, about 40 % of the villagers in Bansbaria do not have access to toilets and routinely defecate in the open, while approximately 50 % use pit latrines. Furthermore, available toilet types differ according to the householder's occupation, as shown in Fig. 7.5. In this figure, a comparison between the two primary village occupations, day laborer and farmer, is shown. Of families that list householder occupations as day laborer and rickshaw puller, the ratio that do not have access to toilets exceeds 50 %, and none of the pit latrines used by such families are water sealed.

In contrast, more than 75 % of the families in Basailbogh use pit latrines, the majority of which are water sealed, although some families use flush toilets linked to septic tanks. Furthermore, there are differences in hand washing practices between two villages, as shown by the accumulative curve in Fig. 7.6.

From answers regarding medical expenses, primarily medicine purchases, it was determined that families in Bansbaria spend relatively high amounts, with 23 % of the respondents answering that they annually spend in excess of 5000 Taka (Tk), which is approximately 75 USD, and 50 % answering that they spend more than 2000 Tk each year. In contrast, in Basailbogh, only 30 % of the villagers pay out more than 2000 Tk on medical expenses, which is significantly less than Bansbaria. People in Bansbaria are more likely to suffer from ailments such as fevers, diarrhea, influenza, and so on. Furthermore, 94 % of the respondents stated that at least one

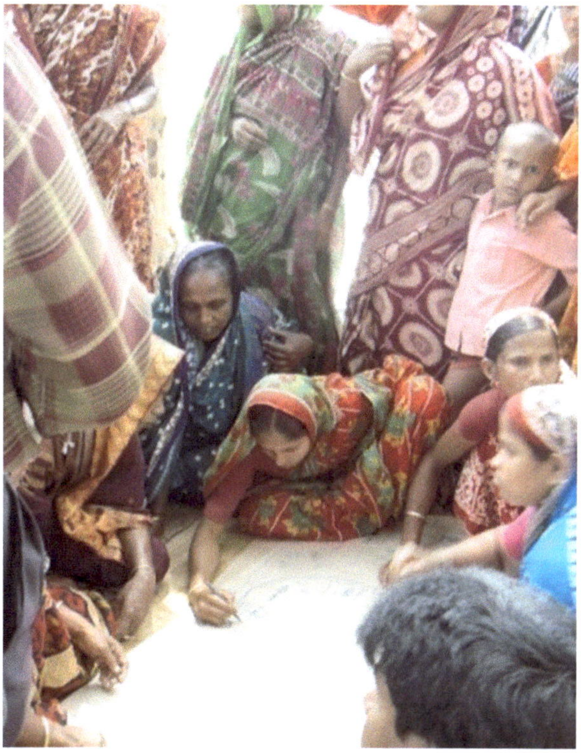

Fig. 7.3 A PRA session in Bansbaria village

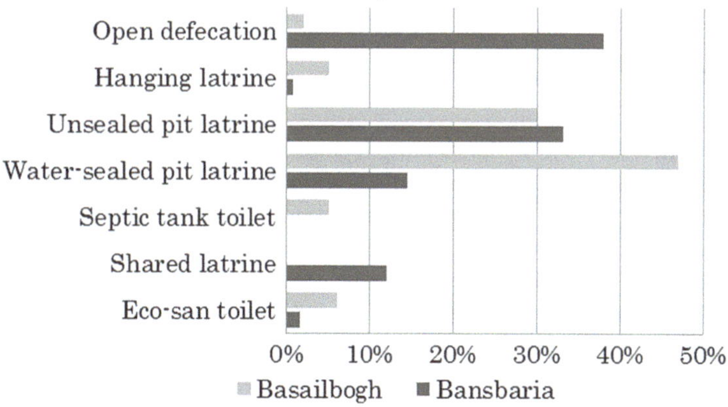

Fig. 7.4 Comparison of current toilet types

member of their family has suffered from disease. This indicates that people in this village suffer from health risk exposure, and that improving sanitation should be a high priority.

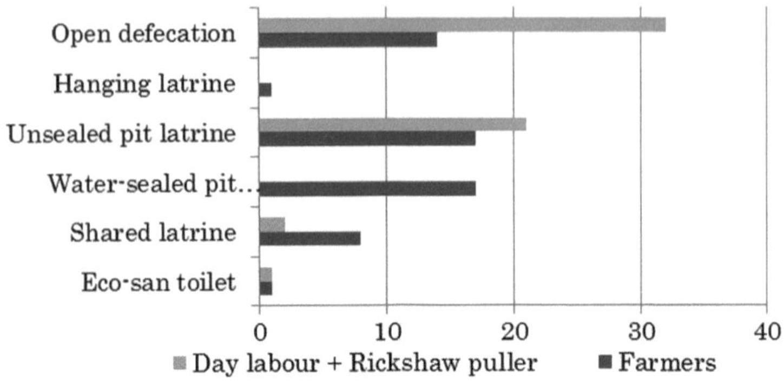

Fig. 7.5 Toilet type comparisons based on occupations in Bansbaria

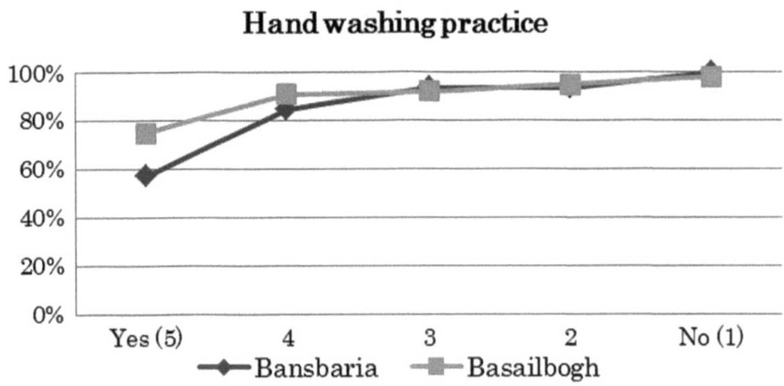

Fig. 7.6 Comparison of hand washing practices (*accumulated curve*)

7.4 Sanitation Awareness

First, the varied levels of understanding regarding the relationship between sanitation, surface water pollution, pond water and water in inundated areas, and disease spread must be noted. For example, as shown in Fig. 7.7, villagers in Bansbaria are significantly less aware that proper sanitation is essential for preventing surface water pollution, and thus for reducing the frequency of disease outbreaks than villagers in Basailbogh. Furthermore, Fig. 7.8 compares sanitation-related consciousness levels between literate and illiterate persons. As can be seen in the figure, there is a significant difference between the two groups in Bansbaria because the awareness level of literate people is notably higher than among illiterate residents. Accordingly, if surface water is to be supplied as a safe drinking water source as part of efforts to mitigate arsenic contamination, it will be necessary to increase villager sanitation awareness. It will also be necessary to educate illiterate persons about the necessity for improving sanitation in order to keep surface water clean.

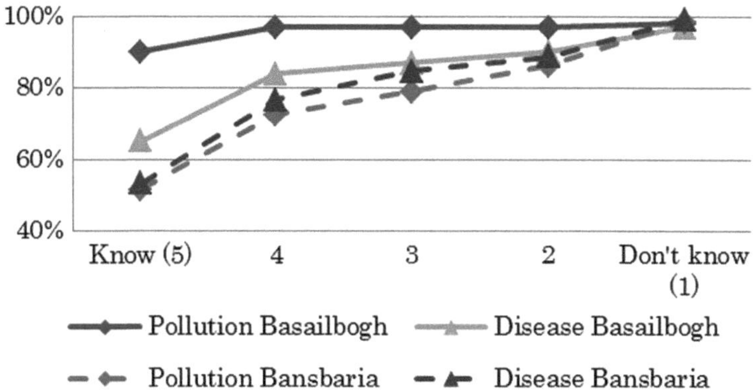

Fig. 7.7 Comparison of awareness of the link between sanitation and water pollution/sanitation and disease (*accumulated curve*)

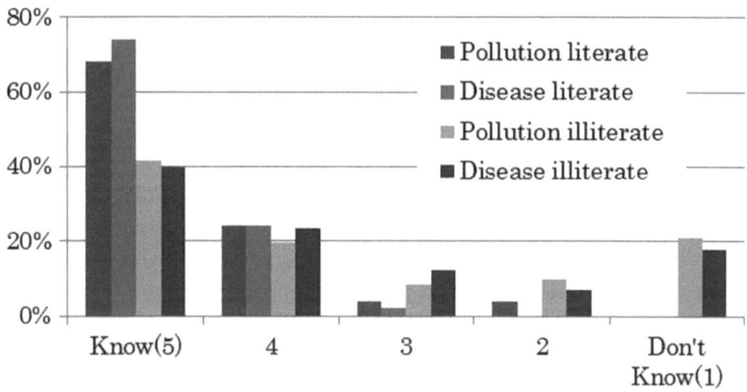

Fig. 7.8 Comparison of awareness of the link between sanitation and water pollution/sanitation and disease incidence between literate and illiterate respondents in Bansbaria village

From the results of the Basailbogh survey, the authors learned that there are a number of people who are using comfortable toilets that are clean and odour free, and who are therefore satisfied with their sanitary arrangements, even if there is no system available to safely and completely dispose of human excreta in their village (Sakai et al. 2009). In Bansbaria, since the number of such toilets is limited compared to Basailbogh, as shown in Fig. 7.4, most of the respondents expressed concerns about sanitation, as shown in Fig. 7.9. However, as in the abovementioned study, there is a similar percentage of comfortable toilet users who are unlikely to express sanitation-related anxieties.

Therefore, it is necessary to share the importance of proper human excreta management with such villagers, and to demonstrate methods for appropriate community-level human excreta management when installing and improving sanitation facilities in Bansbaria village, as most of villagers are anxious to improve their current situation.

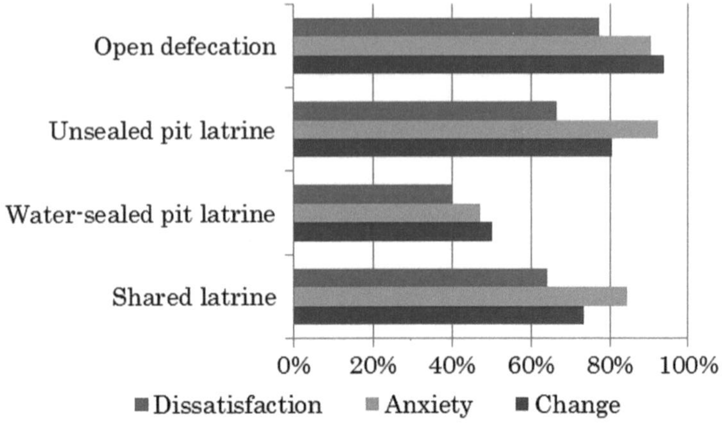

Fig. 7.9 The ratio of "dissatisfaction", "anxiety", and "desire to change current toilet" according to current toilet types in Bansbaria

7.5 Arsenic Contamination Awareness and Safe Water Supply

Figure 7.10 shows comparisons related to drinking water safety, arsenic contamination anxieties, and drinking water satisfaction between residents of the two villages. As can be seen in the figure, the percentages of the residents who answered "safe or not safe" differ depending on the drinking water sources. Yet, in spite of the fact that there are safe water options in the village, the residents of Basailbogh remain anxious about arsenic contamination. Sakai et al. (2009) noted this anxiety and speculated that it resulted from distrust of suppliers caused by the lack of sufficient information regarding the safety of available options. As a result, people in Basailbogh remain unsatisfied with their current drinking water situation.

In Bansbaria, on the other hand, even though the current water safety situation is noticeably inferior to Basailbogh, the ratio of respondents who felt high levels of anxiety was much lower than Basailbogh. Furthermore, while the knowledge and awareness levels are lower in Bansbaria, where 75 % of the respondents think their drinking water is unsafe and arsenic contamination awareness levels are relatively high, the relationship between anxiety levels and drinking water satisfaction (shown in Fig. 7.11) do not indicate corresponding levels of dissatisfaction. In other words, although there is a certain level of anxiety, many villagers are indifferent towards the potential risks and the safety of their drinking water. Regarding the risks posed by arsenic contamination, since the villagers are not fully informed or equipped with ready solutions, the result is cognitive dissonance, and it can be said that they are dealing with this dissonance by considering themselves "satisfied" or as "seeing no need to improve" the current state.

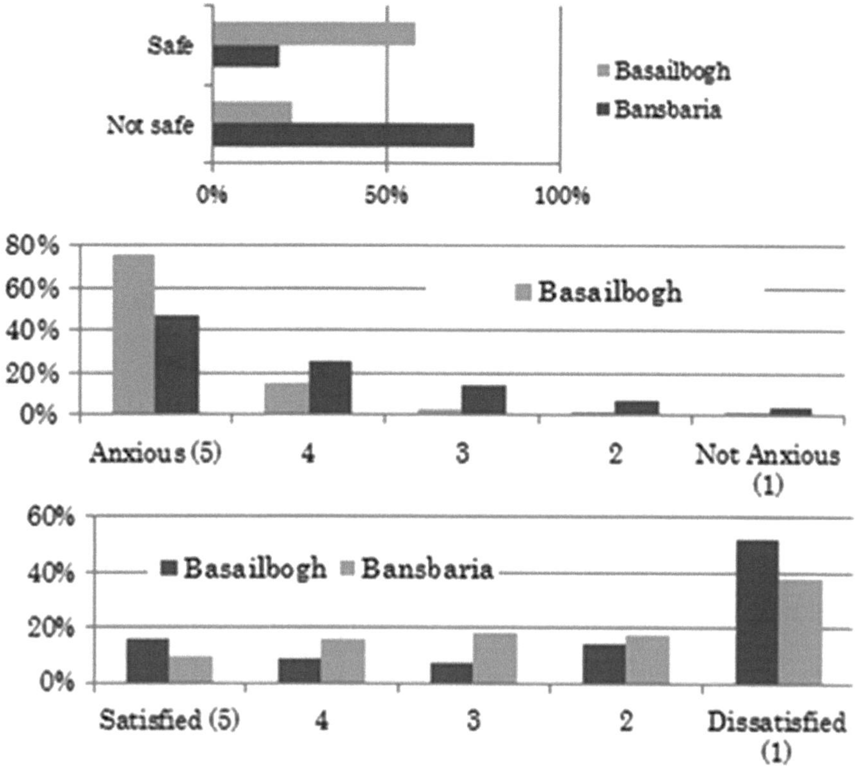

Fig. 7.10 Comparisons between two villages: drinking water safety (*top*), arsenic contamination anxiety (*middle*), and drinking water satisfaction (*bottom*)

Furthermore, since the villagers have very limited experience with support from NGOs or local governments, most have no knowledge of safe water options other than DTW, and see little prospects for resolving the current situation.

7.6 Willingness to Improve Living Environments

Figure 7.12 shows responses to the question, "What worries you right now?" People in Bansbaria have a variety of concerns in their daily life, with arsenic contamination ranking first among them and sanitation and disease close behind. However, the willingness to work on improving their living environment is much lower in Bansbaria than Basailbogh, as shown in Fig. 7.13. One reason for this difference is that, in spite of the various problems and failures that the inhabitants have experienced, Basailbogh has seen a number of improvements over the years that have given the villagers a sense that living environments can be improved and that their living standards are relatively high.

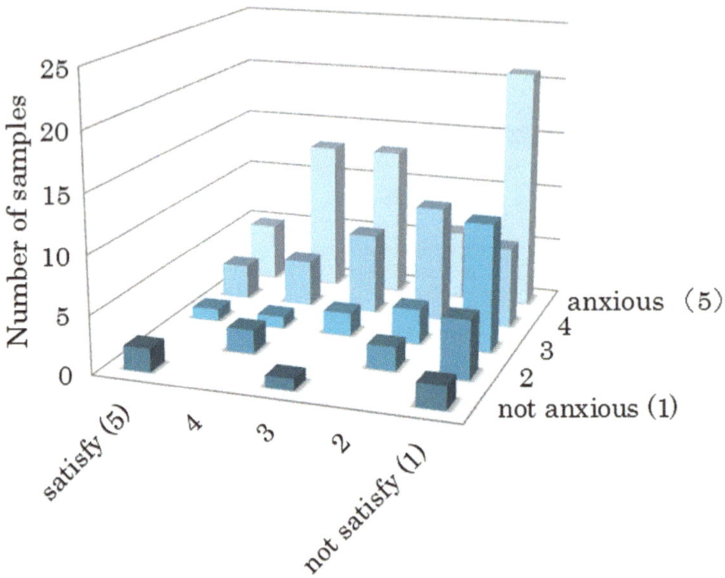

Fig. 7.11 Relationship between anxiety and drinking water satisfaction in Bansbaria Village

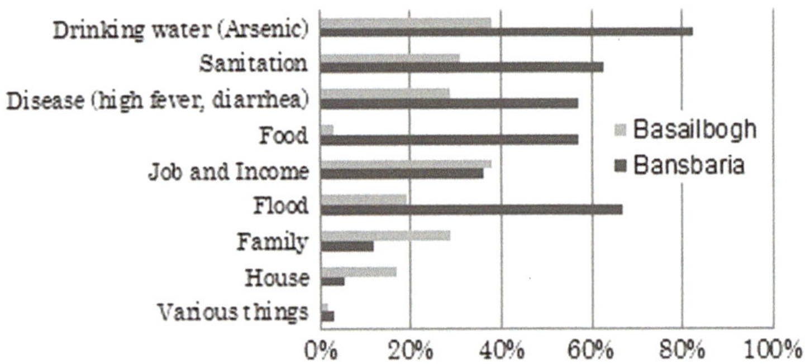

Fig. 7.12 What are you worried about? (Multiple answers were permitted)

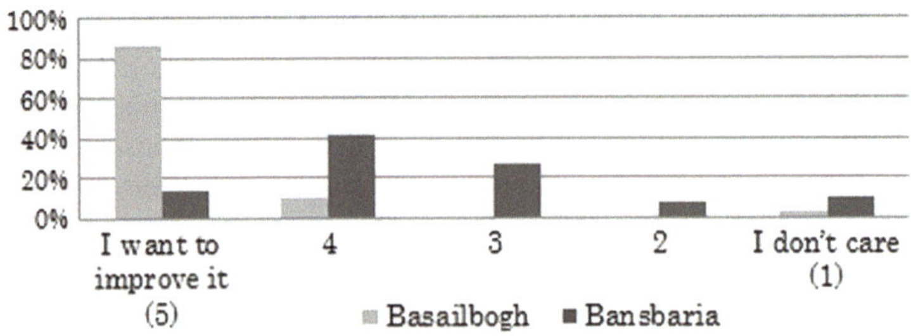

Fig. 7.13 Comparison of willingness to improve living environment

Table 7.1 Relationship between the anxieties in Bansbaria

	Arsenic contamination		
Sanitation(↓)	Anxious	Middle	Not anxious
Anxious	72	13	6
Middle	3	2	5
Not anxious	14	3	3

In contrast, the willingness of Bansbaria villagers to work towards improving their living environment remains undeveloped, even though a considerable portion of the villagers are knowledgeable about sanitation related problems and drinking water safety. From this perspective, cognitive dissonance is considered to be responsible for the Bansbaria villagers' lack of incentive for improving their living environment.

As shown in Table 7.1, the same people in Bansbaria who are anxious about sanitation are likely to be anxious about arsenic contamination. Furthermore, as mentioned earlier, it is known that there is a close relationship between water supply and sanitation and that if surface water is used, then water safety cannot be accomplished without appropriate management of human excreta.

Therefore, an integrated approach is necessary. Along with the process of installing water supply facilities and improving sanitation, increasing awareness and instilling a willingness to improve living environments is essential. Previous results have shown that simply sharing information regarding safe water options and improved sanitation is insufficient. Instead, information sharing and examining different perspectives on how to implement and manage such facilities is considered necessary to inspire unenthusiastic villagers to pursue improvements in their living environment. To achieve such shifts in perspective, community involvement is essential because community-based organization will play an important role in maintaining any installed facilities.

7.7 Conclusions

In this chapter, the authors examined the reasons for the observed differences in awareness and anxiety between two villages in rural areas of Bangladesh that result from the socio-environments of their villages and respective experiences related to sanitation and safe drinking water. The authors believe the following areas need to be addressed if the living environment of Bansbaria is to be improved:

- Instilling willingness to improve the living environment is a key issue given the constraints of limited affluence and low literacy level.
- Development of an integrated methodology for improving living environments through enhancing safe water supply and sanitation while considering the relationship between water supply and sanitation.

- Applying the lessons learnt from Basailbogh, such as addressing the lack of trust between local communities and suppliers and the sense of satisfaction among most people using improved toilets in the absence of a well-developed human excreta management system exists.

Acknowledgment The authors would like to thank Prof. Yoshimi Hagihara, Prof. Kiyoko Hagihara and Mr. Sho Shibata for their kind advice during the formulation of this research, and for their contributions to the questionnaire design. The authors also thank Mr. Qazi Azad-uz-zaman, Mr. Tofayel Ahmed, Mr. Yusouf Ali and the members of related NGOs working in the research areas for their assistance with administration of the questionnaire surveys.

References

Sakai, A., Takahashi, K., Sakamoto, M., Hagihara, Y., & Hagihara, K. (2008). Water supply and sanitation relating risks and social environment in rural areas in Bangladesh. In *4th international conference on water resources and environment research* (pp. 795–806), Adelaide.

Sakai, A., et al. (2009). Diagnosis of sanitation and local people's willingness to improve the living environment in a rural area of Bangladesh. In *2nd international conference on water and flood management* (pp. 45–52). Dhaka.

Shibata, S. et al. (2009). A planning process to install safe water options for arsenic contamination of drinking water in Bangladesh. In *2nd international conference on water and flood management* (pp. 143–150). Dhaka.

Part III
Evaluation of Regional Vulnerability

Chapter 8
Marginal Willingness to Pay for Public Investment Under Urban Environmental Risk: The Case of Municipal Water Use

Kiyoko Hagihara, Chisato Asahi, and Yoshimi Hagihara

Abstract Marginal willingness to pay for public investment under urban environmental risk is considered in this chapter. In particular, we show a model that takes into account a bounded rationality on the ability of risk perception, information situation, and people's threshold acceptance of risk. Then, as one case study of urban environmental risk, the evaluation of risk in municipal water use is shown. In order to investigate the information situation and risk perception of people, a survey is conducted using questionnaires. The survey reveals that risk awareness is a factor in drinking water, and that people change their choice based on risk information. Then, the effects of information on risk and public investment are considered. From some numerical examples, marginal willingness to pay is found to be low in the case of high risk, because of consumers' self-defensive activities. In other words, marginal willingness to pay for public investment is high when there is no or little averting behaviour. Moreover, it is shown that consumers' perception of risk is largely dependent on information on risk, countermeasures taken by public authorities, and overconfidence in private averting goods.

Keywords Marginal willingness to pay • Environmental risk • Municipal water use • Public investment • Bounded rationality

---

K. Hagihara (✉)
School of Sociology, Bukkyo University, 96 Kita-hananobo, Murasakino, Kita-ku, Kyoto 603-8301, Japan
e-mail: hagihara@bukkyo-u.ac.jp

C. Asahi
Division of Urban Policy, Faculty of Urban Liberal Arts, Tokyo Metropolitan University, 1-1 Minami-osawa, Hachioji, Tokyo 192-0397, Japan
e-mail: asahi@tmu.ac.jp

Y. Hagihara
Kyoto University, 264 Suemarucho, Nakagyo-ku, Kyoto 604-0901, Japan
e-mail: y-hagi@gf7.so-net.ne.jp

K. Hagihara, C. Asahi (eds.), *Coping with Regional Vulnerability*, New Frontiers in Regional Science: Asian Perspectives 4, DOI 10.1007/978-4-431-55169-0_8

8.1 Introduction

In the Basic Environmental Plan legislated in 1994 in Japan, the term "environmental risk" is defined as the possibility of adverse effects brought about by the introduction of chemical materials into the environment. In this definition, environmental risk has a very narrow meaning compared with other previous research. For example, environmental risk is regarded as a measure of the probability and severity of adverse effects to health, property, or environment in the national criterion of risk analysis in Canada (Niels 1992).

Not only environmental risk but also general risk has been considered in various research fields such as finance, labour, disaster prevention or mitigation, and health. However, for environmental risk, it seems to us that the definition of risk and the manner of treating risk differ from one research field to another.

The term "urban environmental risk" is defined here as the probability and severity of adverse effects to the urban environment, that is, risks endangering the safe and secure life, a comfortable life, and an affluent life in an urban area (Hagihara 1996).

Consider a water environmental risk. Water pollution in rivers, lakes, and the sea in an urban area might prevent residents from relaxing at such places. What is more important is that such water bodies are sometimes the sources of the municipal water supply, which may be affected by water pollution.

In the municipal water use, risk to consumers of municipal water might arise if they cannot use the amount or quality of water that they want, for example, if there is a shortage of water or the water smells bad. Because of water pollution in urban water bodies, urban residents are unable to live a safe and secure, comfortable, or affluent life in the urban area.

Traditionally, decision theorists classify decisionmaking as occurring in one of three situations: certainty, risk, or uncertainty. The critical aspect that distinguishes these three situations is the decisionmaker's degree of knowledge about the outcome of a particular course of action. If the outcome is known, the decision is said to take place under certainty. Risky decisions occur when more than one outcome is possible from a particular course of action and the probability of each outcome is known, or can be computed actuarially. Uncertain decisions are characterized by multiple outcomes for each alternative action (some of which may be unknown to the decisionmaker), where the probabilities of these outcomes are not known. It is the characteristics of this last type of decision situation which define most consumer purchase decisions – incomplete information, personal preferences, and subjective perceptions.

Risk is regarded as measurable uncertainty, whereas uncertainty is regarded as unmeasurable uncertainty. Consequently, both risky and uncertain situations are called uncertainty situations in this chapter.

The situation in which urban environmental risk can be considered is dependent on the peril (a dangerous event) in question. Assume the following three situations in connection with the relationship between individual preferences and an event.

1. Individuals know only that the event may affect them. The contents and the severity of loss and the probability of the occurrence of loss that the event may cause are not clearly apparent in their preferences.
2. Individuals know the contents and the severity of loss due to the event. These are apparent in their preferences.
3. Individuals know the contents, the severity of loss, and the probability of the occurrence of loss (subjective probability) as a consequence of the event.

Situations 1. and 2. are regarded as those of decisionmaking under uncertainty and 3. is regarded as one of decisionmaking under risk according to the previous classification. In the case of the situation under uncertainty, a traditional theory that apparently does not consider the probability of the occurrence of losses could be applied to estimate the change of environmental risk (compare Hagihara and Hagihara 1990). In contrast, in the case of a risky situation, a theory such as the expected utility theory could be applied to estimate the change in environmental risk (see Hanley et al. 1997).

Environmental risks in a municipal water use are described first. Second, a model for evaluation of environmental risk in municipal water use is setup in Sect. 8.2. Third, empirical application of the model is shown in Sect. 8.3. Finally, a relationship between individual's marginal willingness to pay and public investment is considered in Sect. 8.4.

8.2 Evaluation of Environmental Risk in Municipal Water Use

8.2.1 Environmental Risks in a Municipal Water Use

Water is taken from a river or a lake, treated in a water supply plant, and then distributed to each consumer or household.

Recently, various kinds of risk have arisen in municipal water supplies. Three kinds of environmental peril may occur in the process of municipal water supply. The first is water pollution in rivers, lakes, or groundwater. There is a wide range of factors responsible for such pollution, including the activities of various industries around those water sources and the livelihood of the residents in such areas. Various kinds of chemical and biological materials from economic and social activities flow into rivers, lakes, or groundwater. The pollution may affect not only the ecology around the water source but also the municipal water supply. Recently, nonpoint pollution from farmland and water drainage from urban areas has become a serious problem. Cryptosporidium in public water bodies is one of the major bacteria causing adverse effects on the municipal water supply. Moreover, some chemical substances flowing into public water bodies may affect human health through the municipal water supply. The second environmental peril occurs during the process from intake to treatment. Among other things, trihalomethane is a byproduct of the chlorination treatment process. A model is needed to treat such risks explicitly. The possibilities that nonpoint pollution, Cryptosporidium, trihalomethane, and other chemical substances may affect the municipal water supply are regarded as uncertain and an unknown risk (Hagihara et al. 2000). The third environmental peril occurs during the distribution from the plant to consumers. We do not take the third peril into account in this chapter.

In order to manage these uncertain and unknown risks, municipal water supply authorities have adopted countermeasures, such as reinforcement of control and surveillance, and an advanced water treatment system. However, these measures impose tremendous cost.

On the other hand, the treatment process is undertaken to supply safe water to consumers. However, consumers have recently had little confidence in the treatment process and they are concerned about the quality of tap water. Many consumers are taking averting actions such as drinking bottled water and soft drinks, or using a water purifier. Examples of these averting actions were presented in our previous work (Hagihara and Hagihara 1990).

Consequently, decisionmaking by both municipal water supply authorities and consumers must be included in a framework of the evaluation of risk in the municipal water supply.

8.2.2 Model Based on the Concept of Risk

There is an enormous amount of research which considers risk in willingness to pay for improvements in drinking water quality (see, for example, Crocker et al. 1991; Gurian et al. 2001; Jordan and Elnagheeb 1993; Kask and Shogren 1994; Syme and Williams 1993). As for the framework of the analysis, McConnell and Rosado (2000) show how nonmarginal benefits from higher drinking water quality can be measured by doing a standard welfare analysis, where the parameters for the analysis are obtained through a nested logit model structured according to the defensive inputs available for drinking water treatment. We have already used a logit model to get benefits from improving a waterside area (Hagihara and Hagihara 2004). Crocker et al. (1991) consider the analytical implications of endogenous risk from the viewpoint of the probability and the location of contamination and risk perceptions. Shogren (1990) found reductions in probability were preferred to severity reduction. Whether policy changes the severity of the event or the probability of its occurrence can influence how consumers value a change in the overall risk. So, an analyst must clarify the component of risk that the proposed policy is changing: probability or severity or both (Shogren 1990).

Assume that the individual knows the severity of loss as a consequence of an event and the probability of its occurrence. Suppose the probability is correct, based on the possible information. Therefore, assume the situation to be one under risk rather than uncertainty.

To keep the exposition simple, assume one possible adverse consequence and two states of the world: loss occurs with a given severity, or loss does not occur. Let the variable A be the severity of loss. A takes the value A^* with probability π, and takes the value 0 with probability 1- π (Freeman 1995).

Individual preferences are represented as

$$V = \mathrm{V}(M, P, A) \tag{8.1}$$

where

$$\mathrm{V}(M, P, 0) > \mathrm{V}\left(M, P, A^*\right) \tag{8.2}$$

and M and P represent income and price (both assumed to be constant) and

$$\frac{\partial V}{\partial M} > 0, \quad \frac{\partial^2 V}{\partial M^2} < 0, \quad \frac{\partial V}{\partial A} < 0, \quad \frac{\partial^2 V}{\partial A^2} < 0. \tag{8.3}$$

There may be two measures to prevent risk from the dangerous event that causes loss at the demand side in municipal water use: reduction of the magnitude of loss and reduction of the probability of the occurrence of loss.

Suppose that the severity of loss is determined by both private investment R and government protective investment G:

$$A^* = \mathrm{A}(R, G) \tag{8.4}$$

where $A(0,0) = A^*$, $A_R^* = \partial A^*/\partial R < 0$, *and* $A_G^* = \partial A^*/\partial G < 0$. In the above context, private investment refers to taking averting actions and public investment refers to adopting the advanced treatment system. The individual chooses R, given G, so as to maximize expected utility

$$\mathrm{E}(U) = \pi \mathrm{V}[M - R, \mathrm{A}(R, G)] + (1 - \pi)\mathrm{V}(M - R, 0) \tag{8.5}$$

where π is the possibility of the occurrence of loss.

The first – order conditions include

$$\frac{1}{A_R^*} = \frac{\pi V_A}{\pi V_{M^*} + (1 - \pi)V_{M^0}} \tag{8.6}$$

where V_{M*} is the marginal utility of income evaluated at the level of A associated with the given level of G. By taking the total differential of (8.5), setting it equal to zero, and substituting in the first-order condition for the private investment [Eq. (8.6)], we obtain

$$\frac{dM}{dG} = -\frac{A_G^*}{A_R^*} = \frac{\partial R}{\partial G} \tag{8.7}$$

This means that the individual's marginal willingness to pay for a small increase in government investment is the ratio of the marginal productivities of private investment and public investment in reducing A^*, or the marginal rate of technical substitution between R and G in reducing A^*.

Similarly, suppose that the probability of the occurrence of loss is also determined by both private and public investments as:

$$\pi = \pi(R, G) \tag{8.8}$$

where $\pi(0,0) = \pi^*$, $\pi_R = \partial\pi/\partial R < 0$, *and* $\pi_G = \partial\pi/\partial G < 0$. The individual chooses R so as to maximize the expected utility given by

$$\mathrm{E}(U) = \pi(R,G)\mathrm{V}[M - R, A(R,G)] + [1 - \pi(R,G)]\mathrm{V}(M - R, 0) \tag{8.9}$$

From the first-order condition, we obtain

$$\frac{dM}{dG} = -\frac{\pi_G}{\pi_R} = \frac{\partial R}{\partial G} \tag{8.10}$$

Similarly to the above, the individual's marginal willingness to pay for public investment at the margin is equal to the ratio of the marginal productivities of private and public investments to reduce π or to the marginal rate of technical substitution between R and G, holding A^* constant.

8.2.3 Model Based on the Bounded Rationality

The model described above is basically based on welfare economics. However, it is pointed out that there are many problems in welfare economics, including the Scitovsky and Boadway paradoxes, which demonstrate the logical impossibility of comparing two Pareto-optimal states (see, for example, Just et al. 1982). Modern welfare economics is based upon the Kaldor-Hicks principle of potential compensation (Hanley and Spash 1993). The basic problem is the Hicks-Kaldor criteria use partial equilibrium changes to make statements about general equilibrium (Bowles and Gintis 2000; Hahnel and Albert 1990). Many proposals tried to construct some relevant Pareto-inclusive quasi-orderings. But all these proposed approaches are based on interpersonal welfare comparisons in one form or the another. Defining the Pareto-inclusive test relations which can pass the crucial test of consistency without requiring anything beyond the informational basis characterized by ordinalism and interpersonal noncomparability is by no means an easy task (Suzumura 1999).

It is also argued that traditional welfare theory is in crisis, and a 'quiet revolution' in welfare economics is already under way (Hahnel and Albert 1990). The Walrasian assumptions about preferences and contracting have recently come under scrutiny. Social norms and psychological dispositions extending beyond the selfish motives of *Homo economics* may have an important bearing on outcomes, even in competitive markets. Economists are increasingly concerned with social problems reflecting dimensions of human behaviour and well-being that are not captured by conventional models of 'economic man'. Among these are the management of risky behaviours, and welfare dependency. Economics must become behavioural and more institutional (Bowles and Gintis 2000).

Most modern psychologists and sociologist reject the view of human behaviour envisioned in the traditional economic paradigm as ill suited to defining and addressing the problems that concern them (Hahnel and Albert 1990).

Moreover, the models described above are based on expected utility as a representation of individuals' preferences under uncertainty. There is, however, substantial evidence that individuals' choices frequently violate the expected utility theory. It is reported that there are many examples, such as the Allais paradox, framing effects, and the Ellsberg paradox (for detailed explanation, see, for example, Mas-Colell et al. 1995; Kreps 1990), in which the rationality in decisionmaking that is assumed in the expected utility theory is not realized (see for example, Heap et al. 1992; Kahneman and Tversky 1979; Rubinstein 1998).

An examination of stated preference data and well-behaved preference in the contingent valuation method (see for example, Hanley et al. 1997; Mitchell and Carson 1989) shows no consistency between them, and it is necessary to investigate under which situation people have well-behaved preference and behave in accordance with the axioms of the VNM (von Neumann-Morgenstern) type expected utility theory (Jones-Lee and Loomes 1997). In the VNM-type utility theory, the most important axiom is the independence axiom, which was first proposed by von Neumann and Morgenstern in their book, *Theory of Games and Economic Behavior* (1947), as an incidental result in the theory of games (Mas-Colell et al. 1995).

The model described in Sect. 8.2.2 involves the derivation of the expression for welfare change, and the determination of the first-order conditions for optimization and their substitution into the expression for welfare change. Given the assumptions of the models described here, this substitution allows for the cancelling out of any observable utility terms. Here, the derived observable welfare measures are independent of the particular form of preferences (Freeman 1995).

Let I be some general index of preferences, where the preferences depend on income, prices (implicitly), the probabilities of different states of the world, and the magnitudes of the adverse event in different states:

$$I = \mathrm{f}(M, A, \pi) \tag{8.11}$$

This function has the same expression as the indirect utility function except that it is not based on expected utility theory. So, M denotes income, and prices are included implicitly in M. Assume this function is convex and twice differentiable.

The probability of the occurrence of loss is determined by

$$\pi = \pi(R, G) \tag{8.12}$$

Given the level of G, the first-order condition for the optimum R is

$$\frac{\partial I}{\partial R} = -f_{M*} + f_{\pi}\pi_R \tag{8.13}$$

To find the marginal welfare measure for a policy that reduces π, totally differentiate Eq. (8.11), rearrange terms, and substitute in the first-order condition to obtain

$$dI = f_{M*}dM + (f_{\pi}\pi_R - f_{M*})dR + f_{\pi}\pi_G dG = 0 \tag{8.14}$$

and

$$\frac{dM}{dG} = -\frac{f_{\pi}\pi_G}{f_{M*}} = -\frac{\pi_G}{\pi_R} \tag{8.15}$$

The marginal willingness to pay for publicly supplied risk reduction is equal to the marginal rate of technical substitution between public and private risk (probability of the occurrence of loss) reduction, holding A^* constant.

8.3 Empirical Application of the Model to Valuation of Risk of Municipal Water Use

According to the theory of risk perception (Slovic 1987), risk perception is represented by two dimensions of components: dread risk and unknown risk (degree of perception). Dread risk is dependent on subjective probability and seriousness, and unknown risk (degree of perception) is dependent on the degree of conviction for the above probability and seriousness. This conviction is dependent on the information situation. Hence, risk is governed by the subjective probability, seriousness, conviction for such probability and seriousness, and information situation.

People's knowledge about subjective probability and seriousness is bounded. For example, people may overestimate the quality of data on risk and their ability to perceive risk and mistake their real exposure of risk (Desvousges et al. 1998). Hence, the evaluation of risk may not be appropriate.

In the expected utility theory, the subjective probability is used to represent the possibility of the occurrence of an event. However, if the conviction for this probability differs for each information situation, each possibility based on different information is substantially different. Hence, the possibility of mistaking the risk may increase. Consequently, it is necessary to clarify the information situation and the degree of perception such as conviction, and then to set up the model under the bounded rationality.

Moreover, whether the revealed preference is exactly shown is important. People often take averting behaviour with increasing conviction on the probability of dread risk and seriousness. Their conviction on the dread risk is dependent on the threshold of acceptance of the risk, which differs for each individual.

Consequently, in the case of risk evaluation, it is necessary for the following to be clarified: bounded rationality on the ability of risk perception, the information situation, and people's threshold acceptance of risk.

8.3.1 A Preliminary Survey 1

The first survey questionnaires were distributed to students. The objective of the survey was to investigate what people think of the health risk concerning tap water and to choose substitutes for it under different information situations. The sample size of the first survey was 195 (Asahi and Hagihara 1999).

The questionnaire was designed to ask mainly three questions:

1. What kind of action is taken when they choose drinking water, including substitutable market goods such as soft drinks?
2. Are they concerned about the risk of tap water when they take that action
3. Do they change their action after they are given information about the risk of tap water quality?

The responses to question 1. are shown in Table 8.1. About half of the students drink tap water (48.7 %), whereas most of them (82.1 %) make some other choices such as drinking boiled water, purchasing bottled water or other soft drinks, or drinking tap water treated with a water purifier.

With respect to the relationship between risk perception of tap water and actions, 59.0 % of all students drink something other than tap water because of concern about risks. In contrast, only 5.6 % of students who drink tap water are concerned about risks (see Table 8.2). The differences in the rates of risk consciousness between students who drink tap water and those who select each other choices are significant, as determined by the MacNemar test. Hence, students who drink something other than tap water are strongly conscious of the risk of tap water.

For question 3, we explained the adverse effects of water quality and the advanced treatment process of the municipal water supply system to the students. Then we asked them what actions they would take (see Table 8.3). The result was that more students chose to drink something other than tap water. Table 8.4 shows relationship between actions and perception of risk when additional information was given.

Table 8.1 Choice of drink

	Tap water	Boiled water	Bottled water	Soft drinks	Water treated with purifier	Others
Percentage of total number of students	48.7	59.0	31.3	63.1	29.2	1.5

Table 8.2 Relationship between actions and perception of risk

	Tap water	Boiled water	Bottled water	Soft drinks	Water treated with purifier	Others
Percentage of students making another choice	51.3	41.0	68.7	36.9	70.8	98.5
Percentage of choice without perception of risk	43.1	29.2	10.8	43.1	10.8	0.5
Percentage of choice with perception of risk	5.6	29.7	20.5	20.0	18.5	1.0
Total	100	100	100	100	100	100

Table 8.3 Percentage of choice after additional information

	Tap water	Boiled water	Bottled water	Soft drinks	Water treated with purifier	Others
Before	48.7	59.0	31.3	63.1	29.2	1.5
After (percentage of another choice)	18.5 (81.5)	43.1	44.6	41.0	42.1	2.1

Table 8.4 Relationship between actions and perception of risk when additional information was given

	Tap water	Boiled water	Bottled water	Soft drinks	Water treated with purifier	Others
Percentage of perception of risk of tap water to the number of students taking each action	11.6	50.4	65.6	31.7	63.2	66.7

8.3.2 *Preliminary Survey 2*

Preliminary survey 1 revealed that the perception of the risk of tap water is a factor based on which respondents drink something other than tap water. However, with regard to a misperception of risk because of bounded rationality, it is necessary to know under which type of bounded rationality each respondent would take some kind of action. In order to use the revealed preference data, we must explain the possibility of bounded rationality by examining the information situation and the degree of conviction of the risk level of the quality of tap water.

In the theory of Bayesian decisionmaking, consumers' choices are decided based on both of subjective probability and of conviction of that probability (Matsubara 1992). Moreover, subjective probability and conviction are dependent on the amount of information. Hence, we will investigate the relationship between the information situation and the choice of drinking tap water. First, we explain the difference between each consumer's subjective probability based on the difference between their perception and the standard risk level of the World Health Organization (WHO). The standard risk level is defined as the possibility of suffering from cancer and is legitimated by experts, so it is regarded to be an objective probability in this paper.

We classify respondents whose subjective probability is higher than the WHO level as group A and those whose subjective probability is lower than the WHO level as group B. Then, we investigate the difference between the two groups in the following items: (1) action in the case of drinking tap water; (2) perception of risk when taking a certain action; (3) choice among the substitutes. The second survey questionnaires were also distributed to students. The sample size of the second survey was 151.

First, we asked them where they thought a dangerous event would occur during the process of supplying water from a public water body to the consumer's tap. Most of them were concerned about events both at the public water body and during the treatment process.

Then, we asked them whether they thought the quality of tap water was higher or lower than the standard water quality legitimated by WHO: 34 % of them considered the quality of tap water to be safer than WHO's risk level, that is, the reference point (Kahneman and Tversky 1979), whereas 66.0 % of them thought it was more risky than WHO's level. Moreover, among the latter, the percentage of people who do not drink tap water was 74.5 %. The difference in their consciousness between risky and safe is significant.

According to the above results, people think the risk level of tap water is higher than the objective risk level (the probability determined by experts). However, although the difference between groups A and B is significant concerning drinking tap water, the differences in both perception of risk when taking a certain action and choice among the substitutes are not significant.

The reason they are not significant is considered as follows. The reference point, or WHO's risk level, is not definitely the same as the threshold value of each respondent. In other words, each respondent has a different threshold value of the risk of tap water as his or her own reference point. Hence, if they indeed took action based on their own reference point, our classification of the two groups might have been incorrect. If a respondent who was in group A had a real threshold value that was higher than the reference point and his subjective probability for taking some action was lower than his threshold value, he might have drunk tap water.

In Bayesian decision theory, if the amount of information increases, the subjective probability changes and the degree of conviction concerning the subjective probability becomes high, then the value of the subjective probability will coincide with the objective probability determined by experts in the long run (Matsubara

1992). Hence, if the degree of conviction of the subjective probability is not very high, the subjective probability and the choice based on it may change on being given additional information, as shown by the results of the first survey. Furthermore, we could propose one more reason that their subjective probability is indefinite. That is, although they think that dangerous events occur both at the public water body and during the treatment process, they do not know the exact consequence of the degradation of the quality of the public water body, nor the actual treatment process that is undertaken. Their perception of the risk brought by the event might be ambiguous.

Although, there are some restrictions concerning the revealed preference data obtained, the survey shows that risk awareness is a factor in drinking water, the rate of substitution for the tap water is different depending on the substitute, and people change their choice based on risk information.

8.3.3 Model Setup

It is assumed that a dangerous event occurs only during the treatment process. Moreover, assume first that the probability of suffering from cancer is regarded as a risk of water quality and, second, that the cost of the advanced water treatment process is regarded as a public investment to reduce the probability of suffering from cancer and that the purchasing cost of market goods such as soft drinks is regarded as a private investment.

The general index of preference is defined as

$$\mathrm{I}(M, A, \pi) = \pi(R, G)\mathrm{I}(M - R, A^*) + \{1 - \pi(R, G)\}\mathrm{I}(M - R, 0) \tag{8.16}$$

where A^* is the probability of suffering from cancer (constant), R is the purchasing cost of substitutes for tap water, and G is the investment cost for the advanced water treatment process.

The individual marginal willingness to pay to the investment cost for the advanced water treatment process is equivalent to the marginal rate of substitution between the reduction of the probability of cancer thanks to purchasing substitutes and the reduction of the probability of cancer thanks to public investment.

8.4 Effects of Information on Risk and Public Investment

8.4.1 State of Risk and Public-Private Investment

Preliminary survey 1 revealed that people's willingness to pay is strongly dependent on information about risk. Moreover, their belief in the risk may also be

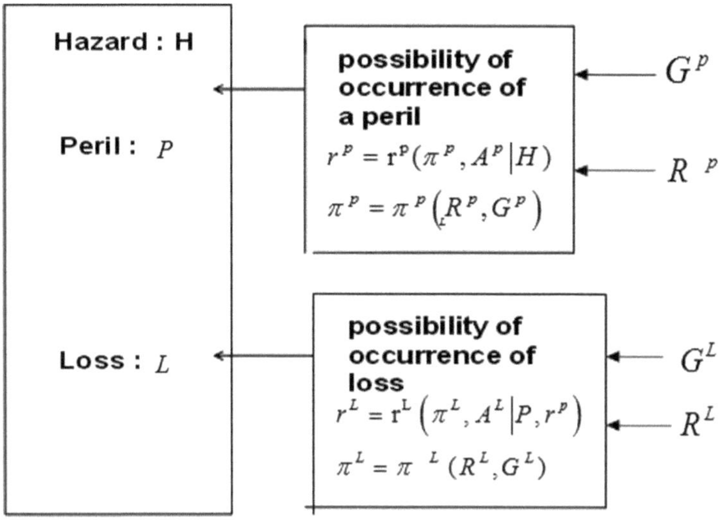

Fig. 8.1 States of risk and public-private investment

dependent on the effect of public investment and other information on the state of the surroundings. Therefore, the above model will be applied to take into account this information.

Figure 8.1 shows the relationship between risk and public-private investment. Hazard refers to the state of nature, climate conditions, and artificial states such as the coverage ratio of the sewerage service and the capacity of the water treatment system. Peril occurs with some probability under the above hazard. We can regard pollution in public water bodies, the incidence of Cryptosporidium, and the incidence of trihalomethane as perils.

Here, G and R represent public and private investments, respectively. Examples of G^P are investment for sewerage service and reinforcement of control and surveillance. Here, superscript P refers to peril. The possibility of the occurrence of peril, π^P, decreases with an increase of G^P, that is, as the coverage ratio of the sewerage service becomes higher and the reinforcement of control and surveillance becomes more strict. The possibility of the occurrence of peril, π^P, also becomes lower as people take action to protect nature through private investment, R^P. That is, the relationship is as follows:

$$\pi^P = \pi^P\left(R^P, G^P\right). \tag{8.17}$$

Consequently, the risk of peril is expressed as

$$r^P = \mathrm{r}^p\left(\pi^P, A^P|H\right). \tag{8.18}$$

On the other hand, examples of G^L, where superscript L refers to loss, are the adoption of an advanced water treatment system and stricter reinforcement of control and surveillance. If an advanced water treatment system is introduced to the water supply plant, that is, G^L increases, the possibility of the occurrence of loss, π^L, decreases. McConnell and Rosado (2000) shows that public investment in water purification or improved wastewater treatment can reduce the incidence of diarrheal diseases. And efficient investment entails some knowledge of the public's willingness to pay for avoiding drinking water potentially contaminated with pathogens (McConnell and Rosado 2000). Moreover, as consumers take averting actions through private investment, R^L, the possibility of the occurrence of loss may again decrease. As above, the relationship is as follows:

$$\pi^L = \mathrm{p}^L\left(R^L, G^L\right) \tag{8.19}$$

Consequently, the risk of loss is expressed as

$$r^L = \mathrm{r}^l\left(\pi^L, A^L|P, r^P\right) \tag{8.20}$$

8.4.2 Factors in Criteria for Decision Making by Consumers

Factors that consumers take into account when they decide their private investment for preventing risk from tap water, that is, their expenditure for averting actions, are as follows:

1. information on water environment, that is, information about the kinds of hazard and peril that exist around the water environment
2. risk perception, that is, the degree of perceptions of the possibility of occurrence of both peril and loss, and severity of loss
3. information on countermeasures taken by government, that is, G^p and G^L, and possible actions for consumers, R^P.

8.4.3 Effects of Information and Risk Perception on Valuation

In order to clarify the effects of subjective perception of risk, and information on risk and public investment, numerical examples are examined for six cases (see Table 8.5).

In order to clarify the effects of subjective perception of risk, and information on risk and public investment, numerical examples are examined for six cases (see Table 8.5).

Table 8.5 Six numerical examples

	Information on risk and public investment	
Subjective perception of risk	low risk or large effect: 1	high risk or small effect: 2
Underestimated risk: I	I-1	I-2
Reasonable evaluation of risk: II	II-1	II-2
Overestimated risk: III	III-1	III-2

In order to calculate the marginal willingness to pay, we assign values to some variables.

1. The standard of water quality is an index by which one can judge whether an event that affects water quality is a dangerous event or not. As an index for health, a risk level of 10^{-5}, which is based on the standard of water quality set by WHO, is selected. This is the risk that one out of one hundred thousand people may suffer from cancer if they drink 2 l of tap water every day for 70 years. Hence, the probability of suffering from cancer in 1 year for a person aged N is $\pi(0,0) = N/70 \times 10^{-5}$. We assume the probability of case 1 is the same as the WHO's risk level, while we assume the probability of case 2 is higher than the WHO's risk level, that is, $N/70 \times 10^{-4}$.
2. Using the total investment cost of the advanced treatment plant by the Tokyo metropolitan government of 27.2 billion yen, a period of durability of 30 years, and a discount rate of 2 %, we can calculate the average annual cost of supplying 2 l of water a day per person as follows:

$$G = \left(\frac{272}{30(1+0.02)^{30}}\right)\left(\frac{730 l \times 365}{520,000,000 l}\right)$$

3. Suppose that an individual substitutes x litres of market goods such as soft drinks because of an awareness of risk. The prices of market goods are $P_m = 200\,\text{yen/l}$, $P_s = 300\,\text{yen/l}$, and $P_j = 200\,\text{yen/l}$, where P_m, P_s, P_j are the prices of bottled water, soft drinks, and a water purifier, respectively. Then the annual private expenditure is

$$R = (P_m + P_s + P_j)/3 \times x \times 365.$$

From 1. and 2.,
$\pi^1(0, G) = N/70 \times 10^{-5} \times (1 - \beta_1)$ in the case of low risk, 1
and
$\pi^2(0, G) = N/70 \times 10^{-4} \times (1 - \beta_2)$ in the case of high risk, 2,
where $\beta_i (i = 1, 2)$ is the ability to reduce trihalomethane, which is assumed to be proportional to the investment cost, that is, $\beta_i = \alpha_i G$ (i =1,2). We assume that the ability to reduce trihalomethane is 60 % in the case of low risk, 1, and 30 % in the case of high risk, 2, respectively. Therefore,

$\pi_G{}^1 = -N/70 \times 10^{-5} \times \alpha_1$ in the case of low risk, 1
and
$\pi_G{}^2 = -N/70 \times 10^{-4} \times \alpha_2$ in the case of high risk, 2.
From 1. and 2.,

$$\pi^i(R,0) = \pi(0,0) \times (1 - \gamma_j), j = I, II, III,$$

where $\gamma_j (j = I, II, III)$ is a percentage of choice of each averting action, and $\gamma_j = \delta_j R (j = I, II, III)$. The value of γ_I is 51.3 %, that is, percentage of students making another choice (see Table 8.2), γ_{II} is 81.5 %, that is, percentage of another choice after additional information given (see Table 8.3), and γ_{III} is 100 % which is a hypothetical percentage and means that people do not drink tap water at all. Therefore,
$\pi_R{}^1 = -N/70 \times 10^{-5} \times \delta_j, j = I, II, III$ in the case of low risk, 1
and
$\pi_R{}^2 = -N/70 \times 10^{-4} \times \delta_j, j = I, II, III$ in the case of high risk, 2.
Consequently, the marginal willingness to pay for the advanced water treatment plant which reduces the probability of suffering from cancer is

$$\pi_G{}^i / \pi_R{}^i = \alpha_i / \delta_j, \quad i = 1, 2; \quad j = I, II, III \quad .$$

The marginal willingness to pay for public investment is determined for each case and shown in Table 8.6. Based on Table 8.6, the effect of information concerning the state of the water environment is as follows. In the case of low risk or large effect of public investment, values of public investment are larger than in the case of high risk or small effect. In other words, the valuation of public investment is low in areas of water pollution in public water bodies or with an effect of small public investment. People take self-defensive action in the case of high risk. These results are obtained under the assumption that consumers can take action taking into account the trade-off between public investment and private investment.

The effect of risk perception on valuation is also obtained from Table 8.6. When consumers underestimate risk, values of public investment are higher than when risk is overestimated. When they feel risk to be high, their valuation of public investment is low. In other words, they take self-defensive action.

Table 8.6 Marginal willingness to pay for public investment

	Information on risk and public investment			
Subjective perception of risk	Low risk or large effect		High risk or small effect	
Underestimated risk: I	I-1	100,923 yen	I-2	50,462 yen
Reasonable evaluation of risk: II	II-1	63,526 yen	II-2	31,763 yen
Overestimated risk: III	III-1	51,774 yen	III-2	25,887 yen

8.5 Concluding Remarks

Urban environmental risk was first defined and the methods of valuation of urban environmental risk were shown with regard to the situation of decisionmaking, that is, decisionmaking under uncertainty and decisionmaking under risk. In particular, in the case of the evaluation of urban environmental risk in a risky decisionmaking situation, bounded rationality is taken into account in the model setup.

Then, the effects of information on risk and public investment were considered. From some numerical examples, marginal willingness to pay was found to be low in the case of high risk, because consumers take self-defensive action. In other words, marginal willingness to pay for public investment is high when there is little or no averting behaviour. Moreover, consumers' perception of risk is predominantly dependent on information on risk, countermeasures by public authorities, and overconfidence in private averting goods. Consequently, risk communication between consumers and public authorities, and an understanding of the exact preferences of consumers by public authorities are needed.

Finally, the above results are based on the preliminary survey, so it is necessary that a survey is undertaken for urban households.

Acknowledgements This paper has originally presented at the second DPRI-IIASA International Symposium on Integrated Disaster Risk Management: Megacity Vulnerability and Resilience, IIASA, Laxenburg, Austria, July 2002. We have benefited from discussion with especially Professor Norio Okada on earlier drafts.

This paper has originally published in *Environment and Planning C: Government and Policy* 2004, volume 22, pages 349–362, Pion LTD, London; www.pion.co.uk and www.envplan.com. We would like to thank the publishers for their permission to use the material here.

References

Asahi, C., & Hagihara, K. (1999). Evaluating environmental risk in drinking water. *Studies in Environmental System, 27*, 395–401 (in Japanese).

Bowles, S., & Gintis, H. (2000). Walrasian economics in retrospect. *Quarterly Journal of Economics, 115*, 1411–1439.

Crocker, T. D., Forster, B. A., & Shogren, J. F. (1991). Valuing potential groundwater protection benefits. *Water Resources Research, 27*, 1–6.

Desvousges, W. H., Reed, J. F., & Spencer, B. H. (1998). *Environmental policy analysis with limited information*. Cheltenham: Edward Elgar.

Freeman, A. M. (1995). Evaluating changes in risk perceptions by revealed preference. In D. W. Bromley (Ed.), *The handbook of environmental economics*. Oxford: Blackwell Publishers.

Gurian, P. L., Small, M. J., Lockwood, J. R., & Schervish, M. J. (2001). Benefit-cost estimation for alternative drinking water maximum contaminant levels. *Water Resources Research, 37*, 2213–2226.

Hagihara, K. (1996). Comprehensive urban economic studies. *Comprehensive Urban Studies, 60*, 95–104 (in Japanese).

Hagihara, K., & Hagihara, Y. (1990). Measuring the benefits of water quality improvement. *Environment and Planning C: Government and Policy, 8*, 195–201.

Hagihara, K., & Hagihara, Y. (2004). The role of environmental valuation in public policymaking: The case of urban waterside area in Japan. *Environment and Planning C: Government and Policy, 22*, 3–13.
Hagihara, Y., et al. (2000). *Planning on rain water environment in urban areas: An approach by the systems analysis*. Tokyo: Keiso Shobo (in Japanese).
Hahnel, R., & Albert, M. (1990). *Quiet revolution in welfare economics*. Princeton: Princeton University Press.
Hanley, N., & Spash, C. L. (1993). *Cost-benefit analysis and the environment*. Cheltenham: Edward Elgar.
Hanley, N., Shogren, J. F., & White, B. (1997). *Environmental economics in theory and practice*. Hampshire: Macmillan Press.
Heap, S. H., Hollis, M., Lyons, B., Sugden, R., & Weale, A. (1992). *The theory of choice- A critical guide*. Oxford: Blackwell Publishers.
Jones-Lee, M. G., & Loomes, G. (1997). Valuing health and safety: Some economic and psychological issues. In R. Nau (Ed.), *The economic and environmental risk and uncertainty- New models and methods*. Dordrecht: Academic.
Jordan, J. F., & Elnagheeb, A. H. (1993). Willingness to pay for improvements in drinking water quality. *Water Resources Research, 29*, 237–245.
Just, R. E., Hueth, D. L., & Schmitz, A. (1982). *Applied welfare economics and public policy*. Englewood Cliffs: Prentice-Hall.
Kahneman, D., & Tversky, A. (1979). Prospect theory: An analysis of decision under risk. *Econometrica, 47*, 263–291.
Kask, S. B., & Shogren, J. F. (1994). Benefit transfer protocol for long-term health risk valuation: A case of surface water contamination. *Water Resources Research, 30*, 2813–2823.
Kreps, D. M. (1990). *A course in microeconomic theory*. Hertfordshire: Harvester Wheatsheaf.
Mas-Colell, A., Whinston, M. D., & Green, J. R. (1995). *Microeconomic theory*. Oxford: Oxford University Press.
Matsubara, N. (1992). *Statistical decision making*. Tokyo: The University of the Air Press (in Japanese).
McConnell, K. E., & Rosado, M. A. (2000). Valuing discrete improvements in drinking water quality through revealed preferences. *Water Resources Research, 36*, 1575–1582.
Mitchell, R., & Carson, R. (1989). *Using surveys to value public goods: Contingent valuation method*. Washington, DC: Resources for the Future.
Niels, C. L. (1992). A national standard for risk analysis. *Risk Abstract, 9*, 1–3.
Rubinstein, A. (1998). *Modelling bounded rationality*. Cambridge, MA: The MIT Press.
Shogren, J. F. (1990). Impact of self-protection and self-insurance on individual response to risk. *Journal of Risk Uncertainty, 3*, 191–204.
Slovic, P. (1987). Perception of risk. *Science, 236*, 286–285.
Suzumura, K. (1999). Paretian welfare judgements and Bergsonian social choice. *The Economic Journal, 109*, 204–220.
Syme, G. J., & Williams, K. D. (1993). The psychology of drinking water quality: An exploratory study. *Water Resources Research, 29*, 4003–4010.

Chapter 9
Economic Valuation for Improving Supply Reliability: Risk Countermeasures for Water Quantity and Quality in Water Supply Systems

Chisato Asahi and Kiyoko Hagihara

Abstract In the supply of water, there is simultaneously a demand for maintenance of supply reliability from the viewpoint of public interest and a demand for efficiency against the backdrop of funding difficulty. In order to examine how these demands can be met through the use of market mechanisms and to identify the optimum replacement investment, maintenance and management, etc. for facilities, a framework must be created that ties together the risks of quantity and quality related to the reliability of supply with decisions concerning investment and the establishment of standards. This chapter uses an economic viewpoint to organize the risks of quantity and quality in and to utility water systems to conduct a valuation of countermeasures for these risks. The first two sections organize the changing nature and methods of understanding the water supply system for utility water and its quantitative and qualitative risks. Next section presents and considers establishment and operation cases of quantitative and qualitative risk countermeasures in current supply systems. Section 9.4 presents the optimization model for evaluating reliability to enable valuation of the supply system under these risks. The welfare measures that should be applied to risk-averse economic agents are organized based on welfare economics theory. The chapter concludes by summarizing critical system design focus points in order to realize a water supply system that is quantitatively and qualitatively reliable and efficient.

Keywords Municipal water supply • Risk management • Welfare measurement under uncertainty • Reliability

---

C. Asahi (✉)
Division of Urban Policy, Faculty of Urban Liberal Arts, Tokyo Metropolitan University, 1-1 Minami-osawa, Hachioji, Tokyo 192-0397, Japan
e-mail: asahi@tmu.ac.jp

K. Hagihara
School of Sociology, Bukkyo University, 96 Kita-hananobo, Murasakino, Kita-ku, Kyoto 603-8301, Japan
e-mail: hagihara@bukkyo-u.ac.jp

K. Hagihara, C. Asahi (eds.), *Coping with Regional Vulnerability*, New Frontiers in Regional Science: Asian Perspectives 4, DOI 10.1007/978-4-431-55169-0_9

9.1 Introduction

Since the 1990s, there has been a continuing trend towards the privatization of government services and public utility works because of the demand for increasingly efficient administration and public finance. In the water utility business, investigations are being conducted into the introduction of various supply systems that have become possible as a result of the following: strengthening business consignment systems for third parties, establishing designated administrator systems for public facilities, and operating local incorporated administrative and other agencies to reform local public enterprise management within broader administrative reforms. However, the other side to the global trend of commercializing public utility works (including water supply) are cases where changes to the supply system have created situations where stable supply was lost (Barlow and Clark 2002). It is also argued that demand for administrative efficiency restricts future investment for the establishment, maintenance, and replacement of social overhead capital. For example, in the water utility business, despite a clearly identified need to replace aging facilities (e.g. pipelines laid in the 1960s), upgrades are not proceeding because of funding difficulties, which are a growing concern.

On the other hand, because utility water and sewage are essential to life, a level of maintenance management must be upheld even if a declining population is reducing the number of users and volume to the same degree as for other infrastructure (Nomura Research Institute 2008). In order to secure this funding, increases in user burden often occur. In the supply of water, there is simultaneously a demand for maintenance of supply reliability from the viewpoint of public interest and a demand for efficiency against the backdrop of funding difficulty. In order to examine how these demands can be met through the use of market mechanisms and to identify the optimum replacement investment, maintenance and management, etc. for facilities, a framework must be created that ties together the risks of quantity and quality related to the reliability of supply with decisions concerning investment and the establishment of standards.

Conscious of these issues, this research uses an economic viewpoint to organize the risks of quantity and quality in and to utility water systems to conduct a valuation of countermeasures for these risks.

The next section organizes the changing nature and methods of understanding the water supply system for utility water and its quantitative and qualitative risks. Section 9.3 presents and considers establishment and operation cases of quantitative and qualitative risk countermeasures in current supply systems. Section 9.4 presents the optimization model for evaluating reliability to enable valuation of the supply system under these risks. The welfare measures that should be applied to risk-averse economic agents are organized based on welfare economics theory. The chapter concludes by summarizing critical system design focus points in order to realize a water supply system that is quantitatively and qualitatively reliable and efficient.

9.2 Risks for Utility Water Supply

9.2.1 *The Changing Nature of Utility Water Supply Risks*

Modern utility water supply in Japan started in Yokohama in 1887. At the time, pollution of drinking water caused frequent outbreaks of cholera, killing tens to hundreds of thousands of people. It was thought that in order to counter this problem, supplying water using steel pipes was essential to protect the sanitation of drinking water; thus, port cities started supplying water, beginning with Yokohama. The planned supply volume in each city was about 70–140 l/person per day, or about 20–40 % of the current volume. Later, because of reasons such as increases in the supply volume and shortages of land sites, water treatment methods were mostly changed from slow sand filtration to rapid filtration, and pipe materials were modified from cast-iron pipes to higher-quality cast-iron pipes and ductile pipes. Regarding the popularization rate of utility water and the number of patients with infectious diseases caused by water treatment equipment, the number of patients with infectious diseases spiked from the latter half of the 1930s and the 1940s because of the war. However, since the 1970s (when the utility water popularization rate exceeded 80 %), the number of patients treated for infectious diseases from water quality has been maintained at a level of several thousand per year.

As the popularization of utility water progressed, it became more difficult to obtain water from non-utility water sources for use in homes and for societal activities; thus, the society was increasingly affected in periods when utility water supply was obstructed. For example, in 1994, a nationwide drought led to reductions or interruptions in the water supply for about 16.4 million people. During fiscal year 1994, out of a total of about 2000 water utility businesses, 72 were involved in water pollution incidents and 120 were affected by odour or taste-related quality issues. Then, in 1995, the Great Hanshin and Awaji Earthquake caused extensive damage to water utility facilities. During the 8–10 weeks it took to recover from this disaster, residents in areas with no working water utility experienced tremendous inconvenience. Moreover, water quality problems caused by water treatment by-products such as trihalomethane have been socially recognized since the 1980s. In 1996, the pathogenic E. coli bacteria O-157 and cryptosporidium caused outbreaks, leading to substantial mistrust in the safety of utility water nationwide. In addition, there was a problem with the so-called environmental hormones ('endocrine disruptors').

In recent years, it has been pointed out that despite the growing need for maintenance or replacement owing to the aging of water utility facilities, it is difficult to perform such investments for reasons relating to technology, systems, and the financial situation (Odanagi 2004). Many of Japan's water utility facilities were constructed concurrently in the 1960s during a period of advanced economic growth. At that time, the popularization rate, supply, and facility capabilities were all broadly increasing. Nationwide, pipes that form the main water utility supply routes stretch for 639,159 km (fiscal year 2011), making up two-thirds of all water utility assets. Of these pipelines, about 47.5 % (i.e. 304,232 km) were laid over 40 years ago. Each year, the ratio of aged pipes is increasing (Japan Water Works Association 2013).

Recently, there have been several supply interruptions caused by large-scale accidents because of aging pipes that have affected civic life and urban activities. Moreover, if a powerful earthquake occurs, the damage to aged facilities may be massive and cause major flooding. Considering this situation, the planned replacement of facilities has been presented as a challenge of the revised edition of Water Supply Vision (Ministry of Health, Labour and Welfare 2008) and the New Water Supply Vision (Ministry of Health, Labour and Welfare 2013), which aims to form the future image of Japan's utility water supply.

Considering this changing nature of utility water supply, objectives for risk countermeasures relating to hygiene have been achieved by realizing universal water supply through water utilities. However, current policy preparations to completely replace non-utility water supply risk the occurrence of supply obstructions.

9.2.2 Risks for the Water Utility Supply System

Odanagi et al. (1997) organizes the risks for the water utility supply system discussed in the previous section according to the system and concepts of risk management. We present these here as an overview of quantitative and qualitative risks and their relationships.

9.2.2.1 Water Utility System

As shown in Fig. 9.1, a water utility system comprises water source and water supply systems. A water source system captures and stores water using water catchment structures, (e.g. dams) and the natural environment (e.g. mountain forests and underground) in order to adjust for temporal rainfall fluctuations. The water supply system takes the amount of water it needs from the water source system and then uses water treatment plants to convert the water quality into clean water. Adjusting for temporal fluctuations of supply and demand using reservoirs, the clean water is then delivered to the consumer through the pipeline.

9.2.2.2 Supply Risks of the Water Utility

In this water utility system, the water source system constantly faces issues related to water quantity and quality because of issues such as droughts, water source quality accidents, eutrophication, and muddy water. In contrast, the water supply system faces potential facility loss caused by problems such as earthquakes.

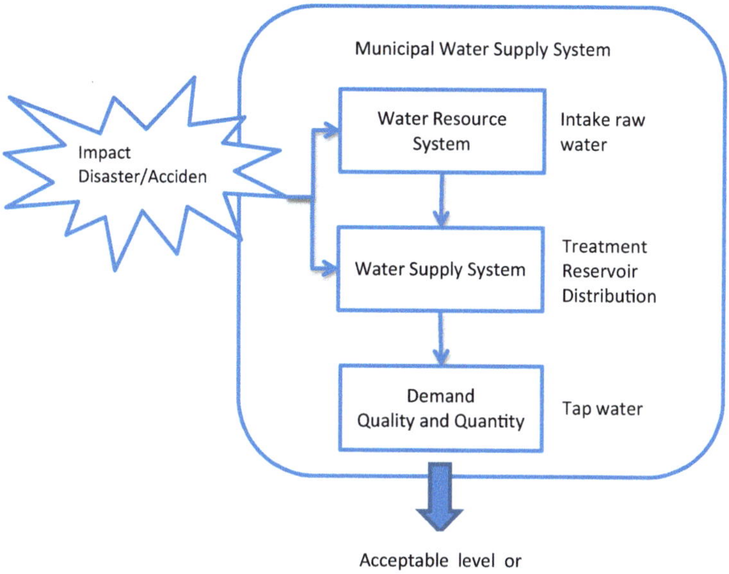

Fig. 9.1 Municipal water supply system (Based on Odanagi et al. 1997)

According to Okada (1985), water utility supply risks are 'water utility risks', which can be understood as 'situations when the water quality or water quantity supplied to the consumer is outside the tolerance range'. As shown in Fig. 9.2, water utility risks are composed of risk phenomena (perils) such as flooding (which directly cause a crisis) and risk circumstances (hazards), which include environmental conditions and factors.

Table 9.1 shows the different categories of perils. For example, water quantity related perils of the water source system are broadly categorized into water quantity shortage (e.g. abnormally low rainfall) and water quality abnormalities (e.g. contamination of poisonous materials). These perils are linked to an inability by the water supply system to take in the required water quantity. Moreover, Table 9.2 is organized from the viewpoint of both the supplier and the consumer and shows the damage that the supplier causes the consumer. Many perils might affect the decision-making of the supplier. On the other hand, considering the decision-making of the consumer, these can be understood within the two risk dimensions of quantitative risks relating to interrupted supply and inadequate pressure and qualitative risks relating to water quality failure. The quantitative and qualitative risks are not necessarily independent of each other, as they have a relationship based on a technological relationship or system. For example, when a drought causes a quantitative risk, it can be assumed that the level of eutrophication becomes severe. Moreover, if a water quality incident is identified in the supply system, damage will not appear as a quality risk but rather as a quantitative one caused by a supply interruption.

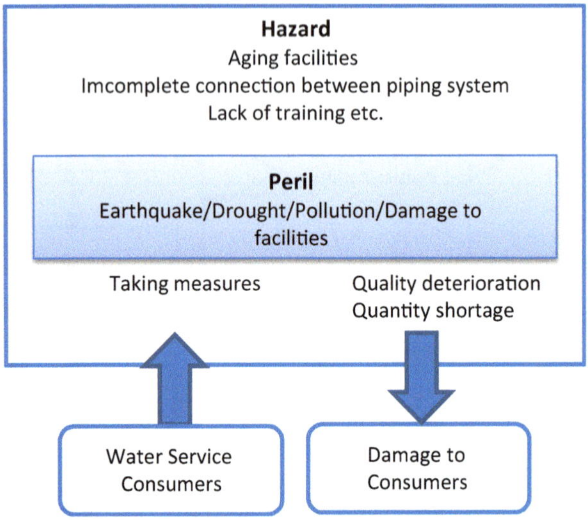

Fig. 9.2 Hazard and peril in municipal water supply system (Based on Okada 1985)

Table 9.1 Peril of water supply system (Based on Odanagi et al. 1997)

	Water resource	Treatment	Reservoir	Distribution
Natural system	Drought	Storm and flood	Storm and flood	Storm and flood
	Eutrophication	Earthquake	Earthquake	Earthquake
	Underground water exhaustion			
	Storm and flood			
	Earthquake			
Human system	Escherichia coli O157	Mis-operation	Mis-operation	Mis-operation
	Cryptosporidium	Terrorism	Terrorism	Terrorism
	Terrorism			
	Pollution with poisonous materials			
Artificial system		Power failure	Power failure	Power failure
		Accidents	Accidents	Accidents

Table 9.2 Risk and damage of water supply system (Based on Odanagi et al. 1997)

Types of hazard		Damage
Water resource	Quantity	Shortage of supply: Cut-off etc.
	Quality	Quality trouble
Treatment		Quality trouble
Distribution		Shortage of supply: Cut-off etc.
Reservoir		

9.3 Risk Countermeasures for the Water Utility System

9.3.1 Current System for Risk Countermeasures

9.3.1.1 Quantitative Risk Countermeasures

Let us consider a drought, which is a typical peril causing quantitative risk. A drought is a situation that requires countermeasures such as water intake restrictions that reduce the quantity of water taken in from rivers when the river flow is expected to not be sustained even if supplemented from water catchment facilities such as dams (Japan Water Works Association 2008). This chapter defines a drought in more practical terms copying Odanagi et al. (1997); namely, 'a drought is a phenomenon due to low precipitation that reduces the flow rate of a river to below the water quantity that provides constant availability'.

Systemically, constant availability is provided by existing, maintenance, and newly developed water quantity. The term used to describe such availability is the 'secured flow rate'.

Under the current system prescribing countermeasures for drought-caused quantitative risks, water utilization plans are proposed that can cope with the equivalent of the worst drought year flow rate in a 10-year period. In other words, a plan standard year is established for the aforesaid river that is able to secure the target flow rate in that year. For the plan standard year, data on the longest possible previous water documentation period are collected (20–30 years or the most recent 10 years if the former is not possible), and a water utilization safe-level valuation is conducted on the recurrence period based on the drought year. As a general rule, the worst drought year within this 10-year period is adopted as the base.

However, Hagihara and Watanabe (1981) point out that because the standard for coping with a drought once in 10 years is conventional, it is administrative custom to establish an adjustment point by considering the disruption of the water-usage order that arises when the safety of water intake is reduced to below this value, or conversely, to the lost opportunity of unused river water that arises when the safety of water intake is increased. Although the worst drought in 10 years is supposed at the time the plan is created, many regions have an occurrence frequency higher than this.

9.3.1.2 Qualitative Risk Countermeasures

A qualitative risk is a crisis relating to the quality of the water, where the peril is contamination of pollutants. Among the various substances that dissolve in water, substances harmful to humans are pollutants, including substances that cause water-borne infectious diseases, acute poisons, and chemicals with carcinogenic properties.

Water quality is managed by water quality standards that prescribe quality requirements for the preparation of water supplied by the water utility. The water quality standard value is established based on the rationale of the 'Guidelines for Drinking-Water Quality' created by WHO and is revised to reflect the latest scientific knowledge. The initial water quality standard, which was prescribed by a Ministry of Health and Welfare ordinance of 1958, had 30 items. Since then, over the course of five revisions, it has been considerably amended. In the 2003 revision, in addition to increasing the water quality standard to 50 items, water quality management target establishment items and required test items were established. Successive revisions have been made. In fiscal year 2008, chloric acid was added as a water quality standard item and heterotrophic bacteria and fipronil were added as water quality management items with established targets. In fiscal year 2009, 1,1-dichloroethylene was removed from water quality standards and repositioned as a water quality management item with a set target. In addition, trans-1,2-dichloroethylene, which was a water quality management item with a set target, was repositioned as a water quality standard item together with cis-1,2-dichloroethylene. As of 2015, the water quality standard has 51 items to which nitrite nitrogen was added in 2014.

The establishment of water quality standards has limits as a countermeasure for quality risks. Asahi and Hagihara (2004) note the following limitations to ex ante countermeasures for quality risks using the water quality standard. Concerning countermeasures for chlorine-resistant pathogenic microorganisms such as cryptosporidium, in order to make chlorine-resistant pathogenic microorganisms a regular water quality standard test item, it is necessary to have large volumes of sample water. However, because this is not practical from a water treatment management viewpoint, they are not included as water quality standard items; furthermore, many water quality incidents have unknown causes. However, since 2007, ex ante risk countermeasures are being conducted—albeit indirectly—by regularly testing for E. coli and anaerobic spore-forming bacteria (indicator bacteria for cryptosporidium) in accordance with 'Guidelines for Cryptosporidium Treatment in Waterworks' (Ministry of Health, Labour, and Welfare 2007).

9.3.2 Issues for Risk Countermeasure Systems

9.3.2.1 Agent for Risk Valuation

The size of the damage supposed for risk countermeasures depends on the level of peril and the state of hazard. However, it can be assumed that the supplier's and consumer's valuations will not always match. Generally, when the supplier is a competitive company, the objective function of decision-making is maximization of profit, and risk for the supplier is related to phenomena that affect profit.

Utility water is considered a cost-decreasing industry. In public interest, it is an industry with economic constraints (e.g. fee regulations) and social ones (e.g. supply obligations). Therefore, risks for the water utility business comprise phenomena that affect profit under the constraint of regulations, including compliance. On the other hand, the objective function of consumer decision-making is utility maximization. For the consumer, risks consist of phenomena that affect the utility obtained from the quantity and quality of the commodity of utility water.

Here, 'phenomena that affect profit under the constraint of regulations' and 'phenomena that affect the utility obtained from the quantity and quality of the commodity that is water' do not necessarily match. For example, in many cases where the supplier conducts risk countermeasures and the damage to the consumer is a 'failure of the supply water quantity or quality to the consumer', valuation uses indicators such as (1) period, (2) scale, and (3) frequency. On the other hand, the consumer evaluates this damage as 'water is unusable' and 'there are restrictions on how the water can be used or it is necessary to change how the water is used'. For examples of risk, we consider earthquakes, droughts, or eutrophication of the water source. The damage caused by an earthquake is of a low frequency, but the scale and period of disruption is comparatively long. In the case of a drought, discontinuation is rare, but the frequency is high and there are many cases where the scale and period are extended. Moreover, in the case of eutrophication of the water source, a mouldy odour is noticed over a long period. With respect to this, the supplier understands the direct damage regarding the scale and the period, but the consumers understand the damage in additional ways, including psychological satisfaction levels that indicate inconvenience or discomfort for daily life.

9.3.2.2 Quantitative Risk Countermeasures

Addressing the disparity between the supplier's and the consumers' valuation for utility water quantitative risks, Hagihara (1995) points out that the water utilization safety level, which indicates quantitative risk, may be biased.

With respect to the water utilization plan proposal, the current system generally adopts the worst drought year in 10 years, and the water utilization safety level is predetermined using the drought recurrence year. In other words, the definition of the water utilization safety level under the current system can be explained as the 'probability of being able to secure the water quantity (secured flow rate) that provides constant availability'. This is called the 'planned safety level'. In this case, a water quantity that provides constant availability is treated as a given, and the responsive behaviour by the water utility agent during the drought is not considered. Because of this, the water utility agent may supply water from substitute sources during a drought to address the issue that the 'water quantity that provides constant availability' cannot be secured, and the consumer may be unaffected by the damage.

In comparison, the 'probability of a state occurring where the supply quantity exceeds the water demand during normal periods' is defined as the 'managed safety level', which is a concept that takes the viewpoint of the consumer. When the supply quantity is lower than the water demand in normal periods, damage occurs;

then, the consumer must respond. In other words, another way of expressing the 'managed safety level' is the 'probability of the consumer not suffering damage'. Hagihara (1995) points out that when a water utilization plan is created under the current system, the 'managed safety level' is not considered although it is the most critical indicator to the consumer.

9.3.2.3 Quality Risk Countermeasures

Addressing the valuation of quality risks by the supplier and the consumer, Asahi and Hagihara (1999) note that the same problem occurs for quality standards, which indicate quality risk.

As mentioned previously, the water quality guideline was established based on the WHO guidelines. However, these standard values have been changed by suppliers based on scientific and epidemiological knowledge. In order to clarify the consumer view of this, Asahi and Hagihara (1999) sought an objective valuation of carcinogenic risk by the consumer by advancing the adoption of standards of carcinogenic rates using total trihalomethane, which is contained in utility water.

According to their research, two groups exist: one that evaluates utility water quality as more dangerous than the values prescribed by the water quality standard, and one that recognizes that the utility water quality is safe. This indicates that a disparity exists from the supplier's evaluation of the water quality standard value. Current risk countermeasures are determined based on risk indicators of only supplier-side water quality standards. However, if risk indicators that consider the valuation of the consumers are used, the possibility exists that the standard and scale of risk countermeasures will differ.

9.4 Observations Based on Economic Theory Related to a Supply System Under Risks

9.4.1 Reliability Model

Up to this section, we investigated the possibilities of discrepancies in risk valuations taken from the respective viewpoints of suppliers and consumers with regard to quantitative and qualitative risk countermeasures. From the viewpoint of policy efficiency, the supplier addresses the respective risks in the supply plan and in the water quality standard. However, as the valuations of these risks are not based on risk preferences of consumers and the willingness-to-pay (WTP) amount, we consider that investment and regulations for these risks may not be economically efficient.

This water utility system is theoretically a cost-decreasing industry that holds characteristics such as public and merit goods. As a general rule, the business is municipally operated and supported by a general budget related to capital investment. To follow systems related to government departments such as obligation to supply, the business conducts facility maintenance and new investment as well as maintenance and replacement investment. Because investment by a public department cannot be evaluated, valuation is conducted by estimating the benefit. With regard to the water utility supply system, in order to consider maintenance and replacement investment that consider risk valuation, it is important to estimate the valuation of the supplier and the consumer toward risk countermeasures. In other words, the cost of the items-at-risk must be evaluated.

However, although the costs necessary to conduct public interest investment and regulations to cope with such risks are generally known, there is insufficient information on the risk preferences of consumers affected by this policy. In order to estimate the benefit of investment and regulations to reduce risk and to determine their optimum level, we require consumer information related to risk preferences. Therefore, we present a model cited from Howe et al. (1993) that finds the optimum level of risk that reflects the risk valuation of the economic agents, that is, both suppliers and consumers. The optimum quantitative and qualitative risk level of a public service also expresses quantitative and qualitative trust of supply, and this model is thus called the 'trust model'(Asahi and Hagihara 2010).

The occurrence probability of damage D related to the quantity and quality of utility water is expressed as follows:

$$P(D) = \int_0^\infty \Phi(D)dD \tag{9.1}$$

Here, Φ(D) expresses the probability density function of the phenomenon. When loss is experienced by the economic agent from this damage L(D), the expected loss is expressed as follows:

$$E[L(D)] = \int_0^\infty L(D)\Phi(D)dD \tag{9.2}$$

As the probability distribution of the risk phenomenon is dependent on the supply system, if the level of investment is incorporated into supply system I, the expected loss is the function of I. Therefore, the decision-making problem of the supplier becomes the investment and minimizing the current value of its expected loss.

$$\min_I \left\{ I + \sum_{t=1}^{T} \frac{E[L(D_t), I]}{(1+r)^t} \right\} \tag{9.3}$$

The discount rate is r. Although the expected loss function expresses the loss in the future, the decision-making of the economic agent is conducted based on a variable in a certain point in time. Then, this decision-making defines the 'standard annual damage: damage event (DE)', which is a concept that expresses a certain time of the expected loss function. Then, if R represents the reliability of the supply system, R can be defined by the event probability of 'standard annual damage (DE)'.

$$R(DE) = 1 - P(DE) \tag{9.4}$$

Because the event probability of standard annual damage (DE) is determined by the investment level relating to the supply system (e.g. damage), it becomes P(DE|I). The expected loss function is as follows:

$$E[L(DE, I)] = P(DE|I) \cdot L(DE) \tag{9.5}$$

When P(DE|I) is a decreasing function with respect to I, R becomes an increasing function with respect to I. If this relationship is clear, it is possible to derive the I required to achieve the optimal level of R.

$$R = R(I) \tag{9.6}$$

$$I = I(R) \tag{9.7}$$

Here, if the duration of investment is n years, the discounted annual investment amount required to achieve R is expressed as follows:

$$A(R) = \frac{r \cdot I(R)}{1 - \left(\frac{1}{1+r}\right)^n} \tag{9.8}$$

Thus, the optimum reliability R* is obtained as the solution to the problem of minimizing the sum of the expected loss for a certain R and the discount annual investment amount that realizes R.

$$\min_R\{A(R) + [1 - R(DE)] \cdot L(DE)\} \tag{9.9}$$

The first order conditions can be obtained using Eq. 9.10. In other words, the level of optimum reliability of the supply system is decided when the decline of the additional expected loss (i.e. the marginal benefit generated from the reliability increase) is equal to the marginal cost of realizing this reliability level.

$$\frac{\partial A(R)}{\partial R} = \frac{dE(L)}{dR} \tag{9.10}$$

9.4.2 *Measurement of Marginal Benefit of Reliability*

According to Eq. 9.10 of the reliability model, it is possible to obtain information on whether the reliability level is efficient by comparing the decline of the expected loss caused by the risk countermeasure (right side) with its marginal cost for a certain reliability level (left side). Moreover, it is possible to test the required level of investment amount against its marginal benefit of reliability. Because the marginal cost of the risk countermeasure on the left side represents the precursory fixed investment for the reliability model, the calculation is simple. On the other hand, the required information for the expected loss decline on the right side (i.e. the marginal benefit of reliability) differs depending on the economic agent's risk behaviour. If the economic agent is risk-neutral, then as for the reliability model, the expected value for the occurrence probability of damage is the expected valuation of loss. However, when the economic agent is risk-averse, a suitable marginal benefit measurement differs depending on the type of risk and the availability of a redistribution market (i.e. an insurance market) for risk. Here, we organize the rationale for the risk-averse economic agent's cost of facing the risk (i.e. the benefits and measurement of risk countermeasures).

9.4.2.1 Benefits of Risk Countermeasures

The economic agent (supplier or consumer) is risk-averse, expressed using utility function U. The economic agent faces a situation where variable x (commodity or service for the consumer or profit for the supplier) differs depending on state i where x occurs. In other words, state i creates the possibility of x_i. The probability of state i is p_i, and when the total is a state of S, the expected utility is as follows.

$$E[U(x)] = \sum_{i=1}^{S} p_i U(x_i) \tag{9.11}$$

Here, $U(x_i)$ expresses cardinal utility. Because the economic agent is risk averse, $U''(x_i) < 0$. At this time, if C represents the amount the economic agent is willing to pay to avert this risk (compensating variation), it is expressed as follows:

$$U\left(\sum_{i=1}^{S} px_i - C\right) = \sum_{i=1}^{S} p_i \cdot U(x_i) \tag{9.12}$$

The approximate value of C is obtained by performing a Taylor expansion on Eq. 9.12 around $\sum_{i=1}^{S} p_i x_i$, which is the expected value of x_i (Boadway and Bruce 1984).

$$C \approx -\frac{1}{2}\frac{U''\left(\sum_{i=1}^{S} p_i x_i\right)}{U'\left(\sum_{i=1}^{S} p_i x_i\right)}\mathrm{var}(x_i);\ \mathrm{var}(x_i) = \sum_{i=1}^{S} p_i\left(x_i - \sum_{i=1}^{S} p_i x_i\right)^2 \tag{9.13}$$

The amount the economic agent is willing to pay for risk aversion depends on the degree to which the utility function is concave with respect to the expected value of the obtained result and dispersion of x_i (i.e. the absolute risk aversion level). The information determined by the WTP amount for risk aversion using these two factors is useful, but it is difficult to estimate. This is because it is difficult to observe the utility function to obtain information for the absolute risk-aversion level of the former. Therefore, welfare measurement is presented for changes in the risk-averse economic agent's risk in a way that includes information on the utility function.

9.4.2.2 Welfare Measurement of the Risk-Averse Economic Agent

The following items are nominated as welfare measurements used for a compensating policy test to address risk: option price, which is a ex ante compensating variable; ex post compensating variable; expected value of the ex post compensating variable; fair-bet point; and expected value of the fair-bet point (for details of the respective measurements, see Just et al. 2004). These measurements are categorized according to the timing of valuation and compensation. The timing of valuation is titled 'ex ante' and 'ex post' decision-making before and after the risk is cancelled, respectively. In addition, the compensating timing is 'ex ante' and 'ex post' compensation before and after the risk is cancelled, respectively. Table 9.3 shows measurement categorization based on valuation and compensation timing.

According to Graham (1981), to address risks based on the created standard, policy compensating tests can be generated from the following measurements: (1) Is the policy cost fixed beforehand? (2) Are the utility functions of the economic agents equal and do the risks being faced have the same quality based on the points expressed by probability and amount? (3) Are the risks being faced by the economic agents collective risks or individual risks? Here, we suppose that with respect to (2), the utility functions and the faced risks of the economic agents are the same quality. Based on this, welfare measurements suitable for policy costs of (1) and types of risks of (3) are organized according to Just, Hueth, and Schmitz (2004).

9.4.2.3 When Contingency Markets Is Not Available

We assume that at the beginning, contingency markets (insurance markets) that redistribute risks is not available. Here, when the cost of policy is fixed and the risk is collective risk, the maximum compensation amount that the risk-averse economic agent is willing to pay is the option price, which is an ex ante compensating variation. In other words, if there are collective risks, risk distribution that satisfies the compensation standard for policy costs does not exist in the first place, because all economic agents face the same state. Therefore, comparing option prices and policy costs is a necessary and sufficient condition for compensation testing.

When the cost of policy is fixed and the risk is individual risk, if the distribution between economic agents in the ex post state is equal to the ex ante occurrence probability distribution that each economic agent faces, in other words, in the case of a purely individual risk, the expected value of the fair-bet point becomes welfare

Table 9.3 Categories of welfare measurement of risk averse consumer

		Compensation	
		Ex ante	Ex post
Valuation	Ex ante	Option price	Expected value of fair bet point
		Precursory compensated variance	Expected value of subsequent compensated variance
	Ex post		Fair bet point
			Subsequent compensated variance

measurement. Because the ratio of economic agents that experience a certain state at a certain time is equal to the occurrence probability of that state (which the respective economic agent faces beforehand), the amount the economic agent is willing to pay becomes the amount that they are willing to pay after the occurrence. The fair-bet point refers to the point when the subsequent expected value of compensating variation is taken, that is, the maximum point. Even when a contingency market is available, risk redistribution for this expected value does not occur. Therefore, comparing the fair-bet point expected value and policy cost is a necessary and sufficient condition for the compensation test.

Next, we consider the case when the policy cost is state-dependent. When the policy cost is state-dependent and the risks are collective risks, the social costs required to suit a state, no matter which state occurs, can be lower than the sum of the compensation amounts. An ex post compensating variation is applicable. If the state-dependent social cost is lower than the ex post compensating variation, the compensation test is satisfied (although it becomes only a sufficient condition).

When the social cost for individual risks is state-dependent, the complexity of welfare measurement increases. The following problem is used to determine the amount that individual j is willing to pay in state i.

$$\max_C \left\{ E\left[\sum\nolimits_j C_{ij} - G_{ij}\right] \middle| \sum\nolimits_j C_{ij} \geq G_{ij} \ \forall i,\ E\left[U_j\left(x_{ij} - C_{ij}, R_1\right)\right] = E\left[U_j\left(x_{ij}, R_0\right)\right] \ \forall j \right\} \tag{9.14}$$

G_j is the state-dependent social cost when reliability state R changes from R_0 to R_1, and the total expected value for state i is calculated for all individuals j. This problem is a nonlinear programming problem for obtaining an efficient risk allocation where the amount of compensation exceeding the social cost is at a maximum. If the excess is positive, the allocation between individuals is made such that at least one is better off without worsening the state of the other people. When a contingency market is not available, acquiring a positive value for the objective function is a necessary and sufficient condition for the compensation test. However, as this is not a prerequisite, even in the case of a compensating system where the compensating variation at a certain state does not exceed the social cost, it may satisfy the potential compensation test. The potential compensation test is not

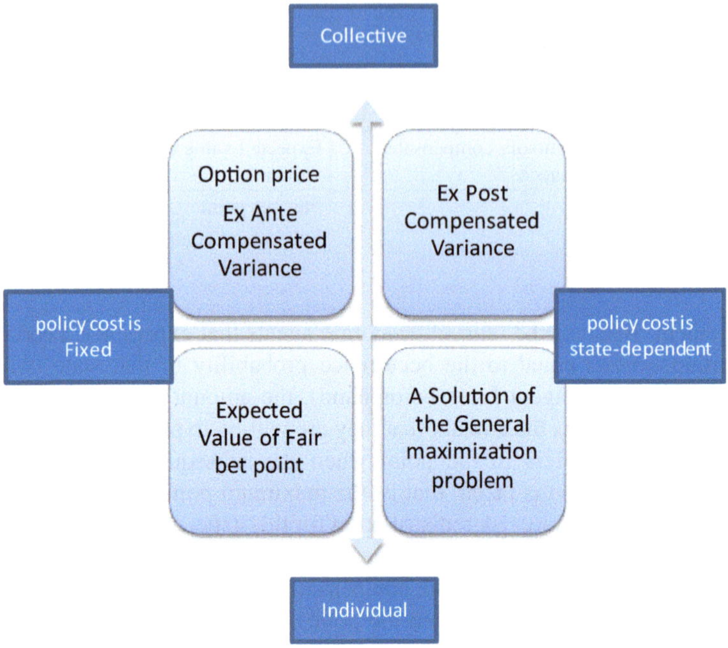

Fig. 9.3 Welfare measurement of risk averse consumer

satisfied when a solution for the general maximization problem does not exist. However, this kind of compensation system is complex, and a large amount of information related to each economic agent's risk behaviour would likely be required. Moreover, the implementation of a public system of paying an uncertain compensation to each economic agent is often not politically acceptable.

Figure 9.3 shows the welfare system that must be applied when a contingency market is not available.

9.4.2.4 When a Contingency Market Is Available

When a contingency market is available, the compensation test is a comparison of the expected value and policy cost of the fair-bet point. When the policy cost is fixed, it suffices to compare the policy cost with the maximum WTP compensation that can be achieved using the contingency market. However, at this time, the maximum WTP compensation becomes the expected value of the fair-bet point. On the other hand, when the policy cost is state-dependent, a condition exists where the total ex ante compensating variation for all states must exceed the social cost. In the case of collective risk, if a contingency market is available that diversifies this risk, testing whether the social cost is smaller than the expected value of the fair-bet point, which is the WTP amount, suffices.

However, in order for the contingency market to be available, considerable transaction and administration costs are incurred because of problems with adverse selection and moral hazards caused by asymmetric information. Therefore, there are many cases when a market will not be formed. In addition, payment of individual and state-specific compensation as well as uncertain compensation after the event may be difficult for practical or political reasons. Therefore, in practice, these measurements may be used when designing a system that absorbs the uncertainty portion of the administrative costs and compensation payment that enables governments to play the role of a contingency market.

9.4.3 Benefit of Reliability Improvement for Water Utility Systems

In order to estimate the benefit of reliability improvement by risk countermeasures for the water utility system using the reliability model of 9.4.1, it is necessary to specify suitable measurements for the type of risk using the categories of 9.4.2.

Table 9.4 lists the risks (perils), damage, and countermeasures of the water utility system using Table 9.1, Miyawaki and Magara (2007), and Japan Water Works Association 2008 as reference sources. The damage in the table is a compilation of damage to the consumer including quantitative supply interruption and reduction damage and qualitative health and odour or taste damage. On the other hand, countermeasures are suppositions planned for by the supplier that provide returns to investment and cost. For example, public investment for earthquake-proofing and replacing facilities as well as establishing backup facilities; operating costs such as management, monitoring, and practice drills; and ex post costs such as emergency water supply or chemical treatment.

When these risk countermeasures are carried out by the supplier or by the government to provide assistance, a valuation is performed using Eq. 9.10 to compare the costs of the countermeasures with the improved reliability benefits of these countermeasures. The improved reliability benefits are estimated using the welfare measurement organized in 9.4.2 2. Although the characteristics of the related risks may be collective or individual, the characteristics of the water utility system (i.e. regional monopolistic supply because of the water source and pipeline) generally imply that the consumer faces collective risk. However, as risks such as infectious diseases or health damage can be a factor for the state of health, individual risk can also be considered on the demand side. Furthermore, when a higher-level government conducts risk countermeasures for multiple water utility business operators, each business entity can be considered to be facing individual risk.

For example, if earthquake-proofing investment is carried out for an earthquake risk in the water supply system, the policy cost is fixed. Because the lowered risk of supply disruption and reduction caused by this investment is collective risk in the

Table 9.4 Damage and countermeasure for water supply system risk

System	Peril		Damage	Countermeasure
Water resource	Quantity	Drought	Cut off	Earthquake resisting
		Underground water exhaustion	Pressure reducing	Multi-systemization
				Emergency water supply
		Storm and flood		
		Earthquake		
	Quality	Eutrophication	Cut off	Multi-systemization
		Escherichia coli O157	Pressure reducing	Emergency water supply
		Cryptosporidium		Use of chemical specialities
		Terrorism	Water-base infection diseases	Monitoring
		Pollution with poisonous materials		
			Health disorder	
Treatment	Quantity	Storm and flood	Cut off	Multi-systemization of incoming power supply system
		Earthquake	Pressure reducing	
		Power failure		Emergency water supply
		Mis-operation		Use of chemical specialties
	Quality	Residual chlorine Disinfection by-product	Cut off Pressure reducing	Increased security of facilities
				Management of water resource
		Terrorism	Health disorder	
			Offensive taste and smell	
Reservoir	Quantity	Storm and flood	Cut off	Emergency water supply
		Earthquake	Pressure reducing	
		Mis-operation		
	Quality	Terrorism	Cut off	Increased security of facilities
			Pressure reducing	
Distribution	Quantity	Storm and flood	Cut off	Multi-systemization of piping system
		Earthquake	Pressure reducing	Multi-systemization of incoming power supply system
		Power failure		
		Mis-operation		Emergency water supply
	Quality	Terrorism	Cut off	Increased security of facilities
		Cross connection		Renewal of lead pipe
		Lead pipe	Pressure reducing	
			Health disorder	
			Offensive	
		Water receiving tank	taste and smell	

target region, it is necessary to estimate the option price, which is the ex ante compensating variation. An example of the quantitative risk countermeasure system shown in 9.3.2 is a policy of changing the standard for the water utilization safety level from a 'planned safety level' to a 'managed safety level'. This is considered taking possession of water sources of multiple systems where the costs involved are fixed; then, the probability of supply disruptions or shortages occurring as part of a response of this policy is collective probability. Therefore, it is possible to obtain information regarding the optimum safety level by comparing the option price, which is the ex ante compensating variation.

Moreover, with respect to setting water quality standards for the qualitative risk countermeasure system, the costs incurred for achieving standards are fixed costs, and because the risk probability of health damage or infections exists in combination with the individual risk according to the standard, information relating to the optimum standard can be obtained by estimating the expected value of the fair-bet point and comparing it with the cost.

9.5 Conclusion

This chapter addresses issues such as utilization of market mechanisms, difficulties with investment for maintenance and replacement, and optimization of maintenance management levels against the backdrop of declining population and efficiency improvements in administration and public finance. From a welfare economics viewpoint, it conducts an examination into the key risks concerning the supply of utility water for which there is pressing urgency for new system design. By organizing a method of understanding the water utility system and risks according to the concept of 'water utility risks', using existing literature, we determined the current risk countermeasure system characteristics through classification into quantitative and qualitative risks. Next, we examined the determination of the system for these risk countermeasures from the point of view of economic agents, and we showed that an efficient system design is not established because the predominant supposition is damage to consumers supposed by the suppliers; thus, the valuation of the damage to consumers is not necessarily reflected directly. Therefore, for a system design that reflects the damage to the economic agents caused by risk, to derive the level of optimum reliability, it is necessary to estimate the potential improvement to the reliability of risk countermeasures using reliability models. To measure this benefit using welfare measurement, we organized differentials depending on whether contingency markets is available, whether policy costs are fixed, and whether the type of risk is collective or individual. We also reviewed several water supply risks and considered applicable measurement types.

For water utility businesses, where stable quantity and quality of supply is essential, economic valuation is required to improve reliability for risk countermeasures to set system design and investment that generates greater efficiency. To achieve this, we suggest a framework that reflects not only an understanding of risks for the supplier but also information relating to the damage and risk preferences of consumers.

On the other hand, there may be challenges in revealing the prerequisites of these measurements that are required to estimate benefits using suitable measurements. First, there are contingency markets with redistribution of risks. When there are numerous water utility business operators, there is a financially-independent principle as a local public enterprise; in this sense, they are risk-averse economic agents. Therefore, although a contingency market must be formed for the risks that water utility business operators face, it is necessary to judge the degree to which the contingency market has been formed using information on the degree to which the general accounting of municipal bodies and the national treasury fulfil the role of absorbing these risks and the participation rate of the water utility insurance. Second, identifying whether a type of risk is collective or individual is difficult. Although, the risks that each economic agent faces are generally independent, 'quasi-individual risks', where collective risks that cannot be ignored are occurring, may result (Just et al. 2004). For example, with respect to contamination of substances that cause health damage, although these are collective risks because supply is conducted through the pipeline, there is a degree of independence in that the actual probability of infection concerns individuals. In such cases, suitable measurement depends on the independence of the damage probability of individuals. Third, there is the degree of dependence on the policy cost state. Although the cost of policy that requires ex post compensation in the form of assistance and insurance is state-dependent, its level is a differential depending on the degree of compensation required based on the level of individual risk. Furthermore, a major challenge concerning the estimation of benefit is how to derive preferences relating to risk and reliability. Because a representative market suitable for estimation is non-existent, estimation frequently uses a method of stated preferences such as CVM (Contingent Valuation Method). In such cases, it is important to pay adequate attention to the characteristics and biases related to risk recognition. The aforementioned challenges are all empirically based challenges, and clarifying each of these risks will be required.

The concept of economic valuation for the improvement of reliability in supply with regard to risk countermeasures can be considered useful for the general supply of not only water utility but also other lifelines such as electric power, gas, and roads that are essential commodities. By examining these challenges, we hope to accumulate knowledge for system design that can achieve the optimum level of reliability and efficiency of supply.

References

Asahi, C., & Hagihara, K. (1999). Evaluating risk in drinking water based on the Choice Research, Environmental information science. *Journal of Environmental Information Science, 13*, 223–226.

Asahi, C., & Hagihara, K. (2004). Ex post risk management of environmental contamination on municipal water. *Studies in Regional Science, 34*(3), 275–287 (in Japanese).

Asahi, C., & Hagihara, K. (2010). Evaluation of policy for supply reliability improvement-the case of municipal water supply system under risk. *Studies in Regional Science, 40*(1), 111–128 (in Japanese).

Barlow, M., & Clark, T. (2002). *Blue gold*. Toronto: Stoddart Publishing.

Boadway, R., & Bruce, N. (1984). *Welfare economics*. Oxford: Basil Blackwell.

Graham, D. A. (1981). Cost-benefit analysis under uncertainty. *American Economic Review, 71* (4), 715–725.

Hagihara, Y. (1995). Water resources and environment, research report of Water Resources Research Center, No. 15, pp. 51–71. Water Resources Research Centre, Kyoto University (in Japanese).

Hagihara, Y., & Watanabe, H. (1981). Model analysis on fair water allocation, Japan Society of Civil Engineers, papers of research meeting on civil engineering planning, pp. 215–220 (in Japanese).

Howe, C. W., Smith, M. G., Bennett, L., Brendecke, C. M., Flack, J. E., Hamm, R. M., et al. (1993). The value of water supply reliability in urban water system. *Journal of Environmental Economics and Management, 26*, 19–30.

Japan Water Works Association. (2008). Outline of water supply (in Japanese).

Japan Water Works Association. (2013). Waterworks statistics (in Japanese).

Just, R. E., Darrell, L. H., & Schmitz, A. (2004). Stochastic welfare economics with applications to agricultural policy analysis. In R. E. Just, L. H. Darrell, & A. Schmitz (Eds.), *The welfare economics of public policy: A practical approach to project and policy evaluation* (pp. 467–526). Cheltenham/Northampton: Edward Elgar Publishing Limited.

Ministry of Health, Labour and Welfare. (2013). New water supply vision (in Japanese).

Ministry of Health, Labour and Welfare, Government of Japan. (2007). The guidelines for cryptosporidium treatment in water supply (in Japanese).

Ministry of Health, Labour and Welfare, Government of Japan. (2008). Water supply vision revised ed. (in Japanese).

Miyawaki, J., & Magara, Y. (2007). For not stopping municipal water services. Tokyo: Jiji Press Ltd. (in Japanese).

Nomura Research Institute. (2008). Seminar, infrastructure in decline of population, Niikei, 11 Dec 2008, p. 23 (in Japanese).

Odanagi, O. (2004). Water resources-systems analysis and planning, Tokyo: Nissuicon Co. Ltd. *29* (1) (in Japanese).

Odanagi, O., Imada, T., Mori, M., & Hagihara, Y. (1997). A study on crisis and risk management of municipal water provision. In *Proceedings of the 10th annual conference*/Society for Risk Analysis, Japan-Section, pp. 98–105.

Okada, N. (1985). An analytical view of disaster risk, Tokyo: Japan Society of Civil Engineers (in Japanese).

Chapter 10
Evaluation of the Vulnerability of Municipal Water Infrastructure

Chisato Asahi and Kiyoko Hagihara

Abstract In this chapter, we examine the welfare effects and their measurement of social overhead capital on a household. In Japan, the decline of population and the structural straits of public finance give rise to difficulties in maintenance and replacement of social overhead capital that have been invested through the period of high economic growth and the following period of the fiscal stimulus over 1990s. The importance of the basic service provided by the infrastructure service, such as public utility service or flood control, is however increasing because of transformation of regional structure caused by population decrease and continually concern for disaster. In order to solve the trade-off between the capacity limit of new public investment and the increasing needs for safety of life provided by social overhead capital, we show a model of welfare evaluation of vulnerability handled by the infrastructure and examine it in case of municipal water supply. First, we show the welfare evaluation model for vulnerability that household face. Second, the welfare measurement of vulnerability is developed especially in the view of theoretically appropriateness of welfare measurement. Third, preliminary inspections show the validity of the application of the model and some required conditions for evaluation.

Keywords Vulnerability • Risk management • Welfare measurement • Social overhead capital • Municipal water supply

C. Asahi (✉)
Division of Urban Policy, Faculty of Urban Liberal Arts, Tokyo Metropolitan University, 1-1 Minami-osawa, Hachioji, Tokyo 192-0397, Japan
e-mail: asahi@tmu.ac.jp

K. Hagihara
School of Sociology, Bukkyo University, 96 Kita-hananobo, Murasakino, Kita-ku, Kyoto 603-8301, Japan
e-mail: hagihara@bukkyo-u.ac.jp

K. Hagihara, C. Asahi (eds.), *Coping with Regional Vulnerability*, New Frontiers in Regional Science: Asian Perspectives 4, DOI 10.1007/978-4-431-55169-0_10

10.1 Introduction

Financial constraints and changes in socioeconomic structure accompanying dwindling populations have led to ongoing reforms toward streamlining and priority-setting within social overhead capital (SOC) development. Public works spending, which increased throughout the 1990s primarily as a means of economic stimulus, has been in continuous decline since the acceleration of administrative and financial reforms of the 2000s. On the other hand, spending has continuously increased on pensions, medical care, and other social security expenses. The funding for these welfare-related policies comes primarily from reductions to public works expenditures, one of various drastic changes in the allocation of annual government spending. Large volumes of SOC constructed in the 1960–1970s are due for renewal, and the financial burden of these maintenance and renewal costs is expected to increase. Large demand for infrastructure is also anticipated in the reconstruction efforts following the March 2011 The Great East Japan Earthquake. In this context, a variety of policy reforms are underway pertaining to the construction of SOC as a stock of public works projects.

The development of SOC has flow effects such as multiplier effects on reflation and employment creation, but also in the long term has stock effects that improve productivity and standard of life (Karaki et al. 2006). It also has income redistribution effects in cases where infrastructure development improves production and standard of life in a way that leads to corrected or increased income gaps between regions or income groups. The stock effects are divided into productivity effects and welfare effects. The productivity effects include, for example, new transportation infrastructure that yields decreased transportation costs and increased cargo volume, leading to greater industrial productivity and greater economic growth. The welfare effects include improvements to sanitation, safety, and comfort that contribute to a higher standard of life for households.

To maintain these effects in the face of declining capacity for investment in SOC, there are three primary challenges to be met. The first such challenge is to reduce the costs of infrastructure development by re-examining construction costs, government procurement, and bidding systems. The second challenge is to prioritize infrastructure development that improves productivity. Sugiura (2003) indicates that Japan's development of SOC in the post-war economic boom period contributed to economic growth by removing the bottlenecks on industrial activity, consistent with the Hirschman hypothesis (Hirschman 1958). On the other hand, in the subsequent periods of stable economic growth and bubble formation/collapse, the mechanisms of economic growth changed from exogenous growth to economies of agglomeration and endogenous growth. It cannot be said, however, that the allocation of SOC development has been revised to effectively suit this change in growth mechanisms, and it remains a challenge to discover the right direction for infrastructure development from a productivity perspective. The third challenge is to prioritize infrastructure development that improves welfare. The standard definition of social overhead capital (SOC) is "capital (infrastructure) that would be significantly undersupplied if its provision were left to private processes" (Director for Infrastructure Planning 2007), but the objectives and effects of this infrastructure are "to provide basic services (security of national territory, life, and property; living environment, health, and hygiene)" and "to preserve and increase socioeconomic vitality" (Japan Society of Civil Engineers 1989). The "provision of basic services" falls under the category of welfare effects. These effects are often examined with respect to concepts such as "civil minimums" or "national minimums".[1] The criteria unrelated to efficiency, such as fairness and sustainability, are also used in deciding the levels at which such infrastructure will be provided. Therefore, in the midst of appeals for budgetary restraint and streamlining, it is necessary to evaluate the multifaceted effects of infrastructure development in setting priorities for such development.

To determine the proper priorities for future SOC development, it is necessary for each of the above challenges to consider a framework that evaluates the effects of the SOC stock from an economic perspective. The present research focuses specifically on the third challenge, examining the welfare effects of the "provision of basic services." I n particular, the purpose of this chapter is to consider a framework for evaluating the benefits of safety and security, i.e. risk management. This chapter will utilize traditional models of decision-making under uncertainty and the findings of work on risk management, as well as examine the applicability to economic analysis of the concept of vulnerability as organized by Asahi (2010).

The second section of this chapter will outline the findings of previous work on risk management and vulnerability vis-à-vis SOC (i.e., infrastructure). The third section will consider household decision-making regarding the use of SOC and utilize a benefit evaluation model to assess the effects of SOC on ameliorating the vulnerability confronted by households. In this, water supply infrastructure, which possesses all of the attributes of a "basic service," will be examined for descriptive purposes. The fourth section will present methods of preliminarily testing a given configuration of the evaluation model, as well as offer points that must be verified in the process of specifying and utilizing the evaluation model. The last section will be a summary.

10.2 Infrastructure: Risk and Vulnerability

10.2.1 Risk Management and Welfare Loss due to Risk

This section will begin with an overview of the work in the field of risk management on risk and household response to risk. From a management perspective, risk

[1] In this chapter, the "civil minimum" is defined as the necessary minimum standard of living in a region. The "national minimum" is defined as the minimum standard of living guaranteed for all citizens of the nation.

is defined in terms of phenomena and their probability of occurrence distributions. Phenomena are described in terms of size, frequency, duration, range, etc. Probability distributions are either known or unknown, and unknown probability distributions are sometimes distinguished by their unpredictable uncertainty. The probability distribution may be known or unknown; if unknown, it may be classified as unpredictable uncertainty.[2] Households have several means to deal with risk, which we shall refer to here as "household-based risk management". In risk management there are concepts of *ex ante* and *ex post*; the former refers to preventive actions taken before the event actually occurs, the latter to countermeasures taken after it does. There are two major types of *ex ante* risk management. One is risk reduction, meaning to directly reduce the chances (probability) of incurring damage from the event; examples include earthquake-proofing buildings and building levees. The second is mitigation, meaning to allay the damage from the event, by means such as purchasing earthquake insurance or building reserve funds. *Ex post* risk management, on the other hand, refers to actions taken in response to damage which has occurred, such as distributing water and foodstuffs.

Households cannot carry out such risk management at optimal levels. The primary reasons are (1) market failures; (2) income distribution; and (3) risk characteristics. First, the risk event or probability distribution may be intrinsically unknown, or there may be an informational asymmetry between the household and the enterprise or government. It is also possible that demand and supply for the insurance or supplies of goods that make risk mitigation possible may diverge from effective levels due to informational asymmetry in the market. These can be categorized as market failures. Second, whether or not actions such as building reserves or investing in preventive medicines or earthquake facilities can be taken depends on income levels. Third, it is known that subjective assessments of risk can diverge from objective assessments due to how risks are perceived. For instance, earthquake countermeasures based on optimistic subjective assessments may be inadequate when viewed from the level of an objective assessment. In addition, when risks are collective risks, or covariate risks, meaning the risk of the risk materializing is not independent of the individual, the risks cannot be pooled, posing limits to the effectiveness of mutual assistance measures based on insurance and social networks. For these reasons, households cannot deal completely with the risks being faced, leaving open the possibility that the risk event may impinge on welfare.

It has traditionally been held to be the role of government agencies to supplement the risk management capabilities lacking in households. In general, the building of infrastructure with the objective of countering risks is intended to have the effect of reducing risk in advance. Since there are restrictions to the resources which can be devoted to building infrastructure, however, the government cannot completely supplement household-based risk management measures. In particular, when viewed from the perspective of the efficiency of resource

[2] The classic definition of Knight (1921) is well known.

allocation, whether to implement risk management policies is assessed based on whether the benefits outweigh the costs.

10.2.2 Vulnerability Valuation

The possibility of substantial losses faced by households thus depends on the level of risk management that can be adopted by the household and publicly. With regard to the former, Siegel and Alwang (1999) note that of the range of available risk management approaches, those which a particular household can implement is dependent on the "assets" it possesses, in the broad sense. By "assets" in the broad sense we assume "assets" as defined by Siegel and Alwang (1999). These "assets" include not only physical capital such as income-generating land, labour, and savings, but also the household's situation as relates to nature, the economy, the society, and politics in enhancing welfare. Specific examples include social capital, social ties, level of education and health, markets and infrastructure condition (roads and water supply), and ability to participate in political decision-making. The degree to which the latter "public risk management" is provided is governed by the cost-effectiveness and absolute criteria (civil minimums). The concept of vulnerability expresses the possibility of damage to household welfare explicitly considering such conditions.

The concept of vulnerability is used in a number of fields, including the poverty issue in developing countries, natural disaster management and food security, and is defined differently in each. Alwang, Siegel and Jorgensen (2001) reviewed the concept of vulnerability across a number of fields and identified the common factors comprising vulnerability. According to them, first a risk exists, and a certain result emerges based on risk management measures taken by the household or government. Household vulnerability is a concept defined vis-à-vis this result, described by the factors governing the risk for the household and the risk management measures that the household can take. The "result" to measure the degree of vulnerability, then, is represented by a "loss in welfare" in comparison to a given benchmark. The benchmark used is normally the minimum level acceptable to society (metrics such as poverty line and civil minimum). Vulnerability can thus be described with three factors. The first is the risk or the event accompanying the risk; the second is the measures taken to manage the risk (household or government); and the third is the ensuing state of affairs represented by the loss in welfare.

Asahi (2010) reviewed the concept of vulnerability and the economic modelling of vulnerability in poverty research by Alwang, Siegel & Jorgensen (2001) and others, and concludes that the effect of the condition of infrastructure in mitigating vulnerability is reflected in its benefits. The concept of vulnerability, not that of risk, is appropriate to introduce in the valuation of the benefits of improving infrastructure for the following two reasons. First, the welfare effect of improving infrastructure as a way to counter risk has a strong character of being a "fundamental service". The interest of households in consuming "basic services" does not lie in the "magnitude and probability of loss to life and living standards" in terms of the overall level of volume and quality of supply, but rather in the "magnitude and probability of the loss beyond minimal levels necessary to maintain life and living standards". Second, although improvements in infrastructure in general have the nature of public goods, they are often not pure public goods but rather to some extent impure-public goods, the nature of which is collective consumption and excludability. Since private sector entities can supply impure-public goods for compensation, risks faced by households are different depending on their income and assets, broadly defined, including such consumability.

The many kinds of risks that are addressed by social infrastructure, therefore, should not be defined merely by the probability of an event occurring and its result. Defining them via the concept of vulnerability, which encompasses the household's capability to deal with risks and minimum supply standards, is more compatible with evaluations of social infrastructure designed to enhance household welfare through stable supply.

10.3 Vulnerability Evaluation Model

10.3.1 Household Water Consumption Decision Problem

Let us consider a model to assess the reduction in vulnerability due to "provision of basic services" through social infrastructure. Let us take the case of water services as a basic service.

Let y be a household's water consumption level. The minimum amount of water the household requires to survive is z, but there is a possibility that a sudden water cut-off due to a natural disaster or malfunctioning equipment, or a reduction in water pressure or cut-off in water from a drought could reduce y to a level below this minimum consumption level of z. Let the probability of harm be π^{a}. π^{a} can be reduced by infrastructure improvements G to the public water facilities, or preventive actions by the household such as purchasing bottled water (private investment R). In other words, we can define π^{a} as Eq. 10.1 with the result that the consumption level is determined by private investment R and public infrastructure level G.

$$\pi^{\mathrm{a}} = \pi^{\mathrm{a}}(R, G, z) \tag{10.1}$$

The household water consumption decision-problem is to determine, given public infrastructure level G and income M, the private investment R in preventive actions which maximizes the preference function I. For the preference function I we use the general preference metric proposed by Freeman (1991)[3]. The consumer decision- problem can now be expressed as follows.

$$Max\ I = f(M, y(R,G), \pi^a) \tag{10.2}$$

$$\pi^a = Prob(y(R,G) < z) \tag{10.3}$$

The household still faces the risk of the actual water consumption level y resulting from optimal choices falling below the minimal consumption level z, the probability of which we denote by π^p. The first-degree condition for the household's optimal investment in water R^* and the probability π^p can be represented as follows.

$$\frac{\partial I}{\partial R} = \frac{\partial f}{\partial y}\frac{\partial y}{\partial R} + \frac{\partial f}{\partial \pi^a}\frac{\partial \pi^a}{\partial R} - \frac{\partial f}{\partial M} = 0 \tag{10.4}$$

$$\pi^p = Prob\big(y\big(R^*, G\big) < z\big) \tag{10.5}$$

In other words, the household determines its optimal private investment in water R^*, given income M and public water infrastructure level G, but since it is not fully aware of the probability π^a, there is a possibility that the actual resulting water consumption level y might fall short of the minimum level z. The a posteriori probability π^p depends on the optimal choice R^*. By "not fully aware of the probability π^a", we mean that the optimal choice for R may have been made to achieve π^a, but viewed at the point after the fact when the consumption level y was reached, they encountered a probability π^p different from π^a. We thus call the situation where if π^a is the a priori probability and π^p the a posteriori probability, the optimal choice R^* such that $\pi^a = \pi^p$ is achieved, a case of perfect foresight. Letting the water consumption level achieved when the optimal investment R^* has been made be y^*, from Eqs. 10.2, 10.4 and 10.5 we can represent y^* as follows.

$$y^* = y\big(M, y\big(R^*, G\big), \pi^p\big(R^*, G, z\big)\big) \tag{10.6}$$

10.3.2 Vulnerability in Water Consumption

One essential part of the concept of vulnerability is to capture welfare by comparing to exogenous benchmarks such as the poverty line or civil minimum, and often to establish these benchmarks we describe dynamic situations. Here we define a household whose water consumption level y^*, based on optimal decisions, is on

---

[3] Freeman (1991) addresses the problem that decisions based on expected utility theory do not conform with actual decision making under risk. He proposes a model to measure changes in wealth outside of expected utility, using general preference index. For examples of applications of this model, see Asahi and Hagihara (1999) and Hagihara and Hagihara (2002). Our model is based on these.

average below the exogenous minimum water consumption level z as vulnerable in water usage. Dercon (2006) proposes a model for measuring vulnerability based on six axioms, which we will employ here.

We represent the measure of shortage v expressing the lack of water vis-à-vis the household's optimal water demand y^* as follows. The measure of shortage v is less than 1 if the optimal water demand y^* is less than the minimum level z.

$$v = \frac{\tilde{y}}{z}\ (0 \le v \le 1),\ \tilde{y} = min\left(y^*, z\right) \tag{10.7}$$

Denoting the measure of shortage in time period t as v_t, if the probability distribution of v over t is θ, then the evaluation function for water consumption V is the expected value for v_t under θ. In other words, if a household's optimal water demand falls short, on average, of the minimum standard for some period of time, meaning that V is less than 1, the household is considered vulnerable as regards water consumption. From Eq. 10.7, V is as follows.

$$V = E[v_t] = V(v_t, \theta) = V\left[\frac{min\left(y^*(t), z(t)\right)}{z(t)}, \theta\right] \tag{10.8}$$

In addition, assuming the household's decision concerning water consumption, from Eq. 10.5 we can write the state evaluation function for water consumption as follows.

$$V = V\left(M(t), y\left(R^*(t), G(t)\right), \pi^p\left(R^*(t), G(t), z(t)\right), z(t), \theta\right) \tag{10.9}$$

10.3.3 Welfare Evaluations of Public Investments for Vulnerability

Let us now consider welfare evaluation for public investments as concerns water consumption levels. In order for households to make the optimal choice for private investment R, from Eq. 10.2 the first-order condition is as given in Eq. 10.4. From the total differential of Eqs. 10.2 and 10.4, the welfare evaluation of public investment (marginal change in infrastructure level) is as follows. In other words, consider the marginal effect of marginal private investment in water on the household's perception of the probability of harm π^a and water consumption level y. The evaluation of public investment is the ratio at which this substitutes for the margin effect of public investment. Since this evaluation is derived from a choice based on the household's a priori probability π^a, depending on the level of y achieved the evaluation of the risk faced (a posteriori probability π^p) is included even less.

$$\frac{dM}{dG} = -\left(\frac{\partial f}{\partial y}\frac{\partial y}{\partial G} + \frac{\partial f}{\partial \pi^a}\frac{\partial \pi^a}{\partial G}\right) \Big/ \left(\frac{\partial f}{\partial y}\frac{\partial y}{\partial R} + \frac{\partial f}{\partial \pi^a}\frac{\partial \pi^a}{\partial R}\right) dI = dz = 0 \quad (10.10)$$

Next let us consider the welfare valuation of vulnerability based on the post facto probability π^P. Since our main purpose is to compare the difference between the a priori and a posteriori probabilities of harm, to simplify matters we will abstract away the existence of the probability distribution of measure of shortage v over t, and consider a static evaluation. From the total differential of the variables in Eq. 10.9, we derive a welfare valuation of the settings for public investment and minimum standards.

$$dV = \frac{\partial V}{\partial M} dM + \left(\frac{\partial V}{\partial y}\frac{\partial y}{\partial R^*} - \frac{\partial V}{\partial \pi^p}\frac{\partial \pi^p}{\partial R^*}\right) dR^* + \left(\frac{\partial V}{\partial y}\frac{\partial y}{\partial G} - \frac{\partial V}{\partial \pi^p}\frac{\partial \pi^p}{\partial G}\right) dG + \left(\frac{\partial V}{\partial \pi^p}\frac{\partial \pi^p}{\partial z} + \frac{\partial V}{\partial z}\right) dz \quad (10.11)$$

$$\frac{dM}{dG} = -\left(\frac{\partial V}{\partial y}\frac{\partial y}{\partial G} + \frac{\partial V}{\partial \pi^p}\frac{\partial \pi^p}{\partial G}\right) \Big/ \frac{\partial V}{\partial M}, \; dV = dz = dR^* = 0 \quad (10.12)$$

$$\frac{dM}{dz} = -\left(\frac{\partial V}{\partial \pi^p}\frac{\partial \pi^p}{\partial z} + \frac{\partial V}{\partial z}\right) \Big/ \frac{\partial V}{\partial M}, \; dV = dR^* = dG = 0 \quad (10.13)$$

Equation 10.12 shows us that there are two routes by which improvements in the level of water infrastructure G exert influence on vulnerability V. The first term in parentheses on the left-hand side is the route by which improvements in infrastructure affect the a priori probability π^a, and thus affect the choice for private investment R. The second term in parentheses is the effect due to the influence of improvements in infrastructure levels on the a posteriori probability π^P, and does not affect consumer decision making. That is, regardless of private investment R, consumers buying bottled water or keeping water on hand, this route works to mitigate the probability that consumption level y will not meet minimum standard z, due to investments such as strengthening water lines against earthquakes. It follows that the second term has the nature of an evaluation of the investment effect to reduce the vulnerability anticipated by water suppliers, while the first term is an evaluation of the same effect revealed by the improvement in consumers revising their risk awareness (a priori probability π^a).

Equation 10.13 shows that the fact that the probability that water consumption will fall short of the minimum level is reduced by relaxing that standard, and the fact that the setting of the minimum standard changes the measure of water shortage, are reflected in the evaluation of the minimum standard.

10.4 Verification of Vulnerability Evaluation Model

10.4.1 The Need for Preliminary Verification

In order to derive an evaluation from the model set out in Sect. 10.3, it is necessary to verify the assumptions of the model and make specifications based on that verification. Next, we will set out the items needed for this and consider preliminary data collection and verification methods (Asahi and Hagihara 2012).

As we can see from the structure of the model, evaluation of vulnerability differs from evaluation of risk countermeasures in that it is the evaluation of the possibility that households suffer losses even if they make optimal decisions before risks materialize—namely the evaluation of ex post risk. In order to study the possibility of suffering losses despite having prepared for risks, we will examine Eq. 10.3. If households do not have perfect foresight regarding water consumption levels, which are the result of their own choices, or regarding such ex post damage probabilities, we use the ex ante private investment function is Eq. 10.13 from Eq. 10.4. However, in the case of perfect foresight, the ex post private investment function Eq. 10.14 can be obtained by assigning Eq. 10.5 to Eq. 10.2.

$$R^* = R(M, y(R, G), \pi^a(R, G, z)) \tag{10.14}$$

$$R^* = R\left(M, y\left(R^*, G\right), \pi^p\left(R^*, G, z\right)\right) \tag{10.15}$$

In the case of perfect foresight, the optimum selections of Eqs. 10.13 and 10.14 are the same, and public investment can be evaluated just through its relationship of substitution with private investment. In other words, households can internalize decision-making on risks through private investment. The effect whereby ex post damage probabilities that households cannot grasp are reduced from out of the value of public investment disappears. By contrast, in the case that households do have perfect foresight regarding damage probabilities, the necessity arises of evaluating public investment from the point of view of vulnerability.

It is widely accepted that households' awareness of risks and the characteristics of decision-making under conditions of risk or uncertainty can result in distortions regarding whether or not there is perfect foresight on damage probability. Accordingly, before making an evaluation with the model, it is necessary to investigate the decision-making characteristics and information of households under risk. Regarding the example of water supply, assuming that maintenance of facilities is carried out by water utilities, it is necessary to conduct an investigation in relation to ex post damage probabilities. That is, what kind of risks can we recognize regarding water quality problems or water outages resulting from water shortages or trouble with water facilities? Are preparations being carried out? Furthermore, what level of actual water outages etc. are being faced?

Below, we will look at water supply as an example. We can assume either Eq. 10.13 or Eq. 10.14, which are applied when evaluating water supply infrastructure improvements, and use the "Performance Index" data based on water utility guidelines and a preliminary questionnaire survey. In other words, we will consider methods for investigating the factors that influence households' preparations regarding risks to municipal water use (private investment).

10.4.2 *Data*

10.4.2.1 Outline of Questionnaire

Using a preliminary questionnaire, we surveyed 290 university students (from Bukkyo University's School of Sociology and Tokyo Metropolitan University) to explore their awareness of the risks and emergency preparedness of the water supply system. To try to ascertain water usage in households consisting of more than one person, we asked the survey respondents questions about the regions in which they were living when they were high school students and about what steps their families had taken at that time in regard to water supply risk and emergency preparedness.

First, we listed five types of risks related to water supply usage and asked them how much they had been aware that each of these risks could materialize in the near future (pre-probability π^a). The five types of risks are as follows: (1) water supply reduction or stoppage due to a water shortage; (2) water stoppage due to an earthquake or to storm and flood damage; (3) water stoppage due to an accident at the facility; (4) water stoppage due to a problem with water quality; and (5) health issues posed by a problem with water quality. The choices for prior probability π^a were "once every 3 years," "once every 5 years," "once every 10 years," "once every 20 years," and "once every 50 years." Next, we asked the respondents what they did to prepare for the above-mentioned five types of risks when the water their family used was for drinking purposes (drinking and cooking). The choices were "bought bottled water," "set aside water in a pot, plastic tank, or other container," and "nothing in particular."

10.4.2.2 Water Supply Performance Indicator

Data on performance indicator for the water supply business have been published on each water utility's website based on the "Water Works Business Guidelines JWWAQ100", which were standardized by the Japan Water Works Association in January 2005 (see Japan Water Works Association 2005)[4]. Because the university students we surveyed were being asked about the situation when they were in high

[4] Japan Water Works Association web address: http://www.jwwa.or.jp/topics/topics_20081126002.html (as of Sept. 30, 2011).

Table 10.1 Sources of data and variables

Variables		Data	Source
Private investment	R	Drought risk	Survey
		Earthquake, wind, and flood damage risk	
		Facility accident risk	
		Water quality accident (water outage) risk	
		Water quality accident (taste, odour, and health) risk	
Infrastructure/ facilities	G	Surplus capacity of resources	Performance indicators
		Self owned resource ratio	
		Drinking water storage volume per population supplied	
		Service reservoir capacity	
		Aging of water treatment facilities	
		Aging of electric and mechanical equipment	
		Aging of mains	
		Mains rehabilitation	
		Capacity for interconnection of raw water	
		Ratio of earthquake-resistant treatment	
		Ratio of earthquake-resistant service reservoir	
		Ratio of earthquake-resistant pipeline	
		Ratio of non utility generation facility	
		Available water volume in an accident	
		Population supplied water in an accident	
Prior probability	πa	Drought risk perception	Survey
		Earthquake, wind, and flood damage risk perception	
		Facility accident risk perception	
		Water quality accident (water outage) risk perception	
		Water quality accident (taste, odour, and health) risk perception	
Posterior probability	πp	Trunk mains failures	Performance indicators
		Number of purification plant accident	
		Number of pipeline failures	
		Number of service pipe failures	
		Restricted water supply	
		Complaints for water quality	
		Hours of water interruption or water turbidity	
Sample characteristics		Types of housing (detached house/apartment/ others)	Survey
		Number of family member	
		Age	
		Gender	
		Experience of water cut-off	
		Experience of offensive taste and odor/health disorder	

school, our study used data from fiscal 2007. The performance indicator data consisted of 137 items classified into six areas: (1) safety, (2) security, (3) survival, (4) environment, (5) oversight, and (6) international. Our analysis focused on the variables corresponding to each of these, rather than on the defining the data items. Table 10.1 gives an overview of the data used in our analysis.

10.4.3 Analysis of the Factors for Choosing Private Investment R

Our research data was prepared by matching up the municipalities in which the survey respondents were living (when they were in high school) with each municipal water works as listed in the performance indicator for fiscal 2007. I n other words, the survey respondents were considered to be sample water users in their resident municipalities, and we analysed the correlations among the variables through a dispersion of data among regions (water supply business units). Performance indicator data for fiscal 2007 were collected from 199 water supply business units nationwide. The survey respondents' former places of residence consisted of 56 municipalities (principally Kyoto City (Kyoto) and Hachioji City (Tokyo), and when these were matched up, our survey group consisted of 33 water supply business units and 141 samples.

We investigated the choice of steps (private investment) to prevent risk in household water consumption using the following procedure.

1. For the risk classification, we used a stepwise regression analysis (ordered logit analysis) to select variables from among all the explanatory variable candidates of pre-probability, post-probability, level of facility equipment, and respondent attributes. Combining all the explanatory variables selected from the risk classes, we determined the explanatory variables to be those estimated in 2. and 3.
2. We conducted a regression analysis (ordered logit analysis) using private investment (R) as the explained variable and using as explanatory variables those selected in 1. from among level of public equipment (*G*), and respondent attributes. However, for pre-probability (π^a), we set all risk classifications as explanatory variables.
3. Changing pre-probability (π^a) in 2. to post-probability (π^p), we ran a similar regression analysis.
4. A comparison of 2. and 3. showed the determinants of private investment (*R*) resulting from an awareness of risk according to probability.

10.4.3.1 Selection of Explanatory Variables

An ordered logit analysis was conducted for each risk category, selecting variables using a stepwise method (variable reduction method at the 10 % level). However, in

Table 10.2 List of explanatory variables

Prior probability: πa	Posterior probability: πp	Infrastructure: G	Sample characteristics
Drought risk perception	Number of service pipe failures	Surplus capacity of resources	Experience of water cut-off
Earthquake, wind, and flood damage risk perception	Complaints for water quality	Drinking water storage volume per population supplied	Experience of offensive taste and odor/ health disorder
Facility accident risk perception	Number of pipeline failures	Ratio of earthquake-resistant pipeline	Age
Water quality accident (water outage) risk perception		Ratio of non-utility generation facility	Gender
Water quality accident (taste, odour, and health) risk perception		Mains rehabilitation	

order to elucidate the relationship between private investment and prior probability variables, the analysis was conducted for all risk categories. Variables used in the estimation are shown in Table 10.2.

10.4.3.2 Regression Analysis Using Prior Probability (π^a) and a Posteriori Probability (π^p)

Regression analyses (ordered logit analysis) were performed on the dependent variable, defined as private investment in each of the risk categories (*R*), the explanatory variable, defined as respondent attributes (Table 10.2), the prior probability (π^a), and public maintenance standards. Then, the prior probability (π^a) is changed to the posterior probability (π^p) and the same regression analysis is performed. The results of the analysis are presented in Table 10.3.

10.4.3.3 Results

The results of the analysis are organized according to risk classification (see Table 10.3).

Water Quality Accident (Water Outage) Risk

In estimating prior probability, the perception of risk of a water quality accident (water outage) is significant and is interpreted as the lower the risk, the higher the

Table 10.3 Results of ordered logit analysis

Variables categories	Variables	Water quality accident (water outage)		Water quality accident (offensive taste and odor/health disorder)		Facilities accident		Earthquake/storm and flood damage		Drought	
		πa	πp	πa	πp	πa	πp	πa	πp	πa	πp
Prior probability: πa	Drought risk perception	0.307		0.206		0.265		0.093		−0.053	
		(0.198)		(0.191)		(0.197)		(0.187)		(0.187)	
	Earthquake, wind, and flood damage risk perception	0.188		0.042		0.260		−0.177		−0.056	
		(0.162)		(0.159)		(0.176)		(0.158)		(0.155)	
	Facility accident risk perception	−0.066		−0.022		−0.243		0.090		0.100	
		(0.205)		(0.206)		(0.216)		(0.198)		(0.203)	
	Water quality accident (water outage) risk perception	−0.466**		−0.351		−0.514**		0.107		−0.010	
		(0.224)		(0.230)		(0.240)		(0.221)		(0.224)	
	Water quality accident (taste, odour, and health) risk perception	−0.218		−0.184		−0.041		−0.189		−0.318	
		(0.201)		(0.202)		(0.201)		(0.194)		(0.199)	

(continued)

Table 10.3 (continued)

Variables categories	Variables	Water quality accident (water outage)		Water quality accident (offensive taste and odor/health disorder)		Facilities accident		Earthquake/storm and flood damage		Drought	
		πa	πp	πa	πp	πa	πp	πa	πp	πa	πp
Posterior probability: πp	Number of service pipe failures		−0.058		−0.124*		−0.082		−0.083		−0.114
			(0.057)		(0.066)		(0.060)		(0.058)		(0.062)
	Complaints for water quality		−2.466***		−1.836**		−1.602*		−1.736		−1.118
			(0.935)		(0.837)		(0.830)		(0.835)		(0.750)
	Number of pipeline failures		−0.086		−0.041		−0.079		−0.032		−0.025
			(0.065)		(0.069)		(0.069)		(0.062)		(0.067)
Infrastructure: G	Surplus capacity of resources	−0.010	0.005	0.005	0.025*	−0.001	0.019	0.005	0.016	0.007	0.022
		(0.013)	(0.014)	(0.012)	(0.014)	(0.013)	(0.014)	(0.012)	(0.013)	(0.013)	(0.014)
	Drinking water storage volume per population supplied	0.002	0.013	−0.000	0.011	−0.000	0.011	0.004	0.010	−0.004	0.003
		(0.009)	(0.009)	(0.009)	(0.009)	(0.010)	(0.009)	(0.009)	(0.009)	(0.009)	(0.009)
	Ratio of earthquake-resistant pipeline	0.009	−0.018	0.020	−0.016	0.056*	0.027	0.128***	0.101	0.039	0.019
		(0.030)	(0.032)	(0.032)	(0.034)	(0.034)	(0.034)	(0.034)	(0.036)	(0.030)	(0.032)
	Ratio of non utility generation facility	0.011	0.028***	0.010	0.029***	0.015	0.033***	0.018**	0.029	0.014	0.030
		(0.009)	(0.010)	(0.009)	(0.010)	(0.010)	(0.011)	(0.009)	(0.010)	(0.009)	(0.010)
	Mains rehabilitation	1.557**	1.315*	0.225	−0.348	0.834	0.067	−0.435	−0.751	0.522	−0.237
		(0.707)	(0.767)	(0.722)	(0.777)	(0.779)	(0.815)	(0.710)	(0.769)	(0.700)	(0.770)

Sample characteristics	Experience of water cut-off	0.034	0.019	−0.172	−0.140	−0.329	−0.541	−1.106***	−0.916	−0.536	−0.380
		(0.423)	(0.395)	(0.429)	(0.402)	(0.418)	(0.397)	(0.424)	(0.403)	(0.419)	(0.387)
	Experience of offensive taste and odor/health disorder	−0.777	−0.579	−0.786	−0.632	−0.796	−0.553	−0.477	−0.461	−0.037	−0.004
		(0.488)	(0.474)	(0.491)	(0.484)	(0.500)	(0.478)	(0.480)	(0.470)	(0.493)	(0.478)
	Age	−0.220	−0.283**	−0.024	−0.072	−0.103	−0.194	−0.078	−0.080	−0.183	−0.129
		(0.143)	(0.140)	(0.148)	(0.140)	(0.120)	(0.138)	(0.087)	(0.087)	(0.130)	(0.116)
	Gender	−0.813**	−0.653*	−0.028	0.011	−0.340	−0.291	−0.474	−0.392	−0.556	−0.389
		(0.373)	(0.372)	(0.376)	(0.374)	(0.388)	(0.379)	(0.362)	(0.367)	(0.365)	(0.359)
Fitting	Number of obs	134.000	136.000	133.000	135.000	134.000	136.000	134.000	136.000	134.000	136.000
	LR chi2(6)	26.760	27.670	12.510	15.790	22.950	22.050	28.000	33.870	17.960	19.540
	Prob > chi2	0.021	0.006	0.566	0.201	0.061	0.037	0.014	0.001	0.209	0.076
	Pseudo R2	0.098	0.100	0.053	0.066	0.096	0.090	0.099	0.117	0.068	0.073

Note: Standard deviations in parentheses
***$p < 0.01$, **$p < 0.05$, *$p < 0.10$

probability of no occurrence of private investment. This estimation conforms with the hypothesis. On the other hand, interpretation of the posterior probability as the higher the number of complaints regarding water quality the higher the probability of no private investment, results in the sign condition not being met. The coefficient of pipeline renewal rate for public maintenance standards is significant, but when the coefficient is high, measures to prevent water quality accidents are become high which also results in the sign condition not being met. For the posterior probability, although the rate of private power generating capacity is significant, the installed capacity rate is prepared for power failures, the interpretation of the relationship between the water outage risk measures by water quality accident is difficult. In terms of respondent attributes including age and sex, women and the elderly were shown to have a low probability of taking preventative measures.

Water Quality Accident (Taste, Odour, and Health) Risk

For estimation of the prior probability, none of the variables were significant and model did not have good fit. For the posterior probability, the proportion of water supply pipe accidents and the rate of complaints regarding water quality were significant, but the sign condition was not satisfied, and the higher the proportion of accidents and complaints the less preventative actions would be taken. For public maintenance standards, "water source margin" and "private power generating capacity" were significant, but each of these involve maintenance for reducing the risk of drought and power failure, respectively. It is difficult to interpret of the impact of drought or power failure, as well as the relationship between taste, odour, and measures to prevent detrimental health effects. Respondent attributes were also not significant.

Facility Accident Risk

For the prior probability estimation, perception of the risk of a water quality accident (water outage) was significant, and the more it is recognized that a water outage due to water quality is unlikely, the lower the probability of taking measures against water outage due to a facility accident. Facility accident risk perception was not significant, but because facility accidents caused by water quality and aging infrastructure are both related to the overall management of water utilities, confidence in the management entity (water utility) was considered as a potential proxy variable. For the posterior probability, since the higher the "rate of complaints regarding water quality" the higher the probability of no private investment, it does not fit the hypothesis. For public maintenance standards, although the earthquake resistance rate of pipelines was significant, the greater the pipeline earthquake resistance the higher the probability of private investment (assuming a facility accident occurs). This does not satisfy the sign condition. The same is true for "private power generating capacity" in the case of a posterior probability: it was not significant for respondent attributes.

Earthquake, Wind, and Flood Damage Risk

Prior probabilities do not affect preventative measures, but the "rate of complaints regarding water quality" was significant. However, since the posterior probability did not satisfy the hypothesis, interpretation is difficult. For public maintenance standards, "earthquake resistance rate of pipelines" "private power generating capacity rate" were significant in both the prior and posterior models, but the signs were positive, in contrast with the hypothesis. For respondent attributes, experience with a water outage was significant, which can be interpreted to mean the less experience with a water outage the less private investment occurs.

Drought Risk

Prior probabilities do not affect preventative measures, but the "accident rate of the water supply pipe" was significant. Since the sign did not fit with the hypothesis, interpretation of this finding is difficult. For public maintenance standards, "private power generating capacity rate" was significant in both the prior and posterior models, but the sign was different from the hypothesis. Enhancement of measures to prevent power outages was shown to be effective against the behaviour of drought risk measures, though it is difficult to interpret this finding as well. No significance was observed for respondent attributes.

According to the above, when viewed by risk classification, the impact of perception of the risk of prior probability risk affecting preventative measures (private investment) is the risk of water quality accidents (water outages) and facility accidents. However, it is inferred that the posterior probability of these risks is not the determining factor of preventative measures. In other words, the determinant of private investment is the prior probability, and households were considered not have foreseen the risk after preventative measures have been taken based on the posterior probability. In addition, for public maintenance standards, the assumption of a good maintenance level of the pipe was a determining factor of preventative measures, and substitution of private investment and maintenance levels did not fit the hypothesis.

10.4.4 Discussion

10.4.4.1 Determinants of Private Investment

Based on the results of the preceding regression analysis, variables other than those that affect the risk of household tap water use (private investment) will be discussed.

First of all, prior probability does not constitute a determining factor for all risk classification categories, and is significant only for water quality accidents (water outages) and facilities accidents. These risks are internal quantitative risks derived from the water supply system. Conversely, earthquakes, flooding and droughts represent external quantitative risks derived from outside the water system. However, water quality accidents (taste, odour, and health) are qualitative risks for which evidence could not obtained in advance of risk perception.

For posterior probability, even though some variables were shown to be significant, they did not satisfy the sign conditions are therefore unlikely to become determining factors of preventative measures. However, for the "rate of complaints regarding water quality" variable there was a high probability of no preventative measures being taken, indicating possible reverse causality in terms of the sign condition.

For public maintenance standards, there were no significant variables among all risk categories, neither the sign condition nor significance were observed, indicating a tendency for the probability of taking preventative measures to increase as maintenance levels increased. There may be three possible explanations for this. First, there exists the possibility of not perceiving the level of public infrastructure capacities. For example, interpreting the contribution of "Mains rehabilitation" to water quality accidents (water outages), "Surplus capacity of resources" in the risk of water quality accidents (taste, odour, and health), the "Ratio of non-utility generation facility" and "facility accidents" to earthquakes, wind, and flood damage risk is difficult, as these factors may not have been the key determinants of protective measures. Secondly, the relationship between private investment and public infrastructure capacity may be complimentary rather than substitutable. "Ratio of earthquake-resistant pipelines" and "Ratio of non-utility generation facility" also relate to the risk of facility accidents, earthquakes and flooding as these measures enhance personal awareness of earthquake resistance and of preventative measures for power outages, which has the possibility of inducing private investment. In such cases, the relationship between private investment and public infrastructure capacity is complementary, and it is necessary to examine the assumptions of an alternative model through a vulnerability evaluation. A third explanation concerns income effects. Although income was not estimated as a variable due to data constraints, when regional income levels are high, there is a possibility for both private investment and maintenance levels to increase.

In accordance with the above, preliminary verification of the determinants of private investment can be captured in characteristics of the following four points.

First, in terms of risk perception being a determining factor of private investment, water quality accidents (water outages) are quantitative risks derived from water supply system risk and facility accident risk. Second, risk perception is a prior probability rather than a posterior probability. Third, in deciding to make a private investment, it is possible that household recognition of public development levels is insufficient. Finally, public infrastructure capacity and private investment may have a complementary relationship.

10.4.4.2 Implications for Vulnerability Valuation Model

From the features described above, applications to a vulnerability valuation model are suggested. These could include taking into account the characteristics of risk perception according to risk category when applying the perception probability of households to an evaluation model. The primary cause of earthquakes, flooding, droughts, and so on may be due to external factors, such as water systems, and since risk perception according to probability was not significant, it is also conceivable that the results provide an estimate of the magnitude of the damage rather than the probability of damage among the components of risk. If this is correct, then the vulnerability valuation model must be calibrated by replacing probability with the degree of damage.

With regard to the second characteristic, this shows that Eq. 10.14 in Sect. 10.4.1 holds, not Eq. 10.15. In other words, if private investment is determined in awareness of the a priori probability (π^a), under that assumption in Eq. 10.4 of Sect. 10.3.3 R^* is too small when compared to the optimal posterior level, indicating the possibility that the evaluation of infrastructure level (Eq. 10.9) is also underestimated. It follows that the evaluation of infrastructure level is best carried out taking into account the evaluation of Eq. 10.11 using the posterior probability (π^p), or in other words vulnerability. This result may be generalized in the case that demand for goods and services requiring probabilistic judgments when households make optimal choices is used in the evaluation of benefits from public investment due to consumer surplus. In other words, although in general benefits assume optimal choices by households, when household choices are made under risk and uncertainty, there is a good possibility that biases in the perception of benefits from policies based on information about those choices will cause them to be underestimated. Benefits from security and comfort which are the object of the "basic services" of infrastructure improvement when substituted by private investment, often appear as risk averting behaviours. For example, under the revealed preference method, household risk averting behaviours and risk allaying behaviours become basic information for evaluating benefits from infrastructure development in water and roads, benefits from improvements designed to contain mountains and water such as dikes, embankments, and forest management, benefits intended to prevent and mitigate disasters such as earthquake-proofing and fire prevention, and improvements and regulations put in place to counter harm to health from environmental pollution. It follows that what is required is not only information about averting behaviours (private investment) based on a priori probabilities, but also the application of a welfare evaluation model for vulnerability that makes explicit the divergence from posterior probabilities.

With regard to the third and fourth characteristics, as they are assumptions in the vulnerability evaluation model, the substitution relationship between the knowledge of public infrastructure level in household risk perception and private investment must be closely scrutinized. In the vulnerability evaluation model, public infrastructure level and income are exogenous variables, but if the substitution and complementarity relationships with private investment are to be dealt with explicitly, it is necessary to define the burden on public investment explicitly as a constraint on income.

10.4.4.3 Significance and Challenges in Preliminary Validation

In this section, we have presented a methodology for validating the assumptions for the vulnerability evaluation model, and shown the validation results based on preliminary gathering of data. We arrived at certain suggestions as shown in previous two sections, but we must again note that due to the data used for validation and methodological constraints, those results are not sufficiently robust to be immediately applied in identification of the vulnerability evaluation model. Specifically, in the preliminary validation we used a methodology involving validating the relationship between a priori and posterior probabilities based on surveys and performance indicator, but due to household averting behaviours there are limits to the explanatory power of a model explaining private investment; nor was the relationship between averting behaviours and risks that could be considered more typical, related to drought, fire, and water, captured. Conceivable constraints include bias in the attributes and age cohorts of the respondents, limitations in the population and sample size due to the availability of operational performance indicator, and limits on the regional representativeness of respondents due to the matching methodology.

For example, due to the low variability in survey respondents (Kyoto City and Hachioji City were frequent), and the limited number of water supply business units with publicly available performance indicator for the matching (199 entities), the sample for analysis was limited to 144 people residing in 33 municipalities, possibly insufficient to obtain meaningful results.

When estimating the vulnerability benefits of infrastructure, it is necessary to carefully choose the regions and risks, estimate Eqs. 10.13 and 10.14 based on unbiased data sets, and identify and correct the evaluation model depending on the results.

10.5 Conclusions

The topic of the welfare effect of the "provision of basic services" via infrastructure improvements, in other words safety and comfort, has received increased attention in the aftermath of the damage from the Great East Japan Earthquake and the subsequent reconstruction work. Notwithstanding the fact that a primary objective of infrastructure improvement is the provision of basic services, we have few insights about methodologies to evaluate its benefits. In this chapter, based on the concept of "vulnerability" that began to be applied broadly to socioeconomic events following 9/11 in America, we presented a model to evaluate economically the impact of ensuring safety and comfort via infrastructure improvement, and demonstrated a methodology for validating its preconditions using the water system as a case study.

The key conclusions from this study are as follows. First, we confirmed that in economic evaluations of infrastructure improvements to deal with risks, we can apply the concept of vulnerability, allowing us to explicitly express the capacity of households to deal with risk, and the absolute minimum level of services to be publicly provided. Second, we proposed a benefit evaluation model for mitigating vulnerability through infrastructure improvements. With the evaluation model for water supply facilities, we demonstrated that the investment effect of reducing vulnerability expected by the supplier, and the effect revealed in the household's choices as it revises its awareness of ex ante risks, are represented separately. Third, we presented a validation methodology for the model, and by means of a trial using data from surveys and water performance indicator showed that we could obtain the results that (1) when applying the evaluation model, the characteristics of risk perception depending on the type of risk must be taken into account; (2) since there is a high likelihood that households will make choices based not on objective posterior probabilities but rather on a priori risk awareness when deciding how to address historical quantitative risks in the water system, it is appropriate to perform the evaluation using a vulnerability evaluation model based on posterior risk awareness; and (3) it is necessary to scrutinize more closely the substitution relationship between knowledge of public investment levels in household perception of risk and private investment.

The following three topics remain from this study. First is to improve the accuracy of the model validation methodology. As discussed in Sect. 10.4.4, due to quantitative and qualitative constraints in the data used for validation, we were unable to obtain robust suggestions for identifying the model, and therefore limited ourselves to proposing a validation methodology. We must do further validation and enhancements in order to identify a model based on household behaviour under risk and uncertainty, using regional and household attributes examined in greater detail. Second is the extension to calculating the evaluation of benefits. To contribute to solving issues related to infrastructure improvement, we must gather empirical insights about evaluation based on validating the model. Third is the important remaining topic, not touched upon in this study, of research into household income constraints and dynamic aspects of vulnerability.

References

Alwang, J., Siegel, P. B., & Jorgensen, S. L. (2001). Vulnerability-a view from different disciplines. Social protection discussion paper series, No. 0115, Social Protection Unit, Human Development Network, Washington D.C.: The World Bank.

Asahi, C. (2010). A study on economic evaluation of vulnerability in infrastructure improvement project. *Journal of Urban Studies, 4*, 127–156 (in Japanese).

Asahi, C., & Hagihara, K. (1999). Evaluating environmental risk in drinking water. *Environmental Systems Research, 27*, 395–401 (in Japanese).

Asahi, C., & Hagihara, K. (2012). Evaluation of the vulnerability of municipal water infrastructures. *Studies in Regional Science, 42*(3), 563–580 (in Japanese).

Dercon, S. (2006). Vulnerability: A micro perspective. In F. Bourguignon, B. Pleskovic, & J. van der Gaag (Eds.), Securing development in an unstable world (pp. 117–145). Washington, DC: The World Bank.

Director General for Policy Planning, Economic and Fiscal Management. (2007). Social capital of Japan 2007 (in Japanese).

Freeman, A. M. (1991). Indirect methods for valuing changes in environmental risks with nonexpected utility preferences. *Journal of Risk and Uncertainty, 4*(2), 153–165.

Hagihara, K., & Hagihara, Y. (2002). Evaluating changes in urban environmental risk based on the model under uncertainty and unknown risk: the case of municipal water use. The second annual meeting of the International Institute for Applied Systems Analysis (IIASA) and the Disaster Prevention Research Institute (DPRI) of Kyoto University on integrated disaster risk management: megacity vulnerability and resilience, IIASA, Laxenburg, Austria, July.

Hirschman, A. (1958). The strategy of economic development, New Haven: Yale University Press.

Japan Water Works Association. (2005). Performance indicator 2003.

Japan Society of Civil Engineers. (1989). *Handbook of civil engineering*. Gihodo: Shuppan (in Japanese).

Karaki, Y., Okuhara, T., Tomari, S., Asahi, C., & Nishihata, T. (2006). The economic effects of public capital stock: measuring productivity effects and welfare effects through urban area classification, PRILIT research reports, Vol. 68, Policy Research Institute for Land, Infrastructure, Transport and Tourism (in Japanese).

Knight, F. H. (1921). *Risk, uncertainty and profit*. New York: Hart, Schaffner and Marx.

Siegel, P., & Alwang, J. (1999). An asset-based approach to social risk management: A conceptual framework. Social protection discussion paper series, No. 9926, 1999, Social Protection Unit, Human Development Network, The World Bank.

Sugiura, S. (2003). *Urban Economics*. Tokyo: Iwanami Shoten (in Japanese).

Chapter 11
Social Environment Analysis Regarding Arsenic-Contaminated Drinking Water in Bangladesh

Yosuke Fukushima, Yoshimi Hagihara, and Kiyoko Hagihara

Abstract Arsenic contamination of drinking water has long been a serious problem in Bangladesh. Many foreign institutions have provided support to Bangladesh in terms of constructing arsenic-free wells, providing arsenic removal equipments and so forth. However, most of them are not accepted by local residents because they cannot understand how to maintain the equipments or their effectiveness for reducing arsenic contamination. Furthermore, they find certain equipment is too inconvenient to use in their daily lives. A survey was conducted in two villages in Bangladesh in order to define the relationship between arsenic contamination in drinking water and their social environment. First, we attempt to analyse residents' satisfaction with the drinking water available to them. Second, we introduce the unhappiness function in our model and finally, we identify alternatives acceptable to the residents by devising a structural model addressing distrust of external support.

Keywords Arsenic contamination • Drinking water • Acceptability • Unhappiness function • MIMIC model

Y. Fukushima
Ministry of Land, Infrastructure, Transport and Tourism, 2-1-3, Kasumigaseki, Chiyoda-ku, Tokyo 100-8918, Japan
e-mail: fukushima.y2ah@gmail.com

Y. Hagihara
Kyoto University, 264 Suemarucho, Nakagyou-ku, Kyoto 604-0901, Japan
e-mail: y-hagi@gf7.so-net.ne.jp

K. Hagihara (✉)
School of Sociology, Bukkyo University, 96 Kita-hananobo, Murasakino, Kita-ku, Kyoto 603-8301, Japan
e-mail: hagihara@bukkyo-u.ac.jp

K. Hagihara, C. Asahi (eds.), *Coping with Regional Vulnerability*, New Frontiers in Regional Science: Asian Perspectives 4, DOI 10.1007/978-4-431-55169-0_11

11.1 Introduction

Current reports from around the world indicate that the ground water in some regions is contaminated by arsenic (Kinnburgh and Smedly 2000). Bangladesh has long been recognised as one of the countries most vulnerable areas to the arsenic contamination of ground water. Bangladesh is one of the world's poorest countries and continues to suffer from many kinds of disasters, such as cyclones, floods, droughts, salt water ingression, infectious diseases, and a low literacy rate (Hagihara et al. 2003). In Bangladesh, ground water from wells built with foreign aid to reduce the risk of infectious diseases is contaminated with arsenic. The problem of arsenic contamination in the country was first discovered in 1993. Scientists recognize that people could develop cancer by drinking arsenic-contaminated water for a long time (Singh et al. 2002). However, people in Bangladesh cannot adopt self-countermeasures against arsenic contamination because their resources are severely stretched by severe poverty and several natural disasters that afflict the country from time to time. Thus, they currently depend on economic and technical aids from other countries and international institutions.

However, some of the arsenic-removal equipment provided by foreign institutions is too inconvenient for daily use. Furthermore, it becomes clear that Bangladesh people accept few of them due to the lack of knowledge to maintain the equipments and to realize their effectiveness for removing arsenic. This chapter focuses on arsenic contamination of drinking water in Bangladesh, which can be regarded as a typical combination of natural disaster and social- environmental disaster.

Given that any acceptable alternative should correspond to the social environment, a survey was conducted in two villages of Bangladesh with the cooperation of a local non-profit organization (NPO) called the Environment and Population Research Centre (EPRC). The purpose of this survey was to reveal the relationship between arsenic contamination and the residents' social environment and to suggest acceptable alternatives to alleviate their unhappiness. In this chapter, the factors affecting satisfaction with drinking water are analysed using the results of this survey. Then, an unhappiness function is proposed. Finally, the latent factors affecting distrust of technical support from other countries are disclosed.

11.2 Social Environment Survey

The field work and survey were carried out with the cooperation of the EPRC. The areas to be investigated were decided depending on the arsenic contamination situation, vulnerability to floods, and the cost of the survey. Two villages, Azimpur and Glora, located in Singair sub-district of Manikganji district, located about 27 km to the west of Dhaka, were selected as the survey areas.

Table 11.1 Features of the surveyed villages

	Azimpur	Glora
Population	4000	1500
Literacy rate	25 %	53 %
The number of tubewell	400	300

Source: UNO (pers. communications)

According to the *Upazilla Nirbahi* Office (UNO or Sub-District Administrative Office) and the Department of Public Health Engineering (DPHE), Azimpur is one of the most seriously arsenic-contaminated areas. It also faces several economic problems. On the other hand, these issues are not as serious in Glora. Table 11.1 shows the features of both villages, and Fig. 11.1 shows the location of Singair as well as the degree of arsenic contamination in the ground water in this area. The tubewell in Table 11.1 is a kind of well which is possibly contaminated by arsenic in Bangladesh.

We conducted the field survey for 6 days in September, 2003. During the surveys, we saw many arsenic-removal filters donated to the villagers of Azimpur by foreign aid agencies and institutions, but the local people used only two of these filters. Furthermore, they were able to correctly maintain only one of these filters. These filters are shared within their community. Figure 11.2 shows an example of an arsenic- removal filter.

In Bangladesh, a sign is painted on each well in order to identify whether the well is contaminated by arsenic ('green' means it is not contaminated, and 'red' means it is contaminated). During the field survey, we found one green well, which is shared within the community. Some people had to walk from their homes for close to 15 min to reach the well and use its arsenic-free water.

Then, we devised a questionnaire based on the KJ Method (Kawakita 1966) and Interpretive Structural Modelling (ISM) (Iida and Okada 1992; Hagihara 2008), and translated it into the local language, Bengali. The questionnaire consisted of 50 items regarding (1) personal data (e.g. sex, number of family members, literacy rate, and so on), (2) how local residents obtain drinking water (e.g. 'carrying drinking water is physically taxing'), (3) current awareness regarding drinking water (e.g. whether local residents are aware that arsenic is injurious to health), (4) possibility of pursuing options to improve drinking water safety (e.g. 'I do not mind bearing the expense to improve drinking water safety'), and (5) local residents concerns in daily life (e.g. worries about maintaining/securing a job). The questionnaire items consisted of 5 large items, each of which include small items, and the number of total items is 50.

The staff of the EPRC conducted the questionnaire survey between September and November, 2003. As a result, 110 samples for Azimpur and 103 samples for Glora were obtained by the questionnaire survey.

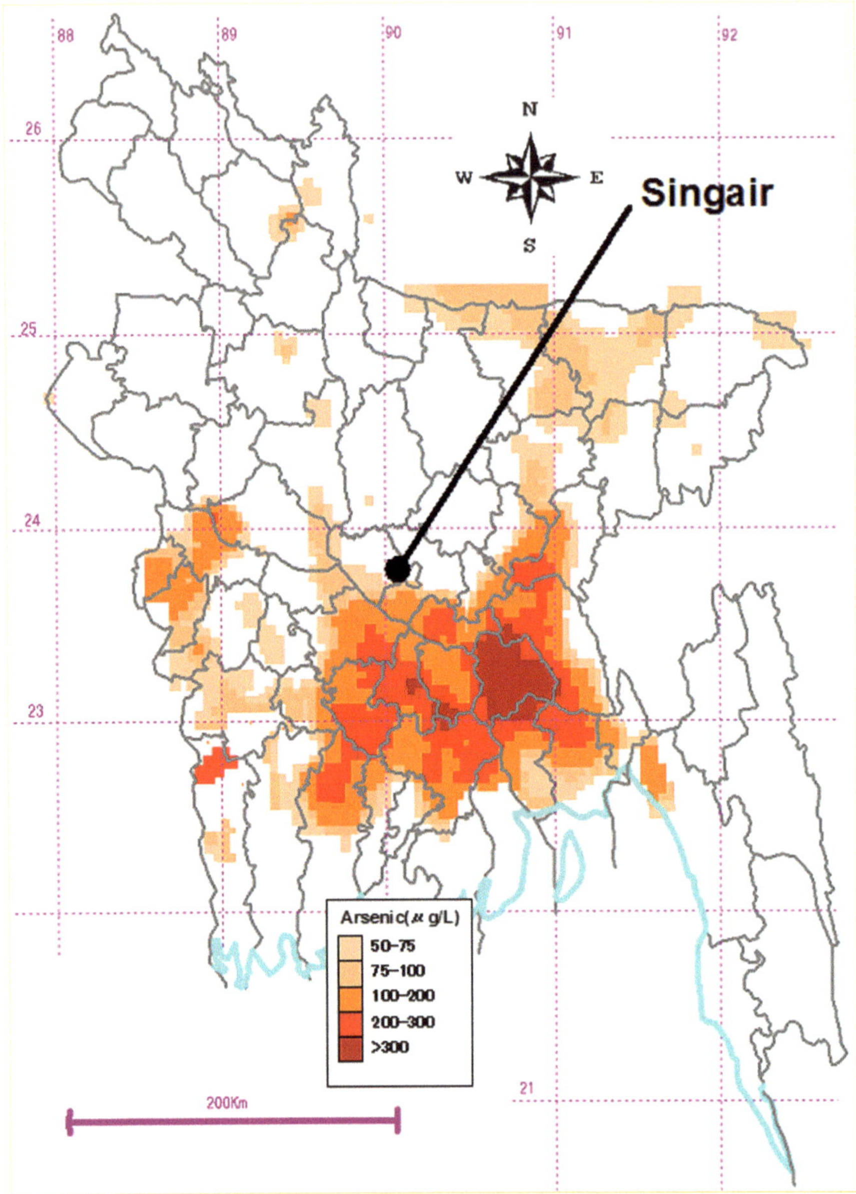

Fig. 11.1 Arsenic contamination in Singair (Hossian 1996)

Fig. 11.2 Arsenic removal filter

11.3 Relationships Between Questionnaire Items

The result of a simple aggregation of samples obtained from the questionnaire surveys indicate comparatively large differences between Azimpur and Glora in terms of the literacy rate, occupation, colours of the signs on the wells, awareness about and actions concerning drinking water use, and ease of obtaining medicine. These results are natural in the sense that the people in Azimpur are poorer than those in Glora, and the ground water in Glora is more contaminated by arsenic. However, the people in both villages have almost the same extent of knowledge and concern about arsenic contamination, and many people are willing to pay either in monetary terms or physically burden to improve the situation. Further, most of the surveyed people are now concerned about many issues including jobs, and earning a steady source of income. The results of the survey show that the community and media are vital sources of information for the people in Azimpur, while those in Glora tend to rely more on the media to obtain information.

In order to analyse the relationships among all items in questionnaire, Cramer's Coefficient of Relationship (Cramer 1946) is applied to all 213 samples obtained from the two villages. Cramer's Coefficient of Relationship which uses χ^2-value to analyse the relationship between two items. Cramer's Coefficient of Relationship, ϕ, is described as below.

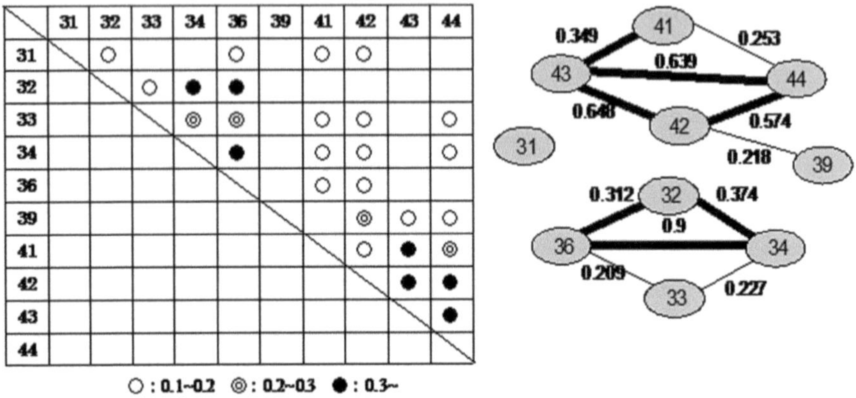

Fig. 11.3 Relationships among items from No. 31 to No. 44

$$\phi = \left[\chi^2/\{N(k-1)\}\right]^{1/2} \tag{11.1}$$

where χ^2 is χ^2-value, N is the sample number, and k is the number of less category, or option in two items. The value of ϕ ranges between 0 and 1.0. The intensity of relationship between two items is interpreted as strong if $\phi \geq 0.3$.

Figure 11.3 shows an example of the representation of relationships among items No. 31 to No. 44 in the questionnaire. These items related to the possibility of pursuing options to improve drinking water safety. In order to apply Cramer's coefficient of relationship, items allowing multiple answers are not considered. The remaining items are as follows. The numbers such as 31 are used as labels in the following.

- {31. Use of well}
- {32. Own well is used for drinking and/or cooking purpose}
- {33. Concern about arsenic contamination}
- {34. Use of some option to mitigate the problems caused by arsenic contamination}
- {36. Use safe well}
- {39. Desire to use a common safe well}
- {41. Do not mind bearing a cost for getting safe water}
- {42. Do not mind bearing an expense (money and/or labour) to improve the safety of drinking water}
- {43. Do not mind bearing an expense (money and/or labour) for the improvement of water quality}
- {44. Do not mind bearing an expense (money and/or labour) for the improvement of water quantity}

In Fig. 11.3, ○ and ◎ denote that ϕ ranges from 0.1 to 0.2 and from 0.2 to 0.3, respectively, while ● denotes that that it exceeds 0.3. In Fig. 11.3, the bold lines represent a strong relationship (above 0.3), and the fine lines represent a moderately strong relationship (above 0.2).

A strong relationship between two items, such as that denoted by ● in Fig. 11.3, implies that the two questions have similar meaning. We will return to these results in the latter part of this chapter.

11.4 Studying People's Satisfaction with Drinking Water

Hayashi's Quantification Theory Type II (Yasuda and Unno 1977) is applied to obtain the main factors regarding people's satisfaction with the drinking water that is currently available in both villages. We model the discriminant functions to evaluate the degree of the satisfaction with the drinking water for each village. We choose item 15, {Satisfied with the currently available drinking water}, as an external criterion in these models. Seven items are chosen as the factors affecting satisfaction. The following steps are used to determine the choice of these items.

1. Select items that have strong relationships with many other items.
2. Select items that probably have some relationship with the satisfaction.
3. Select items having unbiased answers.
4. If two items present a similar idea, select the item that has a broader concept of the two.
5. For step 1., we refer to the result from Sect. 11.3. If an item has many relationships with other items, we can interpret that item as being a pivot.

The following seven items are selected as the factors affecting people's satisfaction with drinking water.

- {3. Literacy}
- {7. Number of family members}
- {17. Carrying drinking water is physically burden}
- {32. Own well is used for drinking and/or cooking purposes}
- {34. Use of some option to mitigate the problems caused by arsenic contamination}
- {41. Do not mind bearing a cost for getting safe water}
- {51a. Worry about arsenic contamination}

The results of the analysis are shown in Table 11.2.

11.4.1 Azimpur

According to the orders of the ranges in Table 11.2, we can select the following as the important factors: {51a}, {7}, {17} and {32}. Using this result, we model the discriminant function of the satisfaction in Azimpur using Eq. 11.2.

Table 11.2 Results of Hayashi's Quantification Theory Type II

			Azimpur			Glora		
j	Number	Category(*k*)	Score(x_{jk})	Range	Order	Score(x_{jk})	Range	Order
1	{3}	1 Yes	0.2034	0.3688	7	−0.0819	0.3758	4
		2 No	−0.1655			0.2939		
2	{7}	1–4	0.5026	0.8967	2	−0.1640	0.3887	3
		2 5,6	−0.0599			−0.0261		
		3 7~	−0.3941			0.2247		
3	{17}	1 Yes	−0.2661	0.8629	3	−0.0469	0.0564	7
		2 No	0.5968			0.0095		
4	{32}	1 Yes	0.3269	0.8327	4	0.0248	0.3131	5
		2 No	−0.5059			−0.2883		
5	{34}	1 Yes	0.2868	0.6976	5	0.5765	1.2131	1
		2 No	−0.4107			−0.6366		
6	{41}	1 Yes	0.0087	0.5272	6	−0.0436	0.2000	6
		2 No	−0.4385			0.1565		
7	{51a}	1 Yes	−0.5861	0.9647	1	−0.6688	1.1074	2
		2 No	0.3787			0.4386		

$$\alpha_i = \sum_{j=1}^{7} \sum_{k=1}^{k_j} \delta_i(jk) x_{jk} \begin{cases} \geq 0.2492\,(\text{sample } i \text{ is satisfied with currently available drinking water}) \\ < 0.2492\,(\text{sample } i \text{ is not satisfied with currently available drinking water}) \end{cases} \tag{11.2}$$

(discriminant rate 78.5 %)
where,

$$\delta_i(jk) = \begin{cases} 1\,(\text{sample } i \text{ responds to category } k \text{ in item } j\,) \\ 0\,(\text{sample } i \text{ does not responds to category } k \text{ in item } j) \end{cases}$$

i is the sample serial number.
j is the item number defined in Table 11.2.
k is the category number of the items defined in Table 11.2, and
The score of (x_{ik}) is shown in Table 11.2.

11.4.2 Glora

For Glora village, we select {34} and {51a} as the important factors. The discriminant function in Glora is modelled using Eq. 11.3.

$$\alpha_i = \sum_{j=1}^{7} \sum_{k=1}^{k_j} \delta_i(jk) x_{jk} \begin{cases} \geq -0.1990\,(\text{sample } i \text{ is satisfied with currently available drinking water}) \\ < -0.1990\,(\text{sample } i \text{ is not satisfied with currently available drinking water}) \end{cases} \tag{11.3}$$

(discriminant rate 83.2 %)

These functions can evaluate the discriminant rate, which indicates whether people are satisfied with the currently available drinking water. In order to improve the degrees of people's satisfaction, the issues referred to by the items having large ranges in Table 11.2 should be tackled. Further discussion about the factors affecting their satisfaction is conducted in Sect. 11.6.

Table 11.3 Items that constitute the unhappiness

j	Items
1	{3. Literacy}
2	{7. Fewer family members}
3	{15. Satisfied with the currently drinking water}
4	{17. Do not feel physical burden to carry drinking water}
5	{51a. Worry about arsenic contamination}
6	{51b. Worry about job and income}
7	{51c. Worry about too many things}
8	{53. Easy access to medicines}

Table 11.4 Scores of items for each axis (Azimpur)

1st axis		2nd axis		3rd axis	
{51a}	−0.083	{51a}	−0.064	{17}	−0.083
{51c}	−0.051	{51b}	−0.026	{3}	−0.065
{53}	−0.030	{53}	−0.011	{15}	−0.042
{3}	−0.017	{3}	−0.007	{53}	−0.010
{7}	0.003	{7}	−0.001	{51a}	−0.008
{51b}	0.053	{17}	−0.001	{51c}	0.018
{17}	0.078	{15}	0.022	{7}	0.044
{15}	0.089	{51c}	0.166	{51b}	0.083

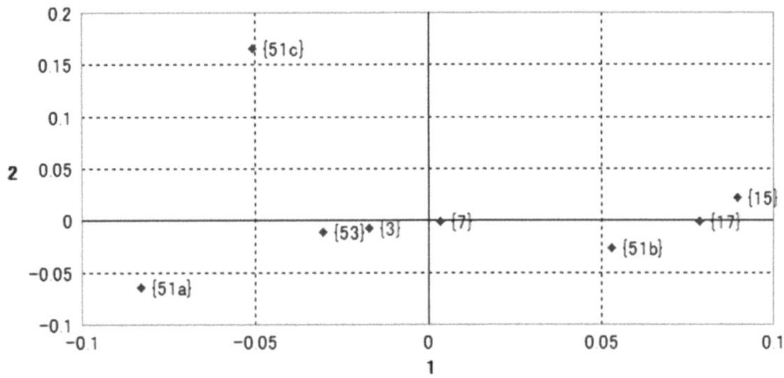

Fig. 11.4 Positions of items in space generated with the 1st and 2nd axes (Azimpur)

Table 11.5 Interpretation of the results of Table 11.4 and contribution ratios

	Azimpur	Glora
1st axis	Satisfaction with water 25.2 %	Worry about and satisfaction with water 28.1 %
2nd axis	Worries about daily life and water 20.8 %	Worries about daily life and water 19.4 %
3rd axis	Rich life 15.0 %	Rich life 15.9 %

11.5 The Unhappiness Function

To evaluate people's unhappiness, we apply Hayashi's Quantification Theory Type III. We refer to Sect. 11.3 to identify the factors that are supposed to constitute the unhappiness are chosen as shown in Table 11.3.

Table 11.4 and Fig. 11.4 show the results of the analysis for Azimpur. Table 11.5 explains the meanings of results for the three axes seen in Table 11.4.

The contribution ratio represents how an axis integrates the data. Therefore the contribution ratio indicates the importance of the axis (Hagihara et al. 1979). Two unhappiness functions are modelled for Azimpur and Glora using Eqs. 11.4 and 11.5 respectively.

For Azimpur

$$D_i = -\frac{25.2}{l_i}\sum_{j=1}^{8}\delta_i(j)x_{1j} + \frac{20.8}{l_i}\sum_{j=1}^{8}\delta_i(j)x_{2j} + \frac{15.0}{l_i}\sum_{j=1}^{8}\delta_i(j)x_{3j} \tag{11.4}$$

For Glora

$$D_i = -\frac{28.1}{l_i}\sum_{j=1}^{8}\delta_i(j)x_{1j} + \frac{19.4}{l_i}\sum_{j=1}^{8}\delta_i(j)x_{2j} - \frac{15.9}{l_i}\sum_{j=1}^{8}\delta_i(j)x_{3j} \tag{11.5}$$

where,

$$\delta_i(j) = \begin{cases} 1\,(\text{sample } i \text{ responds to item } j) \\ 0\,(\text{sample } i \text{ dose not responds to item } j) \end{cases}$$

i is the sample serial number.
j is the item number which is defined in Table 11.3.
l_i is the number of items sample i responds "yes" to the 8 items
x_{rj} is the score of the item j for the rth axis, and the score for Azimpur is shown in Table 11.4.

If a certain person's D_i is positive and the absolute value is large, he/she is relatively unhappy. If D_i is negative and the absolute value is large, he/she is not relatively unhappy.

In order to assess the strength of the contribution of the items towards the unhappiness, the item scores x_j are calculated using Eqs. 11.6 and 11.7. Both Eqs. 11.6 and 11.7 can be obtained by multiplying the contribution rate and score

Table 11.6 Scores combined with contribution rate

j	Items	Azimpur x_j	Glora x_j
1	{3}	−0.69417	0.107076
2	{7}	0.555727	−2.16949
3	{15}	−2.41378	1.299371
4	{17}	−3.22323	0.396022
5	{51a}	0.626394	−1.34232
6	{51b}	−0.61212	−1.09315
7	{51c}	4.995888	4.615948
8	{53}	0.388253	−1.64643
Total contribution rate		60.9 %	63.4 %

for each axis and adding them. Table 11.6 shows the item scores that directly contribute to the unhappiness.

Equations 11.6 and 11.7 represent the definition of x_j

For Azimpur

$$x_j = -25.2x_{1j} + 20.8x_{2j} + 15.0x_{3j} \tag{11.6}$$

For Glora

$$x_j = -28.1x_{1j} + 19.4x_{2j} - 15.9x_{3j} \tag{11.7}$$

where, x_{kj} is the score of item j for the kth axis

According to Table 11.6, addressing the issues posed by items {15} and {17}, which are related to the current drinking water situation, would mitigate the unhappiness in Azimpur. Therefore, in order to mitigate the unhappiness, it is important to improve the current drinking water situation. However, the same logic/action does not apply to Glora. In both villages, the results of the analyses for item {51c} indicate that the people are very unhappy. Thus, we can conclude that other social environmental problems would be also significant issues constituting local people's unhappiness, and mitigating the problems posed by arsenic in drinking water will not be enough to allay their unhappiness.

11.6 Comparative Analysis

11.6.1 Comparison of the Degree of the Satisfaction with the Currently Available Drinking Water

Table 11.7 shows the items that are important toward improving people's satisfaction with drinking water in both villages. It clearly shows that the factors affecting satisfaction with drinking water differ between Azimpur and Glora. Azimpur is characterised by a high literacy rate and small family size, and the people do not to

Table 11.7 Main factors affecting satisfaction with drinking water

Azimpur	Glora
{51a}Worry about arsenic contamination {7} Fewer family members {17} Carrying drinking water is not physically burden {32} Own well is used for drinking and/or cooking purpose	{34} Use some options to mitigate the problems caused by arsenic contamination {51a} Worry about arsenic contamination

mind bearing an expense for procuring safe drinking water. These factors, no doubt, influence their satisfaction. However, Glora is characterised by a low literacy rate and large family size, and the people of this village are unable to bear additional costs to procure safe drinking water. These factors cause the unhappiness with the available drinking water. Notably, the analysis for item 17 {Carrying drinking water is physically burden} shows remarkably different results for the two villages (see Table 11.2). This can be caused by the difference in arsenic contamination in each area. The drinking water sources of Azimpur are strongly contaminated, and people are forced to walk far to get safe water. On the contrary, in Glora, the water is not so contaminated, and people can use their own or their neighbour's wells for drinking water. In both villages, item 51a {Worry about arsenic contamination} is a significant factor contributing to the satisfaction.

11.6.2 Comparison of People's Unhappiness

According to Table 11.6, the responses to item 51c {Worrying about many things} are a strong indication of people's unhappiness in both villages. On the other hand, addressing the issues posed by items related to water, namely item 15 {Satisfied with the currently available drinking water} and item 17 {Carrying drinking water is not physically burden}, can significantly mitigate the unhappiness in Azimpur. However, it appears that addressing item 7 {Fewer family members} can mitigate the unhappiness in Glora.

11.6.3 Scenario Analysis

To consider the changes in satisfaction with drinking water and the unhappiness, we presume a particular scenario. This scenario consists of releasing the people from the physical burden of carrying water and from worrying about arsenic contamination. Therefore, in this scenario, we assume that all samples answer that {Carrying water is not physically burden} for item 17, and {Not worried about arsenic contamination} for item 51a.

Table 11.8 Applied scenario

	Azimpur		Glora	
	Before	After	Before	After
A	29	93	21	23
B	43	4	13	11
Others	35	10	66	66

The extent of satisfaction is evaluated using Eqs. 11.2 and 11.3 for each village. The averages of D_i, which are defined in Eqs. 11.4 and 11.5, are used as the thresholds to classify whether sample i is unhappy for each village. Therefore, we presume that if D_i takes a value above the average, the sample i is unhappy. Conversely, if D_i takes a value below the average, the sample i is not unhappy. The thresholds and equations are fixed throughout the scenario analysis.

Table 11.8 represents the results of applying the scenario. 'A' shows the number of samples who were satisfied with the currently available drinking water and were not unhappy, and 'B' shows the number of samples who were unsatisfied with the currently available drinking water and were unhappy.

The application of this scenario has led to a big change; in Azimpur, group 'B' (unsatisfied and unhappy) changes to group 'A' (satisfied and happy). On the other hand, there is little change in Glora. The results can be explained as follows. In Azimpur, mitigating the issues posed by items 17 and 51a {Carrying water is physically burden} and {Worry about arsenic contamination}, respectively, strongly encourages satisfaction with drinking water (see Table 11.2). Moreover, addressing the issues by items 15 and 17, {Satisfied with currently available drinking water} and {Carrying water is physical burden}, respectively, strongly mitigate the unhappiness (see Table 11.6). On the contrary, in Glora, those two items play smaller roles in people's satisfaction with drinking water. Moreover, the issues indicated by the people's responses to items 15 {Satisfied with currently available drinking water} and 17 {Carrying water is not physical burden} relatively encourage the unhappiness.

According to the results, mitigation of the physical burden and allaying the worry about arsenic problems provide significant results for the case of Azimpur only. It is very important for the people living in Azimpur to get safe drinking water easily. However, this is not so for the people in Glora, where the contamination is not as serious. Consequently, in order to suggest an acceptable alternative to improve their situation, we must consider not a stereotyped way but a harmonized way for each area.

11.7 Analysis of Cognitive Structure for the Foreign Aid

11.7.1 Presupposition for the Analysis and Setting Latent Variables

In this sub-section, we focus only on Azimpur, which faces a more severer situation than Glora, given the higher arsenic contamination and more pressing their economic problems in Azimpur. Foreign technical aid is necessary in order to mitigate the problems posed by arsenic contamination of water. The people in the village is poor to attempt such measures.

Most foreign aid programmes have provided the villages with several types of arsenic-removal equipment. However, most of these are not used regularly by the local people. Therefore, in this section, we consider distrust of foreign aid. To structure the distrust, which is assumed to be a latent variable, we apply the Multiple Indicator Multiple Cause (MIMIC) model (Toyoda 1998; Shimizu et al. 2002).

The following items are selected for modelling the distrust.

- {9. Do not use arsenic-removal devices}
- {10. Are not satisfied with the quantity of drinking water}
- {18. Takes considerable time to fetch drinking water}
- {19. Difficult to fetch drinking water}
- {24. Own well is painted, namely contaminated}
- {26. Knowledge about harmfulness of arsenic}
- {28. Worry about family's health}
- {29. Aware that arsenic can be technically removed}
- {30. Do not want to use additional arsenic-removal option}
- {41. Mind bearing the expense to improve the safety of the drinking water}
- {Note: The definition of 'unhappiness' is defined in Sect. 11.5}

In addition to these items, three latent variables, [anxiety], [dissatisfaction] and [distrust], are assumed.

The first latent variable, [anxiety], can be thought of as a combination of the people's knowledge about the harm that arsenic in drinking water can cause, the situation that own well painted, and the unhappiness. [anxiety] causes the people to worry about the health of their families. Figure 11.5 shows the definition of [anxiety].

The time needed to carry the drinking water and the difficulty of approaching and using a safe well can be thought as contribution to [dissatisfaction]. Further,

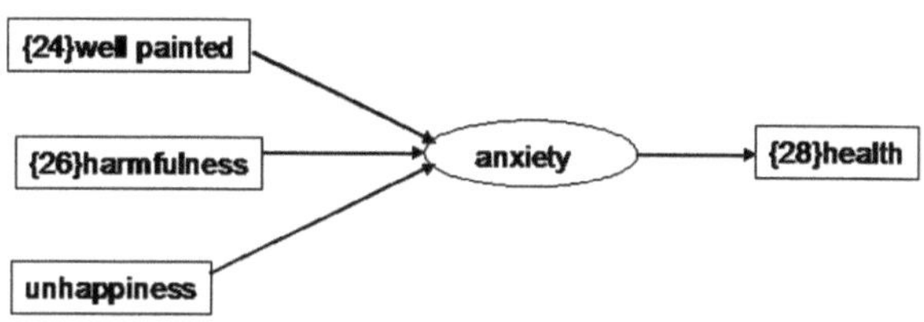

Fig. 11.5 Definition of anxiety

[dissatisfaction] refers to the people's discontent with the quantity of the drinking water. Figure 11.6 shows the definition of [dissatisfaction].

The knowledge of the fact that arsenic can be removed technically may be assumed to cause [distrust], due to which people may not wish to use arsenic-removal devices, do not desire additional arsenic-removal options, and do not wish to bear any expense toward improving the safety of the drinking water. Figure 11.7 shows the definition of [distrust].

11.7.2 Multiple Model of Distrust

We assume that [anxiety] and [dissatisfaction] should cause [distrust]. Therefore, [distrust] is redefined, and structured as shown in Fig. 11.8.

Fig. 11.6 Definition of dissatisfaction

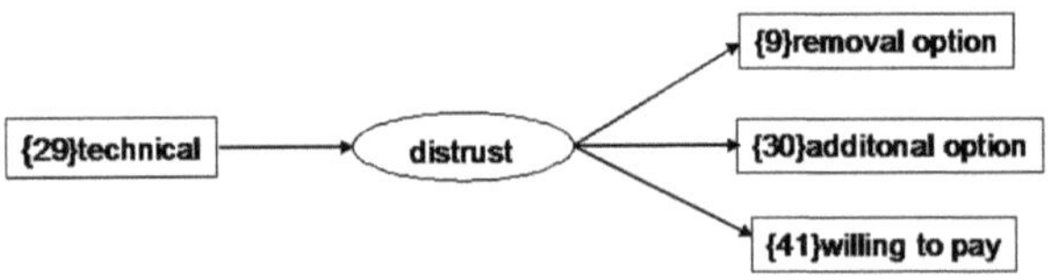

Fig. 11.7 Definition of distrust

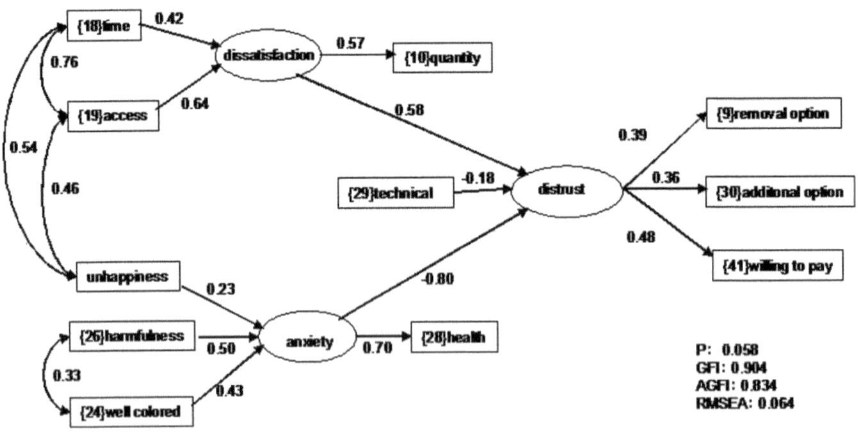

Fig. 11.8 Multiple model for [distrust]

Table 11.9 Criteria for model assessment

Indices	Criteria
P value	Must be above 0.05
GFI	Good if it is above 0.9
AGFI	Good if it is above (GFI-1)
RMSEA	Very good if it is below 0.05. Very bad if it is above 0.1

The coefficients listed beside the recursive passes are the correlation coefficients, and the coefficients listed beside the passes show the strength of the causality. They range from -1 to 1, and if the relationships or causality are positive and strong, they take a value near 1. If they are negative and the relationships are strong, they take a value near -1. If there is no relationship or causality, they take a value near 0. In order to assess the overall model, the values of the significance level (P value), goodness-of-fit index (GFI), adjusted goodness-or-fit index (AGFI), and root-mean-square error of approximation (RMSEA) are applied. Table 11.9 indicates the criteria for these indices. According to Table 11.9, the model in Fig. 11.8 fits the data appropriately.

From Fig. 11.8, we can summarize the result as follows.

1. Knowledge about the harmfulness of arsenic and the labelled wells encourage [anxiety]. While the unhappiness also encourages [anxiety], it has a smaller effect than the above-mentioned knowledge or situation. [anxiety] increased the concern about family health.
2. Difficulties in locating/approaching safe drinking water sources encourages [dissatisfaction], which causes discontent with drinking water quantity.
3. Knowledge about the technical possibility of removing arsenic mitigates [distrust]. However, this effect is relatively small effect. [distrust] leads to non-use of arsenic-removal equipments and lack of desire to use new options to mitigate arsenic contamination and to improve the safety of drinking water by bearing additional expenses.
4. According to the overall model, [dissatisfaction] encourages [distrust]. [anxiety] mitigates [distrust], and [anxiety] has a very strong effect on [distrust].

These results can be interpreted in the following manner: [distrust] should be mitigated by allaying [dissatisfaction] and encouraging [anxiety], so that the local people will accept external support.

11.8 Concluding Remarks

Measures for arsenic removal from drinking water have been provided to Bangladesh by other countries and international organizations. However, the Bangladeshi people have embraced only a few such measures. We studied the possible reasons for this situation and believe that the locals do not accept many of the measures, because the temporal proposals and techniques to reduce the contamination have been promoted by other countries without consideration of local people's traditional ways of life. Thus, the locals are reluctant to become accustomed to these techniques.

Accordingly, the purpose of this chapter was to analyse the relationship between arsenic contamination and the social environment in Bangladesh. First, we conducted interviews with Bangladeshi people with the co-operation of a local NPO. Second, the factors affecting satisfaction with drinking water were analyzed. Third, the unhappiness function was devised. Finally, the results reveal the latent factors affecting distrust of the technical support and arsenic-removal equipment provided by the other countries and foreign aid institutions to Bangladesh.

Consequently, we conclude as follows.

1. The factors constituting satisfaction with the currently available drinking water and unhappiness differ between the surveyed villages of Azimpur and Glora. Therefore, not a stereotyped way but a harmonized way are required to improve the people's lives in these two areas.
2. In order to encourage the let locals of Azimpur to accept external support, it is important to mitigate their distrust in foreign aid. The distrust can be allayed by mitigating the dissatisfaction with the current drinking water situation and encouraging the anxiety about arsenic contamination situation and their health.

In order to suggest acceptable alternatives, we need to investigate the social environment more deeply and analyse the relationships among the problems faced by the local people.

References

Cramer, H. (1946). *Mathematical methods of statistics*. Princeton: Princeton University Press.

Hagihara, Y. (2008). *Kankyo to bousai no doboku keikakugaku (Adaptive system planning methodology for environment risk management)*. Kyoto: Kyoto University Press (in Japanese).

Hagihara, Y., Koizumi, A., Nishizawa, T., & Konda, T. (1979). Anketo chosa wo motonishita mizujyuyo-kouzo narabini sessui ishiki bunseki (Structure of water demand based on questionnaire survey and analysis on consciousness about saving water), Papers of civil engineering, *Eisei Kougaku Kenkyu Toronkai Kouenronbunsyu (Journal of Japan Society of Civil Engineers, Environmental Research), 15*, 188–193 (in Japanese).

Hagihara, Y., Hagihara, K., Hoque, B. A., Yamamura, S., Hatayama, M., Sakamoto, M., & Miyagishima, K. (2003). A study on disaster problems in Bangladesh from natural and social aspects. *The Annuals of the Disaster Prevention Research Institute, Kyoto University*, No. 46B, pp. 15–30.

Hossian, M. (1996). *British geological survey technical report, Graphosman world atlas*. Dhaka: Graphosman.

Iida, Y., & Okada, N. (1992). *Dobokukeikaku system bunseki (A systems analysis of infrastructure planning)*. Tokyo: Morikita Publishing Co. Ltd. (in Japanese).

Kawakita, J. (1966). *Hasso-ho (Method for making ideas)*. Tokyo: Chuko Shinsho (in Japanese).

Kinnburgh, D. G., & Smedly, P. L. (2000). *Arsenic contamination of groundwater in Bangladesh*, Vol. 2, final report, pp. 3–16.

Shimizu, S., Hagihara, K., & Hagihara, Y. (2002). Ninshiki de-ta wo mochiita mizube no kannkyou-hyouka (Environmental valuation on waterside by using cognitive data). *Journal of Japan Society of Hydrology and Water Resources, 15*(2), 152–163 (in Japanese).

Singh, N., Bhattacharga, P., & Jacks, G. (2002). Women and water, the relevance of gender perspective in integrated water resources management in rural India. *ICWRER 2002 Dresden, poster session*

Toyoda, H. (1998). *Kyobunsan kozo bunseki (Covariance analysis: Structural equation modeling)*. Tokyo: Asakura Shoten (in Japanese).

Yasuda, S., & Unno, M. (1977). *Shakai toukeigaku (Social statistics)*. Tokyo: Maruzen (in Japanese).

Chapter 12
The Vulnerability of Toilet Facilities in the Bangladesh Rural Area and Sanitary Improvement by Introduction of the Eco San Toilet

Kunio Takahashi, Akira Sakai, and Tofayel Ahmed

Abstract Bangladesh is a very densely populated agricultural nation and characterized by a tropical monsoon climate and most of the country consists of flat lowlands and a vast delta. Such weather and geographical conditions and vulnerable facilities such as embankment cause severe floods about once in 10 years. On the other hand, the spread of toilets called a 'low-cost Pit Latrine' by the government and international organizations was extended. But with poor management, the Pit Latrine cannot be considered a sustainable and sanitary facility with adverse environmental impact. An appropriate technology for toilets needs to contribute to sanitary improvement and the resource use of human excreta and it should lead to the conservation of the quality of water. The ecological sanitation (henceforth, Eco San) toilet suits the above needs. The present study examined the sanitary improvement effect and benefit evaluation of the Eco San toilet. As the results, the Eco San toilet improves the state of toilets, and reduces household expenditure through mitigation of waterborne diseases. Considering the average medical expenditures, the benefits will be about 2,000 BDT/year/household. And it cannot be emphasized enough that such a result can be achieved through the proper management.

Keywords Ecological sanitation • Sanitary improvement • Resourceful use • Human excreta • Benefit evaluation

---

K. Takahashi (✉)
Japan Association of Drainage and Environment, Shinjuku, Tokyo 162-0067, Japan
e-mail: nqg48559@nifty.com

A. Sakai
University of Marketing and Distribution Sciences, Kobe 651-2188, Japan
e-mail: akira_sakai@red.ids.ac.jp

T. Ahmed
Japan Association of Drainage and Environment Bangladesh Office, Niketon, Gulshan-01, Dhaka 1212, Bangladesh
e-mail: tofaeljade@dhaka.net

K. Hagihara, C. Asahi (eds.), *Coping with Regional Vulnerability*, New Frontiers in Regional Science: Asian Perspectives 4, DOI 10.1007/978-4-431-55169-0_12

12.1 Introduction

Bangladesh, located in South Asia, is a very densely populated agricultural nation that holds 160 million or more people in its 148,000 km^2 area. It is characterized by a tropical monsoon climate and most of the country consists of flat lowlands and a vast delta formed by the Ganges, Brahmaputra, and Meghna rivers. The total area of the river basin of these 3 rivers is 1,730,000 km^2, and the country area constitutes only 8.6 % of the river basin areas.

Such weather and geographical conditions cause severe floods about once in 10 years. Specifically, in the 1998 floods, about 70 % of the country was flooded. Furthermore, regular cyclones occur during the transition from the rainy season to the dry season and vice versa (i.e., April–May, October–November), which can cause extraordinary damage. In a cyclone occurring in May 1991, more than 140,000 people were affected by the high tide. Similarly, the disastrous Cyclone Sidr that occurred in November 2007 remains fresh in people's memories. Furthermore, the loss of land via erosion of the riverbank during the flooding is estimated at 100 km^2 every year (National Plan for Disaster Management 2010) (Figs. 12.1 and 12.2).

On the other hand, the excessive use of chemical fertilizers or agricultural chemicals caused degradation of the quality of soil, and the fall in the groundwater levels caused by the irrigation lead to arsenic contamination.

Subsequently, in the 1980s, to prevent waterborne diseases, wells (tube wells) were promoted by the U.N. (WHO/UNICEF Joint Monitoring Programme (JMP) for Water Supply and Sanitation 2013). However, examination of the arsenic contamination of well water in 1993 revealed that 70 % of the country fell into the contamination zone. Furthermore, although Bangladesh has 64 Districts, contamination is checked in about 63 of them (National Plan for Disaster Management 2010).

The feature of Fig. 12.3 is that almost all water resources are dependent on groundwater. Moreover, this figure does not reflect arsenic contamination.

On the other hand, the provision of toilet facilities has progressed (WHO/UNICEF Joint Monitoring Programme (JMP) for Water Supply and

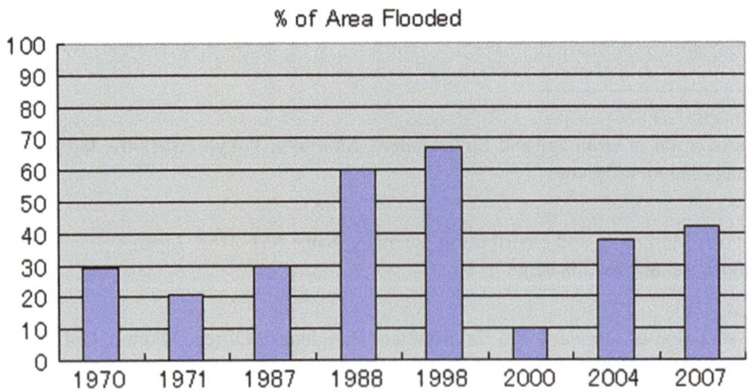

Fig. 12.1 Extent of the area flooded (%)

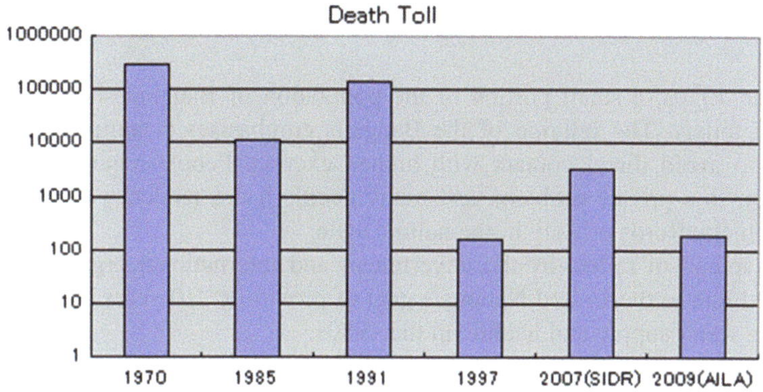

Fig. 12.2 Number of cyclone victims

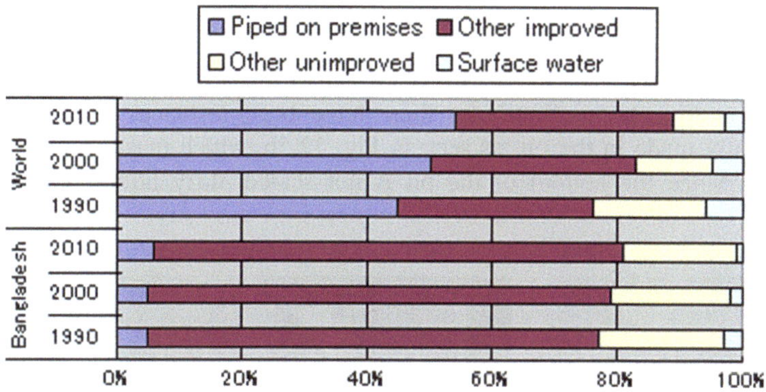

Fig. 12.3 Improved coverage of safe water supply

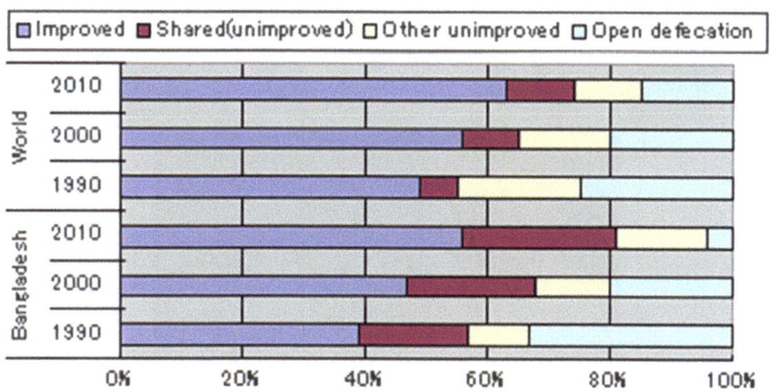

Fig. 12.4 Improved coverage of sanitary toilets

Sanitation, 2013), and it is expected that the government's aim of providing sanitary toilets throughout the entire country by 2015 will be achieved (Fig. 12.4).

12.1.1 Attitudes of Bengalis and Actual Toilet Conditions

Until the 1970s, a small portion of the population of Bangladesh had access to sanitary toilets. The religion of the Bengalis emphasises cleanliness, and urges people to avoid direct contact with human excreta. People regarded the act of excretion as a private problem, and many people had a tendency to do so in the bush, which affords privacy in the natural state.

The spread of toilets by the government and international organizations was extended due to the United Nations's goal of providing "10 years of international drinking water supply and health" in the 1980s.

One type of toilet, called a 'low-cost Pit Latrine', has been the leading choice globally. Similarly, its spread and promotion has been progressing in Bangladesh to date. The substructure of the Pit Latrine lies in the ground, which accumulates faeces in concrete rings. Although the size of the Pit Latrine is based on the number of family members, in order to use it continuously, it is necessary to remove human excreta and dispose of it safely.

However, many such toilets are often damaged, as seen in the housing site depicted in Fig. 12.5a. Similarly, there have been instances where a hole is intentionally made in the pit, as seen in Fig. 12.5b, which makes human excreta flow out. Since the bottom of the pit is not sealed, dirty liquids permeate the

Fig. 12.5 (**a**) Abandoned pit latrine, (**b**) hole was made intentionally, (**c**) a pit submerged in rainwater, (**d**) dumping of the excrement into the waterway by a sweeper

underground and surrounding reservoirs and become a cause of contamination of groundwater and surface water. Further, with a rise in the water levels during floods, pits built by the slope of the Bari (housing site), as depicted in Fig. 12.5c, are almost entirely submerged in rainwater. Furthermore, as shown in Fig. 12.5d, in almost all areas, the emptying of human excreta from the pits is performed by a sweeper; this is not only expensive but also a job hated by most people. Thus, with such poor management, the Pit Latrine cannot be considered a sustainable and sanitary facility with adverse environmental impact (Hosaka et al. 2006).

12.1.2 Sanitary Improvement and Community Development

This section examines the issues in connection with toilets from the viewpoint of sanitary improvement and community development. Figure 12.6 shows the structure of the vicious cycle of poverty, which brings unsanitary situations (Hosaka et al. 2006).

Often, if someone in a poor family falls ill, precious property such as land and livestock is sacrificed to arrange for medical expenses. As a result, the family may fall deeper into the vicious cycle of poverty. Additionally, for children, a loss of education will lead to a loss of future job opportunities, which makes it difficult for them to overcome poverty. Not having a chance to acquire sufficient knowledge about reproductive health in the case of women has an adverse impact on infant mortality rate. Finally, the unsanitary management of human excreta leads to contamination of groundwater and surface water, which has a remarkable adverse effect on the safety of the drinking water sources.

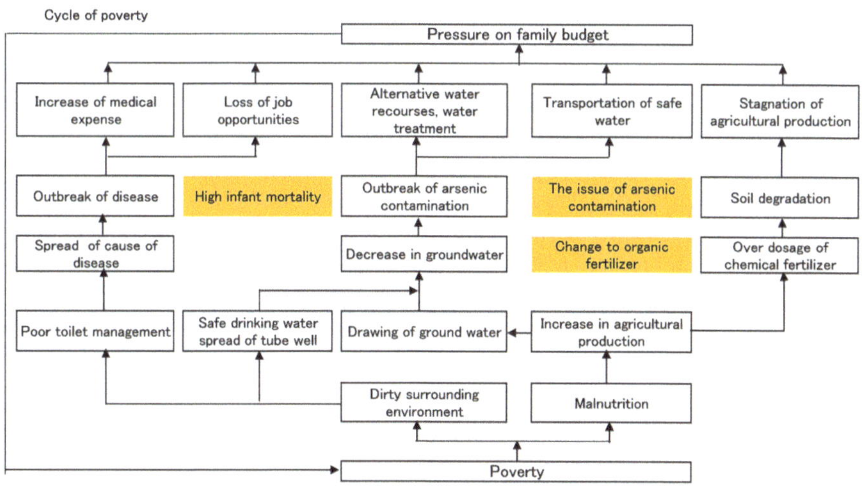

Fig. 12.6 Structure of the vicious cycle of poverty

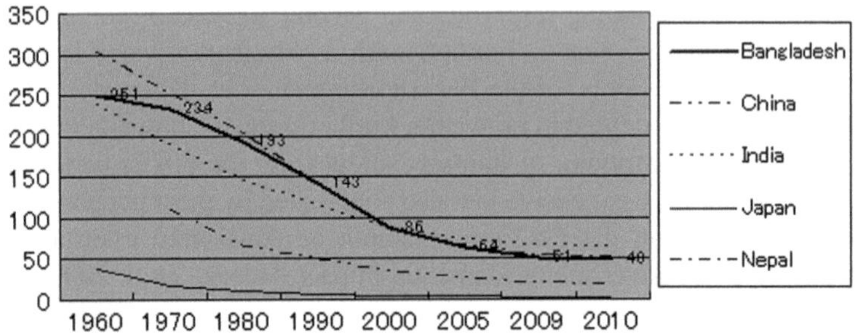

Fig. 12.7 Infant mortality (WHO/UNICEF (JMP) for Water supply and sanitation, 2013)

Additionally, the consumption of chemical fertilizer over many years, to increase agricultural production, reduces the productive capacity of farmland; this, in turn, further compromises the household economy. Thus, poverty poses a challenge for sanitary improvement, and unsanitary condition makes it difficult to overcome poverty. Figure 12.7 shows the infant mortality rate in recent years.

12.1.3 Introduction of the Eco San Toilet as Appropriate Technology

New technology is considered appropriate when the materials, funds, and manpower it requires can be supplied in the area needing it. Furthermore, the technology must suit the peculiarities of that region as well as local needs. Thus, sanitary engineering must not only consider the efficacy of the technology itself, but the residents' preferences and needs. An appropriate technology for toilets needs to contribute to sanitary improvement and the practical use of human excreta. Furthermore, it should lead to the conservation of the quality of surface water, groundwater, reservoir, etc., which are important sources of drinking water. In Japan, such a sanitation system has been established as a social system, although this was done only before a period of high economic growth. It is believed that this will also be applicable to advanced nations, and has been called ecological sanitation (henceforth, Eco San) (Stockholm Environment Institute, Ecological sanitation revised and enlarged edition 2004). Figure 12.8 shows a hypothesized model of the introduction of the Eco San toilet to the structure of the vicious poverty cycle shown in Fig. 12.6 (Takahashi et al. 2009).

As mentioned earlier, the emptying of human excreta from Pit Latrines is performed by a called sweeper, and this work is hated by most of people. However, the Eco San toilet is a technology that is not based on the drawing out of human excreta and transporting it for safe disposal. Instead, it performs sanitary disposal of human excreta and processes it for resourceful use within the toilet itself. An example of a toilet design satisfying the requirements for use in rural areas in Bangladesh has already been developed. This toilet is called the 'urine diverting dry

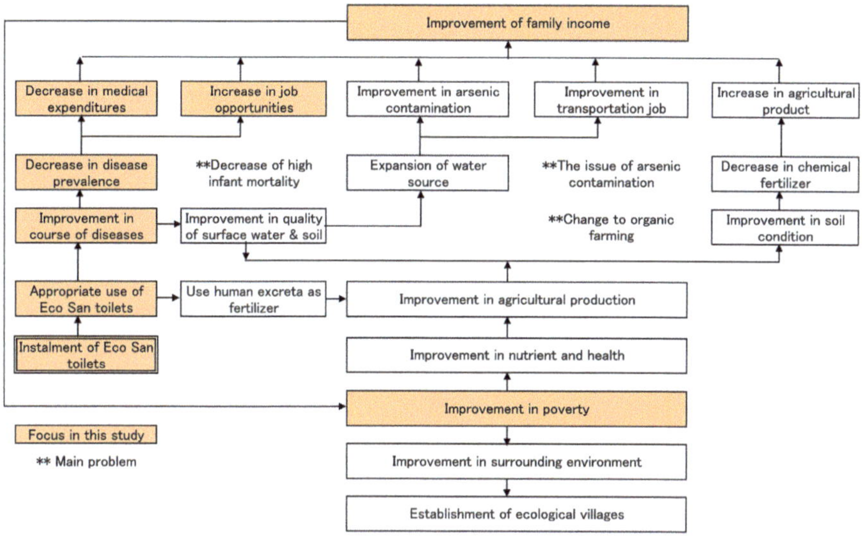

Fig. 12.8 Improvement of the vicious cycle of poverty using the Eco San Toilet

toilet' (UDDT). That is, faeces, which contain pathogenic organisms, and urine, which is rich in nutritive substances as well as pathogenic organisms free, are separated. Urine is diluted with water and sprinkled onto the farmland, and faeces are returned to the farmland after being dried out and stabilized by adding ash.

There are storage tanks to contain the faeces, which are used alternatively. After a tank is filled, it is stored for several months (at least six) to ensure that the process of stabilization is completed. Moreover, people are accustomed to washing the anus with water after defecation. This water is then led to an adjoining bed for evaporation. And the storage tank is waterproof from below, and the toilet seat is higher off the ground (about 70 cm) than the traditional toilet, thereby reducing the possibility of the toilet being affected by flooding. The structure of the toilet and method of using it are shown in Fig. 12.9.

Needless to say, the Eco San toilet is a novel design, thereby requiring its users in Bangladesh to adapt to its use. It is an additional option for toilet facilities here. Therefore, it becomes necessary to define the additional benefits of this toilet, such as how comfortable it is in comparison to the conventional toilet, as well as how it contributes to sanitary improvement and the use of human excreta resourcefully.

The present study examined the sanitary improvement effect of the Eco San toilet, and a benefit evaluation was performed. In Sect. 12.2, the findings of the survey performed in the 2009 fiscal year are presented. Furthermore, in addition to the cross-sectional results of this survey, we describe the results of a longitudinal survey conducted from fiscal years 2011 to 2013 in Sect. 12.3. Both these sections are then summarized in Sect. 12.4.

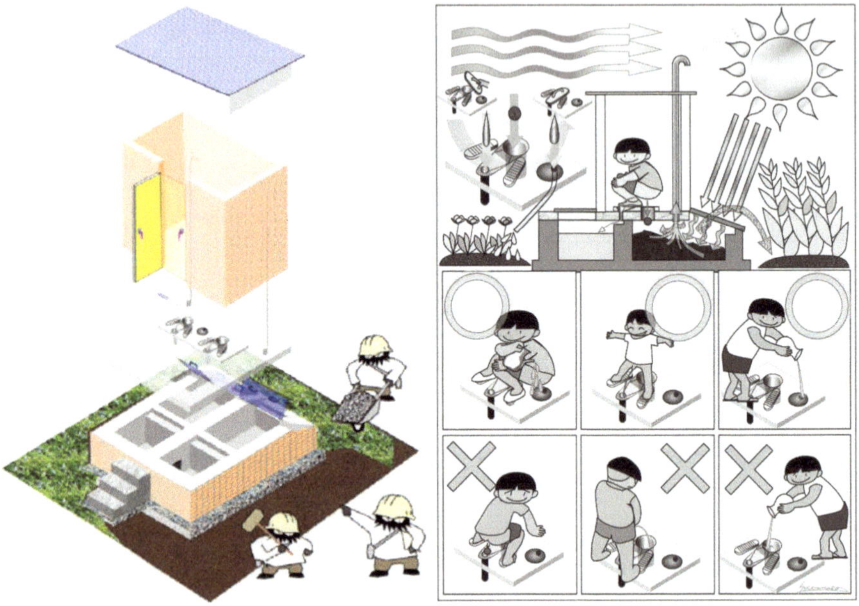

Fig. 12.9 (**a**) Design of Eco San Toilet, (**b**) use of Eco San Toilet

12.2 Sanitary Improvement Effect and Benefit Evaluation (Based on a 2009 Survey)

To investigate the sanitary improvement effect and perform the benefit evaluation on the introduction of the Eco San toilet, we assessed people's attitude towards the toilet after its introduction and measured the disease frequency and medical expenditure of waterborne disease using a questionnaire survey.

The survey areas were Comilla district, Shymnagar, Kolaroa (Satkhira district), Niamotpur (Naogaon district), Keshobpur, and Sharsha (Jessore district) in Bangladesh.

The analysis process and results of the survey in Comilla are presented below. Then, the results of the surveys conducted in Shymnagar, Kolaroa, Niamotpur, Keshobpur, and Shasha, each of which has differing geographical and social characteristics, are presented. Figure 12.10 shows the survey areas and Table 12.1 shows the number of questionnaire responses in these areas.

Fig. 12.10 Survey areas (number of Eco San Toilets)

Table 12.1 Number of questionnaire responses

Area	Users	Others	Total	Total no. of households
Comilla	112	220	332	1,603
Sharsha	20	20	40	512
Keshobpur	37	43	80	495
Niamatpur	40	40	80	520
Kolaroa	19	31	50	492
Shyamnagar	18	32	50	535
Total	246	386	632	4,157

12.2.1 Sanitary Improvement Effect in Comilla

12.2.1.1 Contents of the Questionnaire Survey

Residents are concerned with everyday life, and are the last subjects who can evaluate various living environments surrounding them. For this reason, after the introduction of the Eco San toilet, a questionnaire assessing living environment was performed. The selected questionnaire items were as follows:

- Attributes (e.g., sex, age, literacy rate, occupation, land ownership, household composition)
- Situation of Current Toilet (e.g., type, situation, willingness for improvement, improvement reason)
- Health/Sanitation in Living Conditions (e.g., recognition of infection frequency, pollutant pathways)
- Awareness and Behaviour related to Human Excreta (e.g., management, transportation, permanent disposal)
- Knowledge and Intention to use of Human Excreta (fertilizer and fuel gas use)
- Willingness of payment (e.g., willingness to payment, method of payment)
- Concerns of Daily Life (e.g., medical service/worrying matter of life, satisfaction)

Other toilet users were randomly selected, while the Eco San toilet users were selected and they underwent the interview. These surveys were implemented from June to July 2009, which was about 1 year after the introduction of the Eco San toilet.

12.2.1.2 Sanitary Improvement Effect of the Eco San Toilet

Attributes
As shown in Fig. 12.11, the attributes (including sex, literacy rate, main occupation, and annual income distribution) of Eco San toilet users and users of traditional toilets did not differ substantially.

Situation of Current Toilet
As shown in Fig. 12.12, the differences between Eco San users and other toilet users regarding the current situation (i.e., its dirtiness and smell) of their toilets and their desire to improve said situation is clear. While most of other toilet users wanted to improve their toilet, Eco San toilet users were satisfied with their current toilet.

Awareness and Behaviour Regarding Human Excreta
As shown in Fig. 12.13, a remarkable difference was found with reference to users' awareness and behaviour regarding human excreta. Specifically, the mitigation of aversion awareness was observed in Eco San toilet users. This may be because the

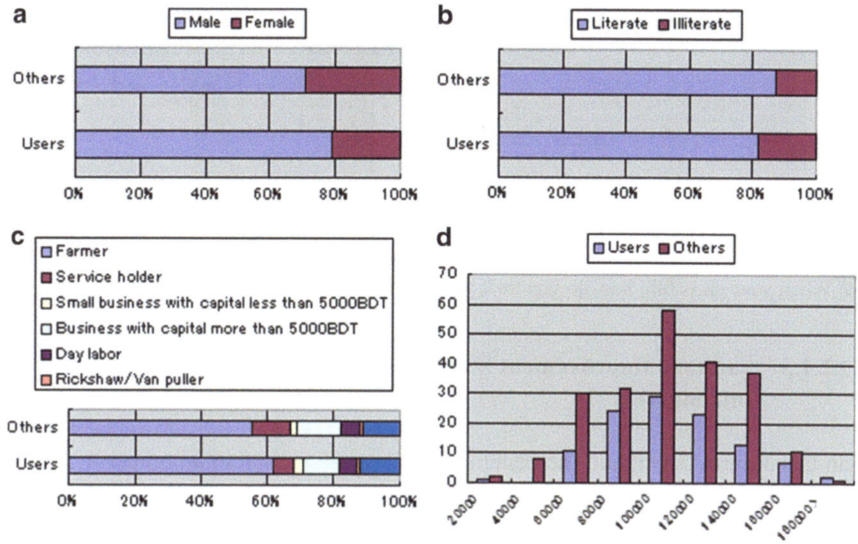

Fig. 12.11 Comparison of Eco San Toilet users and users of other toilets (attributes). (**a**) Sex, (**b**) literacy rate, (**c**) main occupation, (**d**) annual income (in BDT, USD = 78BDT)

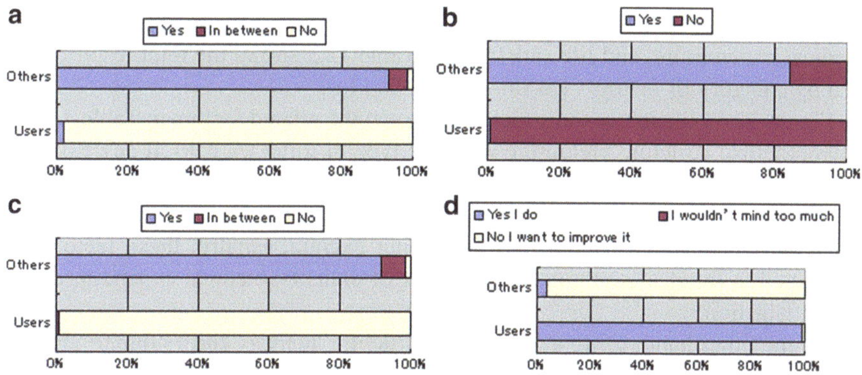

Fig. 12.12 Comparison of Eco San Toilet users and users of other toilets (situation and improvement intention of toilet). (**a**) Dirtiness, (**b**) sometimes unavailable during flood season? (**c**) bad smell, (**d**) do you want to keep using your current toilet?

separation of human excreta (i.e., urine and faeces) and improving the safety of cleaning the toilet by drying the faeces renders it easy to treat. The result shows almost all Eco San toilet users use human excreta as fertilizer.

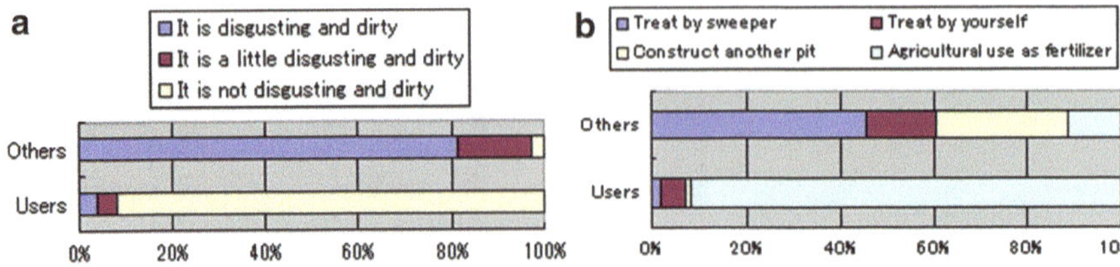

Fig. 12.13 Comparison of Eco San Toilet users and users of other toilets (awareness and behaviour regarding human excreta). (**a**) Do you think it is disgusting and dirty to convey excreta? (**b**) When your pit is full, how do you treat your human excreta?

12.2.1.3 Sanitary Improvement Structure Through Eco San Toilet Introduction

Sanitary improvement and the benefits of toilets are related to a number of factors such as awareness and behaviour regarding toilet use, the situation of the toilet, occurrence of waterborne diseases, medical expenditures, etc. Below, the hypothetical structural model of disease course is shown and identified based on the questionnaire shown previously.

The questionnaire items used for analysis are literacy rate, land ownership, the type of toilet, frequency of toilet cleaning, the state of the toilet (dirtiness, smell), awareness of human excreta, disease frequency of waterborne diseases, annual medical expenditure, and life satisfaction. Table 12.2 shows the total result classified by category of factors (Takahashi et al. 2011a).

Figure 12.14 illustrates the structure between the related factors in Table 12.2 as a hypothetical structure predicting what the type of toilet leads to. It describes the findings of a χ^2 test between these factors. The type of toilet is closely related with the state of the toilet (e.g., its dirtiness and smell) and with the awareness and behaviour related to cleaning and conveying excreta. Finally, these factors are connected to the disease frequency, number of patients, medical expenditure, and life satisfaction.

On the other hand, using a covariance structure analysis allows one to identify the causal relationship between factors (Toyoda 1998). Here, the causal relation between factors is investigated and a covariance structure analysis is applied (Takahashi et al. 2011b). Figure 12.15 shows the results of the multiple indicator model of the covariance structure analysis. The goodness-of-fit of the model is as follows: goodness of fit index, 0.823; adjusted goodness of fit index: 0.723; and root mean square error of approximation, 0.219. The causal relations of various factors can be described though the goodness-of-fit of the model, and the values cited previously indicate that the fit is not optimal. These findings can be interpreted as follows.

There is a close relationship between the type of toilet and the awareness and behaviour of toilet users. The Eco San toilet users declared that they were satisfied with their current toilet and that their aversion awareness for excreta is small. This difference relates to the state of the toilet (i.e., its dirtiness and smell) as well as the

Table 12.2 Totals for variable items

Factors	Category	Num
1. Land ownership	Land owner	292
	No land	24
2. Literacy rate	Literate	271
	Illiterate	46
3. Type of toilet	Eco San	109
	Septic tank	60
	Pit latrine	142
	Hanging latrine	4
4. Dirtiness of toilet	Clean	107
	Between	11
	Dirty	197
5. Smell of toilet	No smell	107
	Between	17
	Bad smell	191
6. Keep toilet	Keep toilet	112
	Improve toilet	203
7. No. of patients	0	88
	1	23
	2	94
	3	103
	≥4	7
8. Medical exp (BDT) *1 USD = 78 BDT	0	88
	1,000	31
	3,000	67
	6,000	79
	>6,000	50
9. Frequency of cleaning	Everyday	78
	Two days	21
	Three days	45
	Week	113
	Rarely	58
10. Dipping dreadful	No dread	108
	Between	8
	Dread	199
11. Convey excreta	No dirty	103
	Between	39
	Dirty	173
12. Dysentery	No dysentery	208
	Dysentery	107
13. Diarrhoea	No diarrhoea	213
	Diarrhoea	102
14. Satisfaction	Satisfaction	132
	Dissatisfaction	185

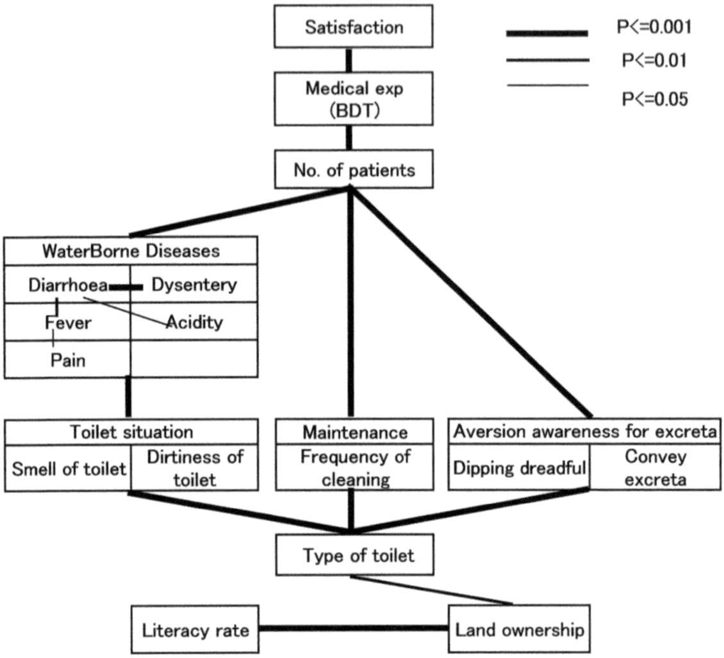

Fig. 12.14 Hypothetical structure between factors

occurrence of waterborne disease. This, in turn, leads to the number of patients and finally medical expenditures.

With reference to users' satisfaction their everyday lives, the results of the multi-dimensional quantification theory II are shown in Fig. 12.16. The external criteria are satisfaction/dissatisfaction with everyday life. This is the result of an analysis of factors selection. The correlation ratio of the analysis result is 0.66, and the distinction efficiency is 87 %; thus, each category score was significantly arranged. The Eco San toilet, as opposed to other toilets, was characterized by less bad smell, comfortable use, less aversion awareness for human excreta, and fewer patients with waterborne diseases.

Summarizing the above, the introduction of the Eco San toilet improves the state of toilets, and contributes to a reduction in household expenditures through mitigation of waterborne diseases and becoming patients. That is, the Eco San toilet contributes to sanitary improvement in rural areas, and thereby to better living. The associative structure noted above shows the context for sanitary improvement benefits evaluation. Of course, the importance of appropriate management of Eco San toilets should be emphasized enough.

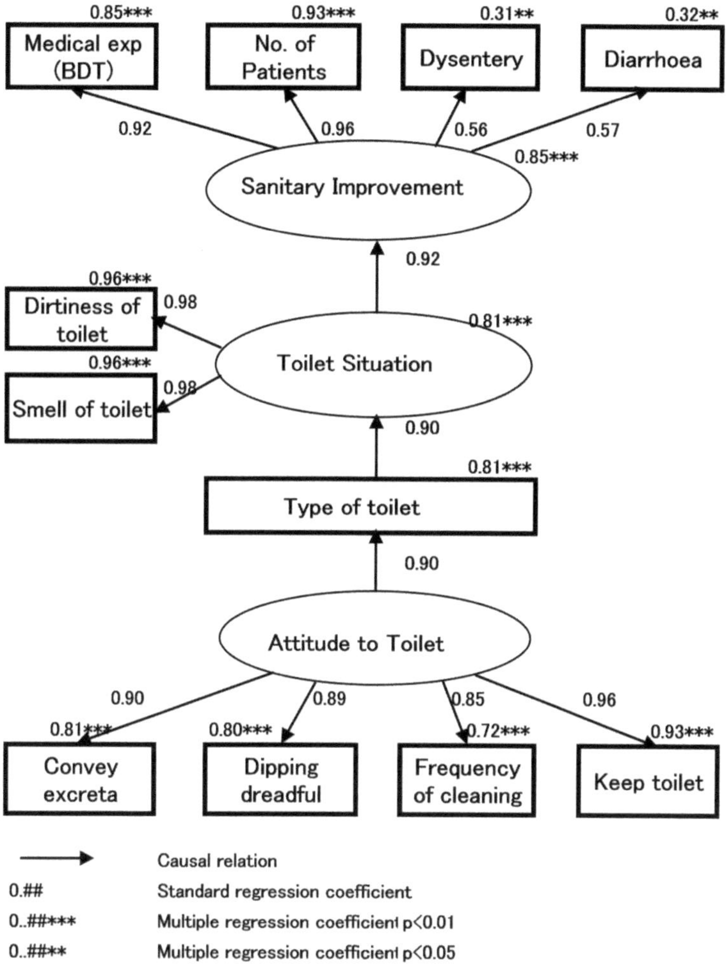

Fig. 12.15 Multiple indicator model

12.2.1.4 Estimation of Sanitary Improvement Benefits

Based on the analysis results, Fig. 12.17 shows the frequency distribution of waterborne diseases, the number of patients with diseases, and the medical expenditures (in BDT, or Bangladesh Taka [1 USD = 78 BDT]) of Eco San toilet users and other toilet users. The figure shows the notable difference between Eco San toilet users and others. If the difference in the expected values of these distributions is classified as benefits, the benefits seen from medical expenditures indicate that use of the Eco San toilet will save around 5,000 BDT/year/household. Since the average annual income per household is about 100,000 BDT, this can reduce most households' expenditures by 5 %. Furthermore, it can be said that the benefit is

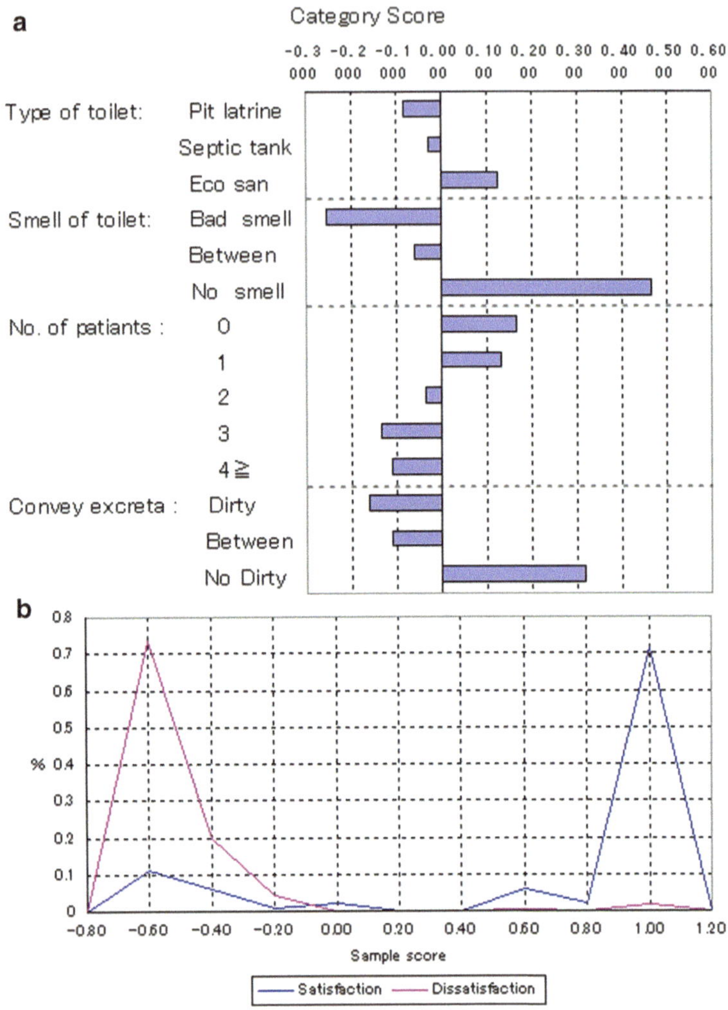

Fig. 12.16 (**a**) Category score of multi-dimensional quantification theory II, (**b**) distinction satisfaction/dissatisfaction group

large in poor families. Moreover, the construction cost of the Eco San toilet is presently about 20,000 BDT, and supposing an annual savings of 5,000 BDT, this expenditure can be recovered within 4 years for each user.

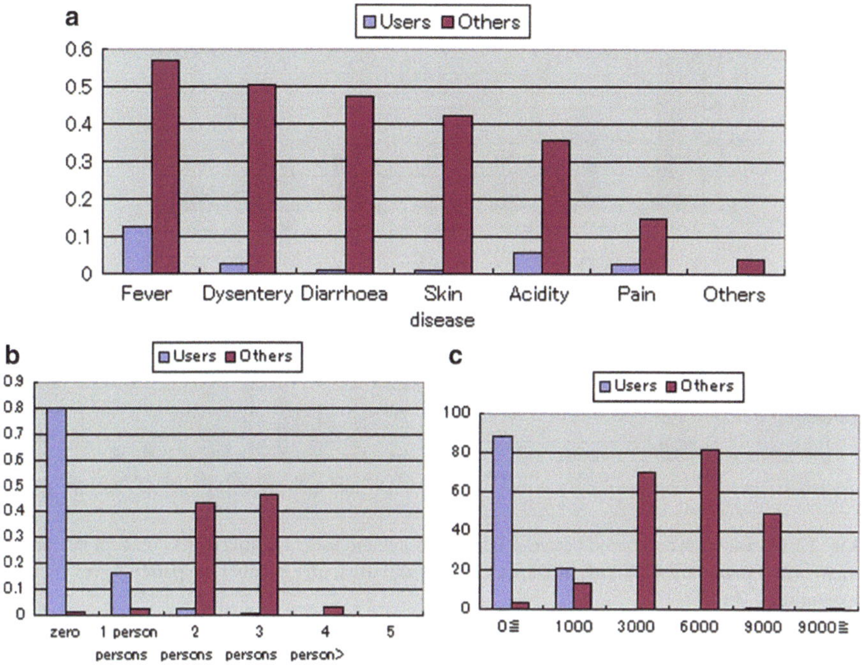

Fig. 12.17 Benefit between Eco San Users and users of other toilets (Comilla area). (**a**) Frequency distribution of waterborne disease, (**b**) number of patients, (**c**) medical expenditures (in BDT)

12.2.2 Results of the Survey in Shyamnagar, Kolaroa, Niamatpur, Keshobpur, and Sharsha Area

The survey contents for the other five areas are similar to those used in the Comilla area. Similarly, the findings are almost equal to those from the Comilla area regarding the state of toilets, the awareness of sanitation and excrement, etc. The difference among the main attributes of these areas is as follows.

The altitudes of the Comilla and Niamatpur areas are comparatively high, with both being around 100 ft. Neither area shows a history of arsenic contamination. Therefore, life in these two areas is somewhat more stable than in other areas.

On the other hand, each of the other 4 areas is contaminated with arsenic and located in the lowlands. Further, Shyamnagar and Kolaroa both see frequent flooding. Especially in the case of Shyamnagar, which suffered the most from Cyclone Aila (which struck at the end of May 2009). Figure 12.18 shows the frequency distribution of waterborne diseases, number of patients, and medical expenditure of Eco San toilet users and users of other toilets. The sanitary improvement benefit in these areas is clear in this figure.

The Eco San toilet is clearly superior to toilets from a sanitary viewpoint. Considering the average medical expenditures across all 6 study areas, the benefits resulting from the mitigation of medical expenditures will be about 2,000

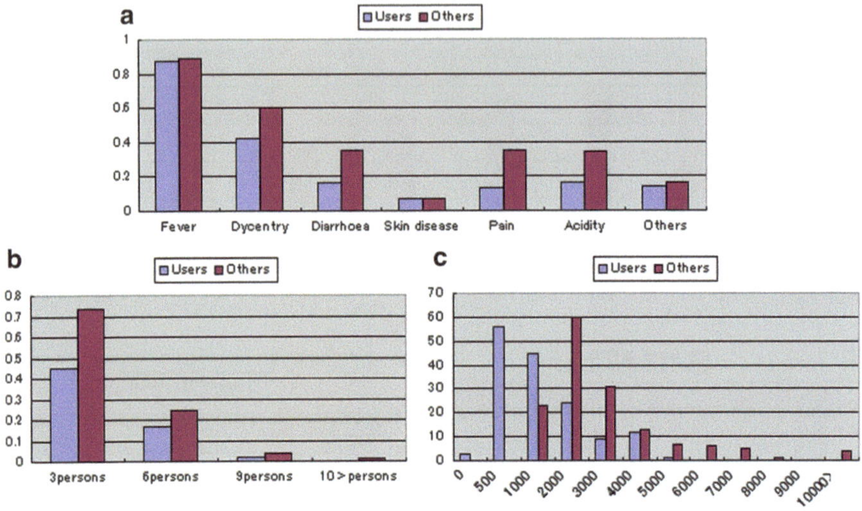

Fig. 12.18 Benefits compared between Eco San users and users of other toilets (except in Comilla area). (**a**) Frequency distribution of waterborne diseases, (**b**) number of patients, (**c**) medical expenditure (in BDT)

BDT/year/household. Of course, it cannot be emphasized enough that such a result can be achieved only through the proper management of the Eco San toilet. Moreover, continuation and verification surveys will be required from now on.

12.3 Sanitary Improvement Effect and Benefit Evaluation (Based on Continuous Survey from 2011 to 2013 Fiscal Year)

12.3.1 Year Transition of Main Indices

The same continuous survey was conducted for the same areas from the 2011 to the 2013 fiscal year. The questionnaire items and the responses (i.e., Eco San toilet users and other toilet users) were the same. The following two indices were observed. Findings revealed a transition in users' awareness towards improving toilets and the aversion of coming into contact with human excreta. The former index evaluates the favourable attitudes towards the toilet, such as acceptance of and satisfaction with the toilet, while the latter index examines the changes in the level of aversion for human excreta, which is relevant to the local culture and customs (Azaduzzaman et al. 2012).

As shown in Fig. 12.19, users exhibited no changes in their awareness of using the toilet. While over 90 % of the Eco San toilet users continued to prefer using the current toilet, the other toilet users desired improvement of their toilet. Further, the

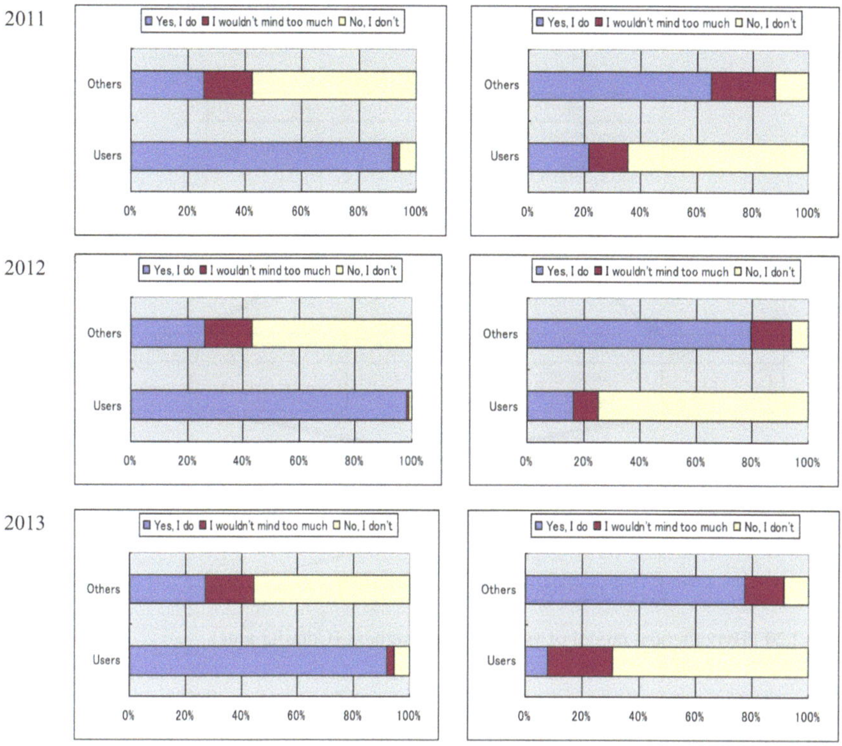

Fig. 12.19 Changes in awareness of toilet improvement and aversion awareness for human excreta

Eco San toilet users showed an improvement in the aversion awareness year by year, while 90 % of the other toilet users continued to avert coming in contact with human excreta. Thus, there was no change in this attitude for the other toilet users.

12.3.2 Sanitary Improvement Structure

Figure 12.20 shows the sanitary improvement hypothetical structure model. The factors for this analysis are literacy rate, land ownership, the type toilet, state of the toilet (dirtiness, smell), evasion awareness to human excreta, toilet cleaning frequency, proper use of toilet, participation in the CBO, frequency of waterborne diseases, no. of patients, annual medical expenditure, and life satisfaction. As in the preceding section, a χ^2 test was performed on these factors, and the findings revealed an associative structure. Specifically, the type of toilet, state of toilet, aversion awareness for human excreta, proper use, and cleaning frequency have a

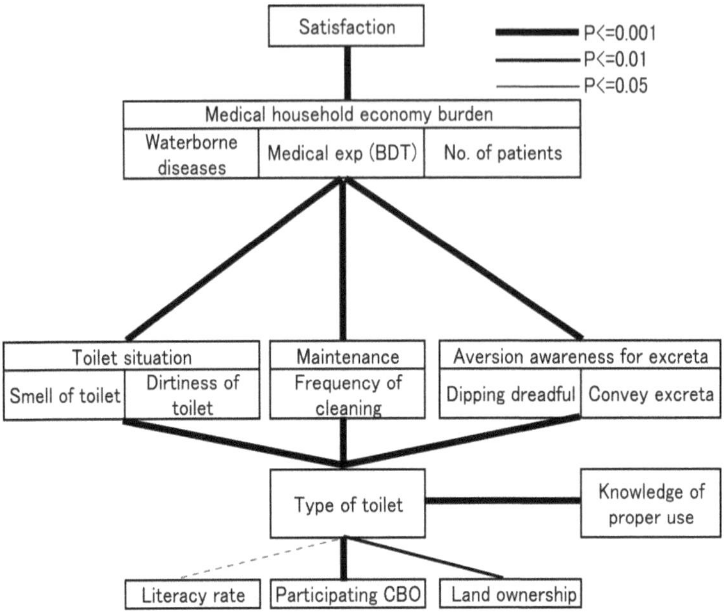

Fig. 12.20 Hypothetical model of sanitary improvement (Comilla area)

close relation with disease frequency and medical expenditure, which in turn contributes to living satisfaction.

12.3.3 Sanitary Improvement Benefits

The disease frequency (number of patients/household/year) and medical expenditure (BDT/household/year) on waterborne disease treatment in the 6 study areas are presented in Figs. 12.21 and 12.22, respectively. The data suggest that the disease frequency and medical expenditure among the Eco San toilet users was less than that among other toilet users. Furthermore, medical expenditures (BDT/household/year) can be expected to be about 2,000 BDT.

Figure 12.23 presents the findings of a survey on medical expenditures for waterborne disease in Bangladesh (World Bank's Water and Sanitation Program, 2007). This figure shows that diarrhoea in particular is the primary cause of medical expenditures, with 60 % of expenditures being related to diarrhoea. These results suggest that if the Eco San toilet is introduced into all households in Bangladesh, and supposing that the sanitary improvement benefit results in 2,000 BDT/household/year, it would lead to 35 % reduction in expenses.

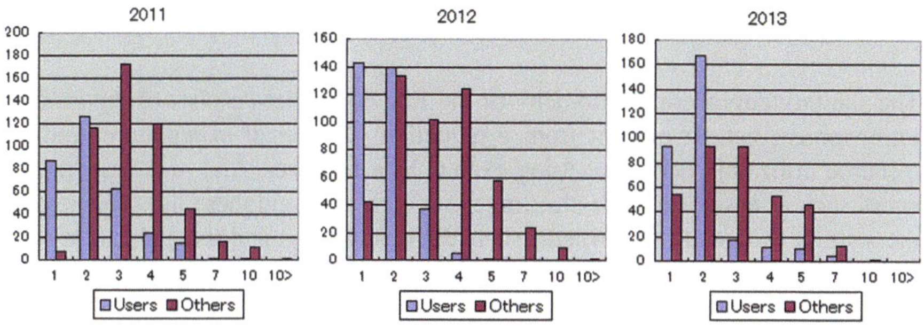

Fig. 12.21 Number of patients (number of patients/household/year)

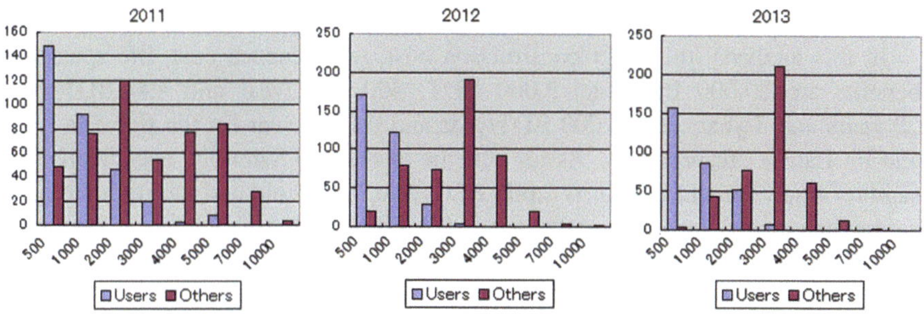

Fig. 12.22 Medical expenditure (in BDT/household/year)

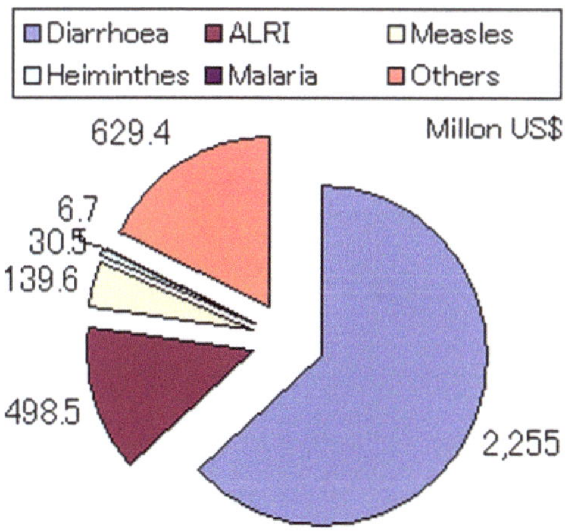

Fig. 12.23 Medical expenditures on waterborne diseases

12.3.4 Benefit Evaluation of the Eco San Toilet

The sanitary engineering rationality of the Eco San toilet consists of the sanitary improvement benefit evident from a reduction of medical expenditure, and the resource utilization benefit of using excreta as a fertilizer. And of course, preferences such as ease of use are also important elements, and these are supported by the state of the toilet (e.g., whether it is dirty or smelly) and its application (e.g., whether it is use easy for children or the aged, possibility of using during floods).

Figure 12.24 shows the construction cost, maintenance cost, and resource utilization and sanitary improvement benefit of the Eco San toilet and Pit Latrine. The Pit Latrine accounts for more than 90 % of the toilets in Bangladesh. Findings reveal that the Eco San toilet construction and maintenance cost will be recovered within 7 years, and that the Eco San toilet will be predominant in about 4 years.

In this analysis, the toilet construction cost, maintenance cost, life span, and benefits are 20,000 BDT and 3,000 BDT, 200 BDT/year and 400 BDT/year, 20 years and 3 years, and 3,000 BDT/year and 0 BDT/year for the Eco San toilet and Pit Latrine, respectively. As for benefits of the Eco San toilet specifically, the sanitary improvement benefit is 2,000 BDT/year/household and the resource utilization benefit of human excreta is 1,000 BDT/year/household (e.g., Takahashi et al. 2009; Parvez et al. 2012b).

Although the effects and benefits of the Eco San toilet are mentioned in detail above, they are subject to the condition that the Eco San toilet is managed appropriately. Figure 12.25 shows the monitoring results related to the proper use of the toilet during the survey period. The monitoring items and the monitoring score were defined as explained in Table 12.3 (Parvez et al. 2012a).

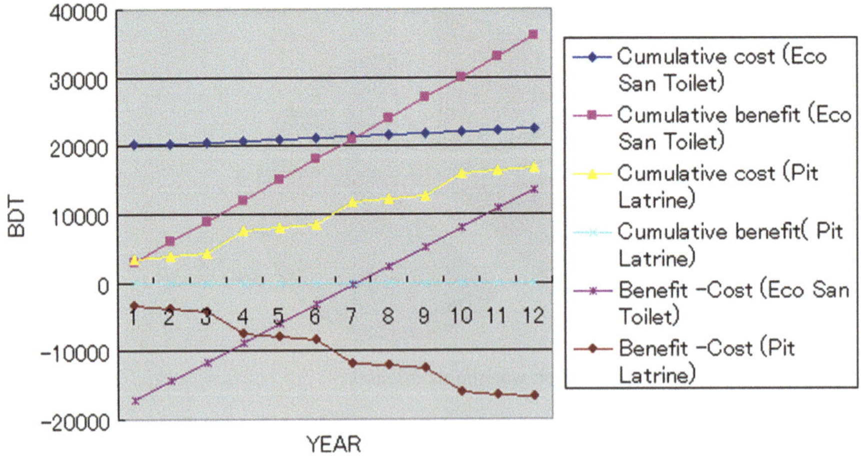

Fig. 12.24 Cost benefit analysis of the Eco San Toilet versus the pit latrine

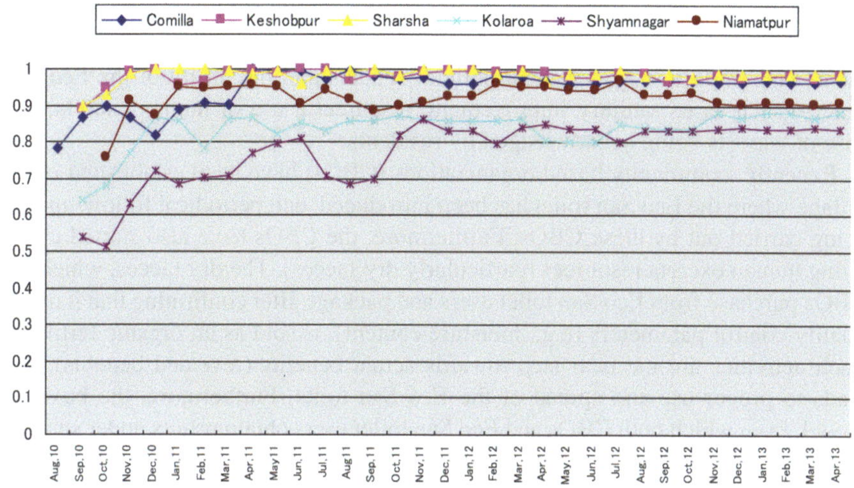

Fig. 12.25 Monitoring score

Table 12.3 Monitoring Items

No.	Monitoring items
1	Cleanliness of the place for urination
2	Cleanliness of the place for defecation
3	Cleanliness of the place for anal washing
4	Availability of a toilet brush
5	Proper adjustment of the urine container
6	Availability of an ash pot and ash
7	Proper placement of the lid over the faeces hole
8	Proper use of ash
9	Presence of water and worms inside the faeces vault
10	Presence of flies in the faeces vault
11	Presence of foul odour
12	Proper use of urine in farm land
13	Condition of the evaporation bed

Monitoring score equation: $S_{Mj} = \sum W_i \times M_{ij}$; S_{Mj}: The monitoring score of the toilet j; M_{ij}: Diagnostic result (0 or 1) of the item i of the toilet j; W_i: Weight (in this case, all were set to 1) of the item i

12.4 Conclusion

The Eco San toilet contributes to sanitary improvement in rural areas, and to better living. The introduction of the Eco San toilet improves the state of toilets, and reduces household expenditure through mitigation of waterborne diseases and resourceful use of human excreta.

Activities to establish Eco San toilets started in 2004, and have now progressed to the stage where the Bangladesh government is aiming at nationwide development. This impact on government policies reflects that the utility of the Eco San toilet – such as its sanitary improvement, resourceful use of human excreta, and ease of use – is being acknowledged by residents.

Recently, community based organizations (CBOs) have been established in the village where the Eco San toilet has been introduced, and periodical follow-ups are being carried out by these CBOs. Furthermore, the CBOs have also started circulating human excreta resources (particularly dry faeces). The dry faeces, which the CBOs purchase from Eco San toilet users and package after confirming that it meets quality control parameters (e.g., moisture content), is sold as an organic fertilizer. Such activities are the next step towards actual benefits (revealed benefits), and leads to proper use and spread of the Eco San toilet. Furthermore, the business model, from which both CBOs and Eco San toilet users obtain profits under suitable management of CBOs, will contribute to the sustainable development of the rural areas of Bangladesh.

References

Azaduzzaman, Q., Mamun, A., & Takahashi, K. (2012, March). Evaluation of Eco San toilet as an appropriate sanitation option for rural areas of Bangladesh in light of sanitary improvement: A case study. ICETCESD, Sylhet, Bangladesh, pp. 181–184.

Flagship Report The Economic Impacts of Inadequate Sanitation on Bangladesh. (2007). World Bank's Water and Sanitation Program (WSP).

Hosaka, K., Takahashi, K., Sakai, A., & Takamura, S. (2006). A study on installation of Eco San Toilet in Bangladesh rural area. *Environmental & Sanitary Engineering Research, 20*(4), 14–23 (in Japanese).

National Plan for Disaster Management. (2010). Government of Bangladesh joint monitoring programme (JMP) for water supply and sanitation, 2013. WHO/UNICEF.

Parvez, M., Azaduzzaman, Q., Mamun, A., & Takahashi, K. (2012a, March). *A case study to evaluate CBO's activity for appropriate management of Eco San toilets in rural areas of Bangladesh*", ICETCESD, Sylhet, Bangladesh, 177–180.

Parvez, M., Azaduzzaman, Q., Mamun, A., Takahashi, K., & Ahmed, T. (2012b, October). *Benefit estimation of human excreta use as fertilizer in agricultural farming*. Paper presented at the International Conference on environmental aspects of Bangladesh & the World, BENJapan, Japan, pp. 134–137.

Stockholm Environment Institute. (2004). Ecological sanitation, revised and enlarged edition.

Takahashi, K., Sakai, A., Hosaka, K., & Takamura, S. (2009). Effect and benefit evaluation of Eco San toilet installation in Bangladesh rural area. *Environmental & Sanitary Engineering Research, 23*(1), 15–24 (in Japanese).

Takahashi, K., Sakai, A., Mamun, A., Tofayel, A., & Azad, Q. (2011a). Sanitary improvement effect and benefit evaluation of Eco San toilet installation in Bangladesh rural area. *Environmental & Sanitary Engineering Research, 25*(4), 17–26 (in Japanese).

Takahashi, K., Sakai, A., Mamun, A., Ahmed, T., & Azaduzzaman, Q. (2011b, October). Sanitary improvement benefit evaluation of Eco San toilet installation in rural areas of Bangladesh. Paper presented at The 4th IWA-ASPIRE.

Toyoda, H. (1998). Covariance structure analysis. *Asakura-Shoten* (in Japanese).

Part IV
Risk Management of Regional Vulnerability

Chapter 13
A Supply System for Municipal Water with Uncertainties

Chisato Asahi and Kiyoko Hagihara

Abstract In this chapter, the system of the provision of municipal water through the market mechanism including some forms of "privatizations" is considered in the view of the normative aspects of resource allocation and welfare under the uncertainty of water quality. First, the first and second fundamental theorems of welfare economics are reviewed and the cases that the market mechanism fails to satisfy the optimal allocation are shown. Second, among those cases, the problems relevant to the municipal water supply under the presence of environmental risk and asymmetric information between water supplier and regional consumer are considered. We review the analytical framework of credence goods and show the advantages of using it when the uncertainty of the water quality is at issue. Third, we assume that some private firms have chance to contract to provide municipal water and examine how the credence of municipal water, that means quality regulation, effects on the price and water quality level using an analytical model. A regulator of municipal water provision should control municipal water supplier by means of some kind of penalties and inspections, which work on the risk attitudes of the supplier and the probabilities of detection under informational asymmetry.

Keywords Municipal water supply • Asymmetric information • Credence goods • Privatization • Public-private partnership

13.1 Introduction

Since the 1990s, there has been a continuing global trend towards the privatization and commodification of water, which specifically refers to private-sector participation in the water utility and bottled drinking water businesses. This chapter

C. Asahi (✉)
Division of Urban Policy, Faculty of Urban Liberal Arts, Tokyo Metropolitan University, 1-1 Minami-osawa, Hachioji, Tokyo 192-0397, Japan
e-mail: asahi@tmu.ac.jp

K. Hagihara
School of Sociology, Bukkyo University, 96 Kita-hananobo, Murasakino, Kita-ku, Kyoto 603-8301, Japan
e-mail: hagihara@bukkyo-u.ac.jp

K. Hagihara, C. Asahi (eds.), *Coping with Regional Vulnerability*, New Frontiers in Regional Science: Asian Perspectives 4, DOI 10.1007/978-4-431-55169-0_13

considers the former classification. There are various levels of private-sector participation in the water utility business, with formats ranging from a fully privatized water authority (e.g. in the UK) to a business consignment to a private enterprise (e.g. in Japan).

For Japan's water utility business, the debate over private-sector participation has been conducted mainly from the viewpoint of reforming the management of local public enterprises as part of structural reforms against the backdrop of forthcoming population decline and financial difficulties. The challenges for management of public enterprises being forwarded through debate are premised on private-sector participation.

The downside to commercialization of public utility works is the occurrence of incidents where commercialization has rocked the entire society, such as the massive power blackouts that occurred in August 2003 in North America and the bankruptcy of Enron Corporation (US), the flag bearer of electric power liberalization. In the water utility business, there have been reports of social problems where privatization has resulted in the deterioration of water quality and soaring water rates. Barlow and Clark (2002) report the background of the water utility system in Bahía Blanca, a city located in the province of Buenos Aires, Argentina. The water system was operated by Azurix Corp., an equity affiliate of Enron Corporation, but Azurix withdrew from the business in 2001 for profitability reasons after suitable water quality and water pressure could not be ensured and the water authority demanded improvements. Moreover, in the UK, 128 counts of water authorities suing the five major water companies for violations such as water leaks and improper treatment of polluted water occurred from 1989 to 1997.

The first contentious issue in the debate on private-sector participation in the water utility business is whether to consider water economic goods that must be efficiently distributed or as goods that is a guarantee of life and a fundamental human right. The standpoint of the latter specifies that support cannot be given to an allocation using market mechanisms. This is because such an allocation follows the Pareto principle and raises the possibility that supply paths may be closed to income earners with a low capability to pay.[1] For the former position (e.g. if water is considered a rare resource that should be distributed using market mechanisms), an examination should be conducted on whether the market mechanism conditions can function effectively based on the characteristics of water as goods supplied (production technology) and demanded (preferences).

For the water utility business in Japan, analysis of resource allocation efficiency is centred on the supply side for the following reasons: management is generally conducted by public municipality enterprises under the Waterworks Act and the Local Public Enterprise Act; there are certain production technology constraints

[1] According to Roemer (1996), 'the question of how a society (organization) allocates scarce resources or goods between individuals with competing needs or requests' is prescribed by the theory of 'distributive justice'. In other words, if consideration includes possible execution, the determination of the resource allocation mechanism expresses the problem of allocation and that which is indivisible.

regarding economies of scale owing to fixed costs; and costs must be reduced through the private sector, as indicated previously.[2] On the other hand, to the extent of the author's knowledge, scant normative analysis exists on how resource allocation efficiency is affected by demand-side requirements, that is, consumer preferences towards water utilities. Analysis related to the demand for water utilities is limited to an empirical viewpoint that contributes to estimates (e.g. demand structure analysis), likely because the normative viewpoint related to resource allocation has not been given much weight.

From the aforementioned background, this chapter organizes the viewpoints examining the demand-side efficiency of resource allocation when supposing the introduction of a market mechanism for a water supply system by a utility. In addition, it presents an analytical framework for how the uncertainty of quality affects efficiency.

The next section takes a general view of the conditions for establishing a fundamental theorem of welfare economics relating to resource allocation of the market economy to provide an understanding of states where efficient resource allocation is not achieved. The third section examines the problems within these states relating to the demand-side characteristics of water utilities. The fourth section uses models of credence goods to examine the points that need to be considered when conducting a system design of supply, focusing in particular on the uncertainty of quality. The last section provides observations on systems that are currently being introduced, and the last section presents a conclusion.

13.2 Examination Using the Fundamental Theorems of Welfare Economics

13.2.1 Conditions of the Fundamental Theorems of Welfare Economics

The fundamental theorems of welfare economics are as follows:

- Fundamental theorem I: 'The economic state will be Pareto efficient as long as competitive equilibrium exists'.
- Fundamental theorem II: 'Supposing that the preferences of all consumers are convex, any Pareto-efficient allocation can be obtained as a competitive equilibrium allocation by suitable reallocation of income by lump-sum tax or subsidy'.

---

[2] For example, according to Japan Water Works Association (2006), issues raised in the background of examination into privatization of the water utility business were upgrade investment accompanying aging, increased costs of wholesale water, and upward pressures on costs because of stricter water quality regulations with no expected increase in demand.

Competitive equilibrium exists by means of a decentralized price mechanism, and the prerequisite for establishing these fundamental theorems of welfare economics is the satisfaction of the following three conditions: (A) perfect competition, (B) market universality, and (C) convex environment. Among these prerequisites, the convex environment is not necessarily essential for fundamental theorem I, because a Pareto-efficient resource allocation is established as long as there is competition equilibrium. In contrast, fundamental theorem II has the prerequisite that the preferences of consumers are convex.[3]

Hagihara, Asahi, and Hagihara (2005) examine the case when these conditions are not respectively satisfied. Based on this, the following three conditions specify why a normative valuation of resource allocation is not achieved using market mechanisms based on the fundamental theorems of welfare economics:

1. It is not possible for each economic agent to act as a price taker (collapse of (A) perfect competition), because production or consumption element are indivisible (collapse of (C) convex environment).
2. There exist economies of scale (collapse of (C) convex environment and thus collapse of (A) perfect competition) by public goods or externalities (collapse of (B) universality of the market) and transaction costs or asymmetric information (collapse of (B) universality of the market).
3. The prerequisites of rationality of choices (completeness, transitivity, continuity, or supposition of the non-saturation of preferences) are not established (collapse of (C) convex environment).

13.2.2 Examination of the Water Utility Market

13.2.2.1 Supply Side

A production technology characteristic of the water utility business is a natural monopoly because of the subadditivity of cost. In other words, because the fixed costs required for supply are large as compared with the market demand scale, market entry or participation does not occur. Inevitably, the industry becomes monopolized. The existence of these indivisible production elements means that the condition for (C) convex environment with regard to production is not satisfied and that the monopoly owing to economies of scale does not satisfy the condition of (A) perfect competition where each economic agent is a price taker. Therefore, a Pareto-efficient resource allocation owing to a price mechanism is not established.

Even in the case where subadditivity of cost exists, explicitly considering the environment where games are repeatedly played between potential participant

[3] Okuno and Suzumura (1988) state that when attempting to realize efficient resource allocation using a decentralized pricing system while relying only on limited information without making sacrifices to this efficiency, the importance of fundamental theorem (2) is the clarification of how strict environmental conditions are required.

companies, contestability theory implies that a Pareto-efficient allocation will be established if several conditions are met (Ihori 1996). The important assumptions here are that economies of scale exist in the cost function and that the withdrawal expense is zero. In other words, the fixed costs must not be sunk costs. In order to collect the fixed costs invested into the production factors when withdrawing from the market, it must be possible to sell these production factors.

These conditions of a contestable market have a normative significance in the point where the conditions are clarified as well as those for perfect competition. The belief that these conditions are actually established in many oligopolistic industries is unrealistic. The water utility business is no exception. Even if there were no system-based barriers to entry, if a company participated by investing capital in production elements such as water treatment facilities or water pipe networks, it would be necessary to verify whether it is possible to withdraw from the business by selling these production factors after obtaining a profit in excess of the participation cost. Thus, even in the context of a contestable market, it is not certain as to whether a Pareto-efficient resource allocation can be established.

On the supply side of the water utility business, resource allocation using market mechanisms does not receive normative support from condition 1 of the previous section. This is why water and electric power utilities have traditionally been public enterprises and regulated companies. Further reasons why the Pareto-efficient allocation is not established concern the characteristics of production technology regarding the indivisibility of the production factors and the fact that the fixed costs are sunk costs.

13.2.2.2 Demand Side

According to conditions 1 to 3 of the previous section, preference characteristics concerning water utilities include the prerequisites of indivisibility of consumption (collapse of (C) convex environment) and that the rationality of choices is not established (collapse of (C) convex environment). For the former prerequisite, water can be considered indivisible consumption goods. For the latter, it is necessary to examine the quantity and quality that determine the preferences for water utilities.

Concerning the quantity of water, as supply obligations are assigned to the water utility company under the current system based on the Waterworks Act, a supply volume is guaranteed that is equal to the sum of the demand function. Therefore, it is possible to perform a rational choice similar to a preference for normal goods.

Concerning quality, supply provided must meet the minimum quality standard stipulated by the Waterworks Act. If consumers consider that satisfactory and consistent water quality is being supplied, they are able to perform a rational choice suited to the desired payment amount. However, in recent years, there has been a growing need to consider the uncertainty of the water supply because of major issues with water from utilities, such as the inflow of pollutants of non-point sources,[4] trihalomethane,[5] cryptosporidia,[6] and endocrine disruptors (Asahi and Hagihara 2000).

When considering these uncertainties, consumer preferences are explained using expected utility theory, which states that if the choice of consumers possesses perfect rationality, preferences and choices will be consistent. It has long been thought that the consumers' choice problems are related to the maximization of expected utility based on perfect rationality. However, in actual choice informed by knowledge of experimental economics and cognitive psychology, there exists a systematic bias contrary to the assumption of perfect rationality in risk-based choice and probability judgement (Hagihara et al. 2005). Furthermore, according to Hagihara et al. (2004), although benefit evaluation is conducted to avoid drinking-water quality risks, it is shown that the risk-evaluated values, that is, marginal willingness-to-pay for public investment (See Chap. 8), differs because of differences in the information environment and risk awareness.

Then, in the case of uncertainty in the water quality, the prerequisites for rationality of consumer choices are not established. That is, the possibility exists that the convex environment relating to preferences will not be satisfied. In this case, it is suggested that efficient resource allocation using market mechanisms shown by the fundamental theorems of welfare economics are not established.

13.3 Uncertainty of Quality

13.3.1 Framework of Analysis for Credence Goods

When there is uncertainty in the quality of goods and welfare analysis that assumes a standard choice based on certainty does not apply, analyses can be divided into the following three main approaches. The first approach is an extension of expected utility theory that explicitly considers bounded rationality in choice decision-making (Hagihara et al. 2005). The second approach is the use of the insurance market to address uncertainties or risks (Asahi and Hagihara 2005). The third approach is an analysis that uses the concept of credence goods that explicitly incorporates the state of information relating to these goods.

The first and second approaches address uncertainty in the state where the consumer has a choice. Currently, there is no uncertainty between the goods and

[4] This is mainly referring to the inflow of pollutants from locations like farms and golf courses.

[5] Trihalomethane is shown to possess carcinogenicity and can be linked to organic material and chlorine in the water through the process of chlorination in water treatment.

[6] Cryptosporidia are waterborne sporozoea that infect humans. They have become a serious global problem in water from utilities, particularly in Europe and North America. In June 1996, a mass infection occurred in Saitama Prefecture. Healthy people recover after experiencing diarrhoeal symptoms for a few days to a few weeks, but people with weakened immune systems, infants, and the elderly can experience prolonged diarrhoea.

the effect realized by its consumption, and uncertainty of quality is expressed using the expected value formed from the effect realized through consumption of the goods. In comparison to these approaches, the third approach of credence goods explicitly deals with situations when the effect of consumption of the goods is not known in advance. This is clearly a situation that conflicts with the condition requiring a preference relationship to satisfy the rational assumption that ensures that this relationship matches the order of the effect. Therefore, when consumption is unavoidable even without judging the quality of goods, if the basis of choice is switched from goods quality to "trust" towards such quality, it is possible to express the uncertainty of quality by the degree of this trust.

The merit of this approach for the first point is that we can recognize its close resemblance to decision-making in choices where the goods quality is uncertain. In other words, in cases where consumption is conducted even if the quality and effect brought about by consumption are uncertain, it is easy to imagine a situation where the basis of choice is not quality itself but rather a substitute that signals quality (e.g. judgement of quality or the supply system). The merit for the second point is that it explicitly addresses trust formed not for the goods itself but for its quality or for a system, and thus, it is easy to examine specific systems.

13.3.2 Definition of Credence Goods

Credence goods are a categorization of goods determined by the state of information related to quality. The first to introduce this concept was Nelson (1970). Darby and Karni (1973) provided the first definitions of credence goods, calling quality understood before purchasing 'search quality', quality understood without cost after purchasing 'experience quality', and quality requiring a high judging cost after purchasing 'credence goods'. Blankart (1987) presents a similar concept and calls these 'trust goods', defined as a service that cannot be evaluated for quality by experience because the quality is incomprehensible. In Anania and Nistico (2004), the concept is expressed as 'credence or trust goods', and they are defined as goods where the product characteristics may remain unknown even after consumption. The term 'trust' considers that consumers choose based on trust in the company because they are unable to judge the characteristics possessed by the product.

Lupton (2005) defines credence goods as goods that cannot have their quality evaluated either by knowing it beforehand or through normal use, and additional large costs are required to evaluate the goods. Dulleck and Kerschbamer (2006), who provide a generalization of the series of analyses of credence goods, offer a definition that further limits the state, describing credence goods as goods and services for which the quality required by the consumer is better known by experts rather than by the actual consumers.

13.3.3 Advance Research

Assumptions common to the analyses of credence goods were first described by Darby and Karni (1973). These definitions clarify that there is asymmetric information where suppliers, 'experts' in their field, have superior information than consumers, that is, 'customers'. As a result, because suppliers may act improperly with respect to quality and rates, efficiency is lost through undertreatment, overtreatment, and overcharging.

Concerning the quality uncertainty problem in the context of private-sector participation in a public-sector field, Blankart (1987) analyses the case where uncertainty of quality is regarded from both sides: the 'production technology' side and the 'consumption technology' one. In this chapter, the problem of quality uncertainty is considered consumption technology and stated as follows: in the case where information on a certain goods' quality cannot be obtained, it is substituted by supply rules for the goods. Therefore, utility is not determined by the goodness or badness of quality but rather based on whether the quantity of 'care', which is measured by the observability of prescribed rules, is appropriate in the production process. Because compliance with the supply rules is generally a constraining condition of production, a private agent does not necessarily have an advantage regarding this point, and its advantages with respect to efficiency are probably not that large. Therefore, among public services, while privatization has advantages for search and experience goods, there is little evidence that it has merits for trust goods.

However, asymmetric information in the case of credence goods does not necessarily lead to inefficiency, and if some conditions can be satisfied, it is possible to have efficient allocation by market mechanisms (Dulleck and Kerschbamer 2006). In such a case, examination is focused on the effectiveness of the supply rules that serve as a substitute for the information and trust lacking in credence goods. Marette et al. (2000) analyses the relationship between the safety of the goods and the supply system. Furthermore, Anania and Nistico (2004) analyses whether food safety regulations can be obtained as substitutes for information and trust.

13.3.4 Credence Goods and the Convex Environment

Blankart (1987) states that revealed preferences in market mechanisms do not guarantee efficiency when there is uncertainty in quality. When viewing credence goods from a convex environment (See Sect. 13.2.1), because it is a state where the relationship of choice and effect is unknown, it is not possible to determine the ordering that satisfies the preference relationship as the prerequisite of convex preferences towards quality. On the other hand, if there were a substitute that forms trust towards quality, it may be possible to determine such a preference relationship. Therefore, analysis of credence goods is possible when this analysis relates not to the quality of the goods itself but to the substitute that forms trust towards quality.

13.4 Analysis Relating to Quality of the Water

13.4.1 Quality in Choice of Water

The judgement that determines whether goods are credence goods is based on whether search or experience can be used to determine quality. With respect to the goods attributes of water, while the consumer may be able to judge quality at the time of choice for taste and smell, judgement cannot likely be made beforehand when it is impossible to perceive such matters as the existence of contamination by pathogens or chemical substances or the effects that these might have on health. In this case, the dependence when the choice is made is thought to be information relating to the supply agent and the method that is considered to affect water quality, such as whether it is supplied by a water utility business entity governed by regulations relating to water quality, whether it is well water, the state of the source water, and in the case of a high-rise building, whether it is a storage tank system or a directly connected one. According to a questionnaire survey conducted by Asahi and Hagihara (2000), consumers were shown to choose consumption but were unable to judge water quality with respect to health risks based on normal search and experience. Therefore, water is considered a credence good for which choices on quality are conducted based on trust toward the supply system.

The condition for establishing a credence goods market is generally the existence of a substitute for the information and trust that the consumer lacks. The items raised as substitutes generally include company reputation, quality assurance by a third party organization, and regulations. The substitutes for the quality of the water can be considered the water utility business entity and the supply system. Here, we assume that a substitute for trust is the degree of compliance with water quality standards.

13.4.2 Model

The model shown in Anania and Nistico (2004) related to the quality of food is applied here to examine how demand, supply, and equilibrium of water are affected by the level of trust toward the quality of the goods.

13.4.2.1 Prerequisites

In water supply, regulations are imposed to ensure a constant standard of water quality is met. There are two types of markets: the H-type market in a region that requires advanced technology to meet constant water quality (e.g. with poor source water or where pollution is terrible), and the L-type market that can easily meet water standards using normal technology. Hypothetically, private enterprise can freely participate in either market. The company supposed here is not necessarily limited to an agent that carries out the entire water business. It can also be a business operator entrusted with maintenance and management of existing facilities to meet water quality standards or parts related to water quality testing, and the operator is referred to hereafter as the company or supplier. To establish subadditivity in water production, one company out of multiple suppliers is assumed to provide a regionally monopolistic supply for each region. Potential suppliers have a difference in technological capacity to ensure water quality. H-type companies possess the technological capability to supply to H-type markets, and L-type companies have the ability to supply only to L-type markets, as they do not possess advanced technology.

When a company participates in an H-type market, a regulator performs tests based on water quality standards to determine whether the company possesses the technology to meet the water quality standard; however, testing performance is not perfect. In other words, the possibility of an L-type company participating in an H-type market cannot be excluded, and there is the possibility that an L-type company participates in an H-type market and that water with quality meeting H-type market regulations is not supplied.

The consumers (or the regulator as the agent for the consumers) of the H-type market knows that an L-type company may participate, but they are unable to make an observation about the company type that supplies in their region. In this case, the water quality standard monitored by consumers is a substitute for the type of company, which cannot be observed.

13.4.2.2 Settings

The supply of water must meet a constant water quality standard. Let us assume that the total number of potential suppliers is N, the number of H-type companies is N_H, and the number of L-type companies that possess only normal technology is N_L.

$$N = N_H + N_L \tag{13.1}$$

Let us also assume that the production cost of water is higher when advanced technology is used compared with normal technology. As mentioned previously, we assume that the possibility exists that an L-type company participates in an H-type market. Let us assume that water supplied to the H-type market is checked using testing to determine whether the water meets water standards, and that the probability that this testing identifies an L-type company is λ.

Here, we express λ as test performance, because the test targets all companies. We set the possible values of λ to $(0 \leq \lambda \leq 1)$, and we assume $\lambda = 1$ if the test performance is perfect.

When testing reveals that the water quality standard is not met, a penalty ρ is charged. Moreover, as it is compulsory to supply water quality that meets the standard, the withdrawal of the L-type company is unavoidable.

13.4.2.3 Decision-Making Related to the Participation of an L-Type Company

The aspect providing an incentive for an L-type company to participate in an H-type market is when the expected profit by supplying in the H-type market exceeds the profit by supplying in the L-type market. However, when participating in an H-type market, the risk exists that testing leads to market withdrawal and a penalty. Therefore, the expected profit function is dependent on the company's own risk-aversion level and on the penalty.

$$\pi_{nc} = \pi[\Pi_{nc}] \tag{13.2}$$

$$\pi_c^i = \pi[E(\Pi_c), \varphi_i] \quad i = \{1, \cdots, N_L\} \tag{13.3}$$

Index nc expresses not concealing or feigning technology and index c expresses concealing (feigning) technology. Π expresses profit, and φ_i expresses the risk-aversion level of company i. Risk-aversion level φ_i is different for each company and a monotonic increasing in risk aversion.

The expected profit in the case where technology is feigned, $E(\Pi_c)$, is determined for the combination of testing detection probability λ and penalty ρ. In other words, when falsification is detected by probability λ, the production cost cannot be collected, because production by advanced technology is enforced and penalty ρ is suffered. On the other hand, if falsification is not detected by probability $(1 - \lambda)$, it is possible to sell water at a water quality that does not meet the regulatory value in the H-type market at price P_H. Therefore, participation in the H-type market is decided by comparing the profit obtained from producing in the L-type market with the expected profit of producing in the H-type market. Then, because detection probability λ, penalty ρ, and price P_H in the H-type market are exogenous for the L-type company, the size of the expected profit is determined by the risk-aversion level φ_i. In other words, $\varphi_i < \varphi^*$ for the risk aversion level φ^* as defined in Eq. 13.4. Thus, when $\pi_{nc} < \pi_c^i$, the L-type company has an incentive to participate in the H-type market.

$$\pi[\Pi_{nc}] = \pi^*\left[E(\Pi_c), \varphi^*\right] \tag{13.4}$$

Here, among the number of L-type companies participating in H-type markets, if undetected by testing, n_{LH}^c remains in the H-type market, whereas if detected, n_{LE}^c is forced to withdraw.

$$\begin{aligned} N_L &= n_L^{nc}(\varphi_i > \varphi^*) + N_L^c(\varphi_i < \varphi^*) \\ N_L^c &= (1-\lambda)N_L^c + \lambda N_L^c = n_{LH}^c + n_{LE}^c \end{aligned} \tag{13.5}$$

13.4.2.4 Types of Suppliers in Each Market

When an L-type company behaves as described previously, the number of suppliers in the H-type market is expressed using Eq. 13.6. On the other hand, the number of suppliers in the L-type market is expressed using n_L^{nc}, and all are L-type companies.

$$N_H + n_{LH}^c \tag{13.6}$$

13.4.2.5 Reliability of Regulation for Water Quality

When we assume that the reliability of the regulation of water quality is θ, θ is expressed by the portion comprised by the L-type companies remaining in the market that participate in the L-type market. On the other hand, the portion comprised by the L-type companies that participate in the H-type market is expressed by $(1-\theta)$. We assume $\partial\theta/\partial\lambda > 0$, $\partial\theta/\partial\varphi^* > 0$, and $\partial\theta/\partial\rho > 0$.

When $\theta = 1$, because an L-type product is not supplied to an H-type market, regulation of water quality is a perfect substitute for trust of water quality. Here, $n_{LH}^c = (1-\lambda)N_L^c$ is 0, and detection probability λ is 1 or penalty ρ (which is the determining cause of the risk-aversion level φ^* that decides N_L^c) is high. On the other hand, when $\theta = 0$, regulation of water quality completely fails to form trust in the quality of the water. Because this is a state where the L-type company that aims to supply the L-type market is completely absent, it is considered to be a state where penalty ρ, which is the determining cause of the risk-aversion level, is extremely low.

$$\begin{aligned} &\theta = \frac{n_L^{nc}}{n_L}, \quad 1-\theta = \frac{n_{LH}^c}{n_L} \\ &n_L = n_L^{nc} + n_{LH}^c \end{aligned} \tag{13.7}$$

13.4.2.6 Decision-Making of the Company

The total cost function TC of the L-type company and the H-type company is specified in Eq. 13.8. α and β are the parameters for fixed and variable cost, respectively, and we assume $\alpha > 0$ and $\beta > 0$. In either market, we assume a natural monopoly of supply to establish subadditivity of cost. When the production volume can be divided into m items, the production volume by subadditivity satisfies Eq. 13.9.

$$TC_j = \alpha_j + \beta_j q_j^2 \quad j = \{L, H\} \tag{13.8}$$

$$0 < q_j < \sqrt{\frac{\alpha_j}{m\beta_j}} \quad j = \{L, H\} \quad m \geq 1 \tag{13.9}$$

Figure 13.1 shows the cost function. For simplification, we assume m = 1. While producing, the average cost (AC) is decreasing, and we assume that fixed costs are sunk and that fixed costs do not affect the decision-making of production. Because marginal cost (MC) exceeds average variable cost (AVC), we assume that the price imposed by the supply company is not a monopoly price but a price formed using marginal cost.

Because of the profit maximization condition, the inverse supply function for the H-type and L-type markets is as follows. Here, we set β_H and β_L for the marginal cost parameters.

$$MC_H = P_H = 2\beta_H q_H \tag{13.10}$$

$$MC_L = P_L = 2\beta_L q_L \tag{13.11}$$

At this time, because we can expect revenue from P_H, assuming the L-type company has not been revealed by testing, the volume supplied by the L-type company in the H-type market is determined by setting P_L and the expected price in the H-type market equal. In other words, it is determined by satisfying Eq. 13.12.

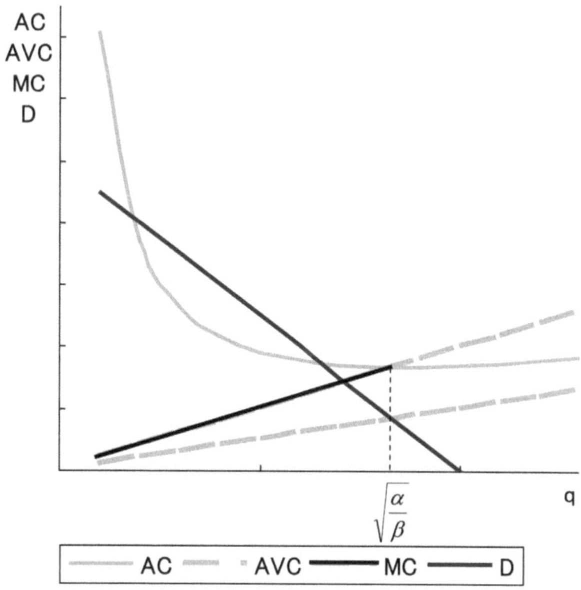

Fig. 13.1 Cost function

Thus, the supply quantity when the L-type company supplies in the H-type market is shown by Eq. 13.13.

$$P_L = 2\beta_L q_L = (1-\lambda)P_H + \lambda \cdot 0 \tag{13.12}$$

$$q_L = \frac{(1-\lambda)P_H}{2\beta_L} \tag{13.13}$$

13.4.2.7 Supply in Regions

Let us consider the supply functions directly faced by regions belonging in each market.

First, we assume the potential supply volume for the L-type market is $n_L^{nc} q_L$, because there is n_L^{nc} L-type companies participating. Eq. 13.14 shows the adjustment to an expression that uses reliability θ. Here, the supply volume that each region in an L-type market faces is the expected value generated using the ratio n_L^{nc}/N of the number of companies remaining in the L-type market, because it is the supply volume for one of n_L^{nc} companies remaining in the market. Here, when standardized using $N = 1$, Eq. 13.14 expresses the supply of the region belonging in the L-type market.

$$Q_L = n_L^{nc} q_L = n_L^{nc}\left(\frac{P_L}{2\beta_L}\right) = \theta n_L\left(\frac{P_L}{2\beta_L}\right) \tag{13.14}$$

The potential supply volume of the H-type market is formed from the supply of the L-type and H-type companies, and Eq. 13.15 shows this expression when reliability θ is used. Similar to the L-type market, the supply volume that a region belonging to the H-type market faces can be expressed using the expected value dependent on the portion of the entire number of suppliers N, which are companies participating in the H-type market. In other words, Eq. 13.15, which expresses the supply of the region belonging to the H-type market, is a weighted average of the possibility of supply by H-type company N_H (standardized using $N = 1$) and by L-type company n_{LH}^c (standardized using $N = 1$).

$$Q_H = N_H q_H + n_{LH}^c q_L = N_H\left(\frac{P_H}{2\beta_H}\right) + (1-\theta)n_L\left(\frac{(1-\lambda)P_H}{2\beta_L}\right) \tag{13.15}$$

13.4.2.8 Demand for the Region

Depending on their region, consumers belong in either the H-type market or the L-type one, and they consume the supply provided by suppliers to their local region.

The inverse demand function of the L-type market is shown in Eq. 13.16. Here, a demand curve drops down on the right, for which γ_L and δ_L are positive parameters, and the marginal benefit is declining.

$$P_L = \gamma_L - \delta_L Q_L \tag{13.16}$$

On the other hand, in the H-type market, the possibility exists that an L-type company supplies water that does not meet quality standards. In this case, because the fact that the L-type company is participating in the H-type market is already known by the consumer, marginal benefit δ_H declines in proportion to the possibility of the L-type company's participation. Therefore, the inverse demand function is expressed by Eq. 13.17. γ_H and δ_H are positive parameters.

$$P_H = \gamma_H - \delta_H Q_H - \frac{(1-\theta)n_L}{N_H} \delta_H Q_H \tag{13.17}$$

Here, $(1-\theta)n_L/N_H$ expresses the portion of supply of the L-type product in the H-type market. Considering the relationship between reliability and demand, even when $\theta = 0$ (i.e. zero regulation reliability), intention to pay for the H-type product exists. In other words, even when the possibility exists that a company with inferior technological capabilities is the supplier in a region with bad pollution, there is intention to pay for the price formation of the H-type market.

This assumption is not realistic in the case of normal goods. In cases where the reliability is lower than a certain standard θ^*, it is supposed that the H-type market is not established (Anania and Nistico 2004). However, because we assume a regional monopoly, because there is no substitute supply source of water for the consumer, even while knowing the possibility of production by the L-type company, there is an willingness to pay for supply in the H-type market, and the H-type market is thus established.

13.4.3 Market Equilibrium

13.4.3.1 Equilibrium

Next, we derive the equilibrium from the supply and demand functions of the region and examine its relationship with the reliability of water quality θ.

Using Eqs. 13.14, 13.15, 13.16 and 13.17, the price and production volume for the equilibrium of the L-type and H-type markets are expressed by Eqs. 13.18, 13.19, 13.20 and 13.21. In the L-type market, the equilibrium is determined by the number of potential suppliers, a marginal cost parameter (β_L), parameters for consumer demand (γ_L, δ_L), and reliability (θ) of water quality regulation. For an H-type market, the effect of testing detection probability (λ) is added to these parameters.

$$P_L = \frac{2\beta_L \gamma_L}{2\beta_L + \theta n_L \delta_L} \tag{13.18}$$

$$Q_L = \frac{\theta n_L \gamma_L}{2\beta_L + \theta n_L \delta_L} \tag{13.19}$$

$$P_H = \frac{2N_H \beta_L \beta_H \gamma_H}{(1-\theta)^2(1-\lambda)n_L^2 \beta_H \delta_H + N_H^2 \beta_L \delta_H + N_H[(1-\theta)n_L \beta_L \delta_H + \beta_H(2\beta_L + (1-\theta)(1-\lambda)n_L \delta_H)]} \tag{13.20}$$

$$Q_H = \frac{N_H \gamma_H[(1-\theta)(1-\lambda)n_L \beta_H + N_H \beta_L]}{(1-\theta)^2(1-\lambda)n_L^2 \beta_H \delta_H + N_H^2 \beta_L \delta_H + N_H[(1-\theta)n_L \beta_L \delta_H + \beta_H(2\beta_L + (1-\theta)(1-\lambda)n_L \delta_H)]} \tag{13.21}$$

13.4.3.2 Change in Reliability θ

Let us examine how the equilibrium price and quantity changes when reliability changes. When Eqs. 13.18, 13.19, 13.20 and 13.21 are differentiated by reliability θ, Eqs. 13.22, 13.23, 13.24 and 13.25 are obtained.

$$\frac{\partial P_L}{\partial \theta} = -\frac{2n_L \beta_L \gamma_L \delta_L}{(2\beta_L + \theta n_L \delta_L)^2} < 0 \tag{13.22}$$

$$\frac{\partial Q_L}{\partial \theta} = \frac{2n_L \beta_L \gamma_L}{(2\beta_L + \theta n_L \delta_L)^2} > 0 \tag{13.23}$$

$$\frac{\partial P_H}{\partial \theta} = -\frac{[2n_L N_H \beta_H \{-2(\theta-1)(\lambda-1)n_L \beta_H + N_H((\lambda-1)\beta_H - \beta_L)\}\beta_L \gamma_H \delta_H]}{\left[-(\theta-1)^2(\lambda-1)n_L^2 \beta_H \delta_H + N_H^2 \beta_L \delta_H + N_H\{-(\theta-1)n_L \beta_L \delta_H + \beta_H(2\beta_L + (\theta-1)(\lambda-1)n_L \delta_H)\}\right]^2} > 0 \tag{13.24}$$

$$\frac{\partial Q_H}{\partial \theta} = \frac{\left[n_L N_H \gamma_H \left\{(\theta-1)^2(\lambda-1)^2 n_L^2 \beta_H^2 \delta_H + N_H^2 \beta_L^2 \delta_H + 2(\lambda-1)N_H \beta_H \beta_L(\beta_H + (\theta-1)n_L \delta_H)\right\}\right]}{\left[-(\theta-1)^2(\lambda-1)n_L^2 \beta_H \delta_H + N_H^2 \beta_L \delta_H + N_H\{-(\theta-1)n_L \beta_L \delta_H + \beta_H(2\beta_L + (\theta-1)(\lambda-1)n_L \delta_H)\}\right]^2} \tag{13.25}$$

When reliability increases, the number of L-type companies remaining in the L-type market increase. In addition, because the expected value of the supply volume that the L-type region faces increases, the price in the L-type market declines and the supply volume increases. On the other hand, in the H-type region, when reliability increases, the price increases, but it does not determine an increase or decrease in the supply volume. When the number of L-type companies participating in the H-type market decreases, the price rises because the expected value of the supply volume that the H-type region faces shrinks. The supply volume trend relies on the size of the parameters β_H and δ_H, which determine the inclination of the supply and demand curve.

Perfect Substitute Case (θ = 1)

When θ = 1, that is, when the water standard becomes a perfect substitute of information for water quality, the equilibrium of the two markets is as shown by Eqs. 13.26, 13.27, 13.28 and 13.29. Factors determining the price and production volume for equilibrium are understood by the following parameters: number of potential suppliers, marginal cost, and marginal benefit.

$$P_L = \frac{2\beta_L \gamma L}{2\beta_L + n_L \delta_L} \tag{13.26}$$

$$Q_L = \frac{n_L \gamma_L}{2\beta_L + n_L \delta_L} \tag{13.27}$$

$$P_H = \frac{2\beta_H \gamma_H}{2\beta_H + N_H \delta_H} \tag{13.28}$$

$$Q_H = \frac{N_H \gamma_H}{2\beta_H + N_H \delta_H} \tag{13.29}$$

No Substitute Case (θ = 0)

When θ = 0, that is, when the water quality standard cannot be trusted, the equilibrium is as shown by Eqs. 13.30, 13.31, 13.32 and 13.33. This is a state where penalty ρ is too low for the risk-aversion level. While the L-type companies supplying in the L-type market no longer exist, the possibility of L-type companies participating in the H-type market is at a maximum. On the other hand, for the H-type market, in addition to the parameter in the case of a perfect substitute, the equilibrium is decided by test detection probability λ.

$$P_L = \gamma_L \tag{13.30}$$

$$Q_L = 0 \tag{13.31}$$

$$P_H = \frac{2N_H \beta_L \beta_H \gamma_H}{(1-\lambda)n_L^2 \beta_H \delta_H + N_H^2 \beta_L \delta_H + N_h \left[n_L \beta_L \delta_H + \beta_H \left(2\beta_L + (1-\lambda) n_L \delta_H\right]\right.} \tag{13.32}$$

$$Q_H = \frac{N_H \gamma_H [(1-\lambda) n_L \beta_H + N_H \beta_L]}{(1-\lambda)n_L^2 \beta_H \delta_H + N_H^2 \beta_L \delta_H + N_H [n_L \beta_L \delta_H + \beta_H (2\beta_L + (1-\lambda) n_L \delta_H)]} \tag{13.33}$$

The equilibrium for the L-type and H-type regions with changing θ is shown in Fig. 13.2.

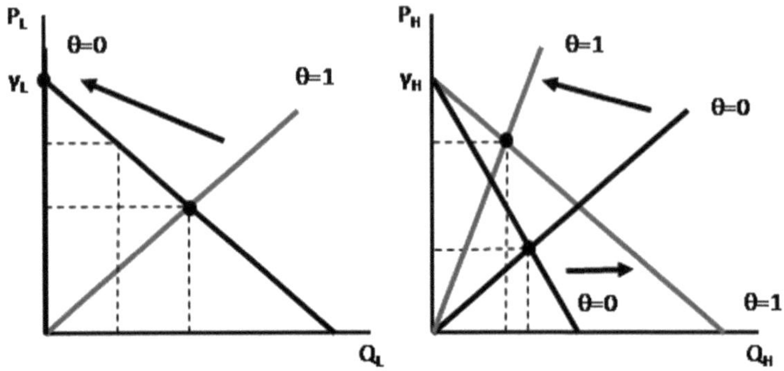

Fig. 13.2 Comparative statics of changing θ

13.4.3.3 Changes to Testing Detection Probability λ

Next, we examine how the equilibrium price and volume change for the testing detection probability. Because the effect of detection probability λ exists only in the H-type regions, Eqs. 13.26 and 13.27 are obtained by differentiating Eqs. 13.20 and 13.21.

$$\frac{\partial P_H}{\partial \lambda} = -\frac{\left[2(\theta-1)n_L N_H\{-(\theta-1)n_L + N_H\}\beta_H^2\beta_L\gamma_H\delta_H\right]}{\left[-(\theta-1)^2(\lambda-1)n_L^2\beta_H\delta_H + N_H^2\beta_L\delta_H + N_H\{-(\theta-1)n_L\beta_L\delta_H + \beta_H(2\beta_L + (\theta-1)(\lambda-1)n_L\delta_H)\}\right]^2} > 0 \tag{13.34}$$

$$\frac{\partial Q_H}{\partial \lambda} = \frac{\left[2(\theta-1)n_L N_H^2\beta_H^2\beta_L\gamma_H\right]}{\left[-(\theta-1)^2(\lambda-1)n_L^2\beta_H\delta_H + N_H^2\beta_L\delta_H + N_H\{-(\theta-1)n_L\beta_L\delta_H + \beta_H(2\beta_L + (\theta-1)(\lambda-1)n_L\delta_H)\}\right]^2} < 0 \tag{13.35}$$

Because the number of L-type companies that have the potential to participate in the H-type market decreases as the testing performance increases, the price increases because the expected value of the supply volume that the H-type region faces shrinks. The supply volume decreases, because unlike the reliability, the demand curve does not change. The equilibrium for the H-type region when λ changes is shown in Fig. 13.3.

13.4.4 Discussion

With respect to preference for water quality when consumers can choose the water utility under quality uncertainty, we examine how the price and volume of a market equilibrium is affected by the reliability of water quality regulation by assuming credence goods (where regulation is the substitute signal for water quality).

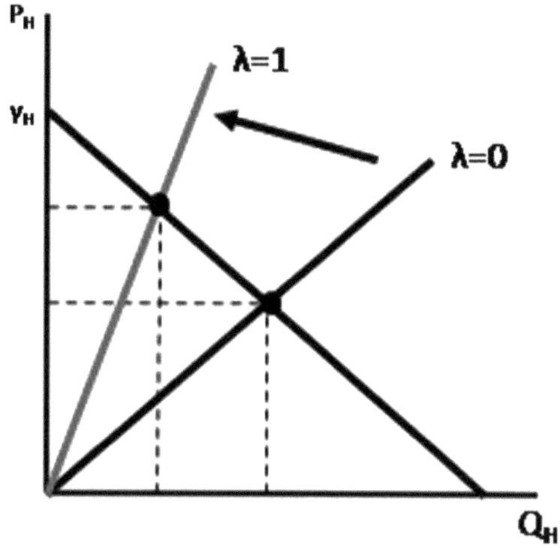

Fig. 13.3 Comparative statics of changing λ

In the L-type region (L-type market), in which the technology required for supply is low, the equilibrium price and volume are determined in relationship with reliability, which is the degree to which regulation acts as a signal for water quality. In the H-type region (H-type market), in which technology required for supply is high, in addition to reliability, equilibrium price and volume rely on the testing performance (detection probability) to detect supply that does not meet the quality standard.

Under circumstances where regulations are a perfect substitute signal for quality and there is certainty that water of a quality below the regulation standard is not supplied regardless of the state of the source water, a market is established in either region. However, in cases where regulation does not become the signal and there is the possibility that the actual water quality is below the standard value, supply in the L-type region is stopped.

The equilibrium for the L-type region is shown when increased reliability is accompanied by a decline in price. This is because when there is an increase in reliability, the expected value of the supply volume increases because the number of suppliers who intend to participate in the L-type region increases. On the other hand, in H-type regions, an increase in reliability is accompanied by an increase in price, and this is considered because the expected value of the supply volume decreases as a result in the decline of L-type companies wishing to participate in an H-type region.

Moreover, by establishing this model, the equilibrium in the L-type region does not exhibit the probability of detecting unmet regulations through testing.

This shows that the behaviour of the L-type company is not affected by testing performance and that even if the testing had a perfect detection probability of 1, the L-type company would have an incentive to participate in the H-type market based on their own risk-aversion level. However, as withdrawal from the market is certain if the detection probability is 1, the expected profit is zero (or a loss when a penalty is considered), and supposing a company possessing a low risk-aversion level that corresponds to this profile is unlikely. On the other hand, equilibrium in an H-type region is explicitly affected by detection probability, and like reliability, an increase in the detection probability is accompanied by a decrease in companies wishing to participate; thus, the price rises.

Because of the aforementioned result, how regulations become a signal for the consumer to judge water quality is shown to depend on whether an L-type company will supply to an H-type region and to what extent the supplying company is excluded through testing when it participates in this H-type region. By establishing a model, the first factor concerning decision-making for participation is determined by the penalty at the time of detection and the company's own risk-aversion level. The second factor concerning the detection probability is determined by the testing performance.

Therefore, a valid means for realizing a situation where the water quality regulation can be a perfect substitute for quality is the establishment of a penalty level that considers a company's risk-aversion level and the guarantee of testing whether a company has the capability to comply with the water quality standards. In order to design such a supply system, it is necessary to have information relating to a company's risk-aversion level and technology. Penalties fulfil a critical role, because they are a factor controlling the incentive to participate in a market with technological capability higher than the existing capabilities even if the testing performance is not perfect.

13.5 Examination of Private Sector Management Methods

Public-Private Partnership has been promoted in water utility business in Japan: business consignments, Private Finance Initiative (PFI), a third-party consignment system, a designated administrator system, and local incorporated administrative agencies. Table 13.1 shows the categories of PPP and the number of cases introduced in water utility business (Ministry of Health, Labour and Welfare 2014). We review earlier eight cases based on a survey conducted by Japan Water Works Association (2006) and organized in Asahi and Hagihara (2008). The cases consists of one PFI case, six third-party consignment system cases, and one designated administrator system involving the introduction of maintenance and operation of

Table 13.1 PPP in water utility business as of 2012 April, based on Ministry of Health, Labour and Welfare (2014)

PPP categories	Introduced
Private Finance Initiative: PFI	11
Third-party consignment system	145
Concession	0
Consignment of operations	Almost all of water business utilities

water purification facilities related to the supply of quality water. The implications based on the analysis of the previous section are studied in those cases.

According to the analysis in the previous section, reliability θ related to water quality is an increasing function of risk-aversion level φ_i of the supply company and penalty ρ. By introducing private sector management methods, we can determine how these factors are considered in the outsourcer's process of deciding the outsourcee. Table 13.2 identifies the relevant items in the eight case surveys. Here, for descriptive purposes, risk-aversion level φ_i of the supply company is included in the survey item 'important points of consideration for selection of outsourcees', penalty ρ is included in the expected value of potential costs prescribed when the outsourcee is allocated risks such as nonfulfillment of obligations, business closure, or failure to meet performance standards by business operators in the survey item 'outsourcees risk allocation', and testing performance (detection probability) λ is included in the survey item 'confirmation of status of work fulfilment'.

Concerning the 'important points of consideration for selection of outsourcees', points concerning technology and performance are the most prevalent. However as seen in items such as business base, financial stability, and financial fundamentals (cases 2, 5, and 8), importance is placed on the stability of the management base, which has a direct link to the risk-aversion level of the outsourcee company. When using a bidding process, we can assume that fundamental consideration is given to risk-aversion level, because management matters are also scrutinized as part of the examination of qualifications.

However, the optimal risk-aversion level φ^* depends on the expected profit based on the analysis of the previous section (See Eq. 13.3). In other words, it is decided by the level of penalty ρ, which is the potential cost. When reviewing the prescription of risk allocation, in most cases (except cases 2 and 3), increases in costs arising from noncompliance with necessary standards are included in the outsourcee's allocation.

As this is reflected in the expected profit, it contributes to the establishment of a risk-aversion level that generates an incentive to participate in the market that is suited to the supply capability. However, because penalty ρ strictly means an imposed penalty for meeting non-conformance to the required level, there must be a framework where the penalty related to nonfulfillment of contract is explicitly considered at the time of participation in the market.

Concerning the methods for confirming the status of nonfulfillment of commissioned work with relation to testing performance λ, most cases required a work design document, completion report, and submission of daily and monthly reports as a regular check. In the PFI in case 1, a check system conducted by third-party monitoring is shown. This validity affects the incentive for an L-type company to join an H-type market, and therefore quantitative analysis on the status of fulfilment and check systems are required.

Table 13.2 Case studies of private sector management system, based on Asahi and Hagihara (2008)

No.	Outsourcer	Project	Method	Outsourcees	Important points of consideration for selection of outsourcees Risk-aversion level□ψ	Outsourcees risk allocation Penalty ρ	Confirmation of status of work fulfillment Testing performance λ
1	Matsuyama City Local Public Enterprise	The construction of filtering treatment facility at Kakitsubata purification plant	PFI(DBO)	Matsuyama Safety Water Corporation	Items of Technical investigation of the proposal: Sufficient results of design, maintenance and management of water facilities Construction Plan, Maintenance and management Plan, and Risk management of the membrane filtering facility	Plan change risk Neighbouring complaints risk Law/Tax system change risk Permission delay risk Price risk before operation Accident risk Performance risk Facility renewal risk Default risk of operator	Monitoring of maintenance and management operation by a third party

2	Hobetsu Town Simple Water-supply System	Outsourcing of Maintenance and Management of Water Facilities in Hogetsu Town	Outsourcing to a third party	H limited Company	an Ex-official of Hobetsu Town in the outsourcee having been set up a collaborative framework in emergency	Seeing 5crisis levels of water quality, an operator (agent) incurs lower level 1–2.	Implementation report of Annual/Monthly business plan Daily report Report on performance of water quality inspection Records on chemical take in/out Records on failures and accidents Plug opening/closing list
3	Ohta City Waterworks Bureau	Outsourcing of Maintenance and Management of Purification Plant in Ohta City	Outsourcing to a third party	Meidensha Corporation	Technological capacity in proposal	Contract risk Financial risk Labour management risk Political/Economic risk	Monthly meeting Check list for active plan of annual business plan Performance index of facilities criteria Production cost criteria Performance Index of JWWA Q 100

(continued)

Table 13.2 (continued)

No.	Outsourcer	Project	Method	Outsourcees	Important points of consideration for selection of outsourcees Risk-aversion level□ψ	Outsourcees risk allocation Penalty ρ	Confirmation of status of work fulfillment Testing performance λ
4	Minami-Ashigara City Water and Sewerage Division	Outsourcing of Maintenance and Management of Water Facilities in Minami-Ashigara City	Outsourcing to a third party	Toshiba South-Metropolitan Area Branch	delivery results of facilities to the Ohta City Technological capacity Qualified management technician	Human error risk (e.g. inappropriate operation)	Business implementation plan Daily report on maintenance of maintenance Annual/monthly report Accident and warning report
5	Miyoshi City Waterworks Bureau	Outsourcing of Maintenance and Management of Purification Plant in Miyoshi City	Outsourcing to a third party	Japan Water Corporation	Sufficient results of operation and management Ability of quick response Ability of efficient operation and management Sufficient financial basis for business continuity	N.A.	Inspection by completion reports of annual/monthly plan Monthly meeting

6	Tabuse-Hirao Water Supply Authority	Outsourcing of Maintenance and Management of Purification Plant	Outsourcing to a third party	Super Water Corporation	Technical capacity of operation and management Ability of preventing accidents Ability of security of facilities	Contract risk Liability risk for damages to third party Neighbouring complaints risk Accident risk Production and operating cost increasing risk Environmental protection risk Price risk	Annual/monthly report Report operation matters at morning meeting
7	Satsuma-Sendai City Waterworks Bureau	Outsourcing of Operation and Management of Maruyama Purification Plant	Outsourcing to a third party	Tsukishima Technology Maintenance Service Corporation	Sufficient results of water facilities as an outsourcee Nearby facilities in emergency Technological capacity Technological proposal from the regional viewpoint	Accident risk	Daily operation report Daily check list report
8	Takayama City Waterworks Department	Outsourcing of Management of Facilities in Takayama City Waterworks, Iwataki Simple Water-supply system, and Takayama Simple Water-supply system	Designated Manager System	Takayama-City Tube Equipment Group	Technological capacity in proposal Sufficient financial basis Well informed about local water	Price/interest rate risk Operational cost risk Facility damage risk Quality non-fulfillment risk Liability risk for damages to third party	Daily/Monthly report Annual business report Specific reports or oral explanations if required

13.6 Conclusion

Considering private sector participation in the water utility business, this study conducted three main investigations.

First, we arranged the points of view considered to determine the efficiency of resource allocation when assuming the introduction of a market mechanism for a water supply utility system. For conditions established by the fundamental theorems of welfare economics, on the water supply utility side, we established that the convex environment collapses because of the indivisibility of production elements and sunk fixed costs, preventing the realization of efficient resource allocation. On the demand side, the collapse of the convex environment relating to preferences owing to the uncertainty of the water quality showed the possibility that efficient resource allocation cannot be guaranteed.

Second, in cases of uncertainty in water quality, we assumed an analysis framework using the concept of credence goods and showed the merits of this analysis through simple review. Because it is highly likely that the assumption of a convex environment is not met, choices under conditions of uncertainty require an interpretation with reservations for the implications on welfare for analysis performed by alleviating these prerequisites for descriptive purposes. The analysis framework of credence goods enabled a definition of preferences not as certain preferences for quality of uncertain goods but as those for goods and a status that is expressed as a substitute for evaluating quality. Therefore, while close to preferences and choices under actual uncertainty, this framework is able to explicitly handle a system that becomes a substitute.

Third, we regarded the water quality of the virtual utility as a credence good and performed analysis for when the market mechanism is introduced in a shape that allows private business to participate in receiving orders in a regional monopoly. The preference for water quality was investigated to determine the type of effect generated on the price and volume of market equilibrium using the reliability of regulation of water control based on credence goods with respect to water quality. To realize an ideal system for the consumer where regulation becomes a signal for judging water quality, it was shown that it is effective to establish a level for penalties that considers the company's risk-aversion level and improves the performance of testing in relation to its capability to comply with the water standard. Furthermore, such action requires information relating to the company's risk-aversion level and technology.

For future research based on the analysis of this chapter, the following subjects can be considered. First, in terms of models, we suppose a natural monopoly of supply using the subadditivity of cost. However, economies of scale are not necessarily applicable to all water utility businesses, and many cases of diminishing returns in relation to scale have been shown for municipally owned businesses in particular (Nakayama 2003). Therefore, it is necessary to perform analysis according to economies of scale and the possibility of price formation. We also obtained the implication that the reliability of regulation for water quality depends on the company's risk-aversion level, penalties, and the detection probability. However, this has been given exogenously and independently. It is hoped that these correlations will be explicitly included in a model. Furthermore, concerning private participation in the water utility business, considering the current situation of investigations into the suitability of specific systems (e.g. consignment of operations, PFIs, third-party consignment systems, and concessions), reviewing cases where these systems have been introduced to quantitatively investigate demand-side implications with respect to these supply systems (i.e. uncertainty of quality), is likely useful.

References

Anania, G., & Nistico, R. (2004). Public regulation as a substitute for trust in quality food markets: What if the trust substitute cannot be fully trusted? *Journal of Institutional and Theoretical Economics, 160*(4), 681–701.

Asahi, C., & Hagihara, K. (2000). A study on evaluating environmental risks in drinking water. *Regional Studies, 30*(1), 167–181 (in Japanese).

Asahi, C., & Hagihara, K. (2005). Ex post risk management of environmental contamination on municipal water. *Regional Studies, 34*(3), 275–287 (in Japanese).

Asahi, C., & Hagihara, K. (2008). A provision system for municipal water with uncertainties. *Studies in Regional Science, 38*(2), 415–434 (in Japanese).

Barlow, M., & Clark, T. (2002). *BLUE GOLD*, Toronto: Stoddart Publishing.

Blankart, C. B. (1987). Limits to privatization. *European Economic Review, 31*, 346–351.

Darby, M., & Karni, E. (1973). Free competition and the optimal amount of fraud. *Journal of Law and Economics, 16*, 67–88.

Dulleck, U., & Kerschbamer, R. (2006). On doctors, mechanics, and computer specialists: The economics of credence goods. *Journal of Economic Literature, 44*(1), 5–42.

Hagihara, K., Asahi, C., & Hagihara, Y. (2004). Marginal willingness to pay for public investment under urban environmental risk: The case of municipal water use. *Environment and Planning C: Government and Policy, 22*(3), 349–362.

Hagihara, K., Asahi, C., & Hagihara, Y. (2005). A study on system design for drinking water supply under risk. Symposium of planning for water and green environment, Kyoto.

Ihori, T. (1996). *Theory of public economics*. Tokyo: Yuhikaku (in Japanese).

Japan Water Works Association. (2006). Introduction of private sector management methods in waterworks (in Japanese).

Lupton, S. (2005). Shared quality uncertainty and the introduction of indeterminate goods. *Cambridge Journal of Economics, 29*(3), 399–421.

Marette, S., Bureau, J. C., & Gozlan, E. (2000). Product safety provision and consumers' information. *Australian Economic Papers, 39*(4), 426–441.

Ministry of Health, Labour and Welfare. (2014). PPP/PFI in water utility business (in Japanese).

Nakayama, T. (2003). *Efficiency analysis of waterworks in Japan*, Tokyo: Taga Shuppan (in Japanese)

Nelson, P. (1970). Information and consumer behavior. *Journal of Political Economy, 78*, 311–329.

Okuno, M., & Suzumura, K. (1988). *Micro economics*. Tokyo: Iwanami Shoten (in Japanese).

Roemer, J. E. (1996). Theories of distributive justice. Harvard University Press (trans: Kitani, S., & Kawamoto, T. in 2001).

Chapter 14
Community Level Planning for Arsenic Contaminated Drinking Water in Bangladesh

Sho Shibata, Kiyoko Hagihara, Yoshimi Hagihara, and Akira Sakai

Abstract In Bangladesh, arsenic contaminated drinking water has become a serious problem. Since the revelation of arsenic contamination, various water supply options have been installed to supply drinking water. However, some of them have already been abandoned. These options are not acceptable for local residents without consideration of their social environments and support after introduction. Purpose of this chapter aims to plan acceptable processes to install water supply options for the disaster of arsenic contaminated drinking water considering local social environment in Bangladesh.

Keywords Arsenic contaminated • Water supply options • Acceptability • Stress • Multi-criteria

14.1 Introduction

Recently, arsenic contamination of drinking water has become a serious problem in rural areas of Bangladesh. Arsenic causes skin disease, cancers, and in the worst case, death. Bangladesh has suffered many kinds of disasters such as cyclones, floods, droughts, and damage caused by salt. Therefore, Bangladesh is considered a fragile area for arsenic problems. In 1993, arsenic contamination was

S. Shibata
International Operations, NEWJEC Inc., 2-3-20 Honjo-Higashi, Kita-ku
Osaka 531-0074, Japan
e-mail: shibatasu@newjec.co.jp

K. Hagihara (✉)
School of Sociology, Bukkyo University, 96 Kita-hananobo, Murasakino, Kita-ku
Kyoto 603-8301, Japan
e-mail: hagihara@bukkyo-u.ac.jp

Y. Hagihara
Kyoto University, 264 Suemarucho, Nakagyou-ku, Kyoto 604-0901, Japan
e-mail: y-hagi@gf7.so-net.ne.jp

A. Sakai
University of Marketing and Distribution Sciences, Kobe 651-2188, Japan
e-mail: akira_sakai@red.umds.ac.jp

K. Hagihara, C. Asahi (eds.), *Coping with Regional Vulnerability*, New Frontiers in Regional Science: Asian Perspectives 4, DOI 10.1007/978-4-431-55169-0_14

first found in the ground water from shallow tube-wells (STWs) in Bangladesh. Arsenic affection covers 70 % areas of Bangladesh and 30 % of STWs used as drinking water are exceeded the standard of arsenic concentration in Bangladesh. Since the revelation of arsenic contamination, various water supply options have been installed to supply drinking water. However some of them have already been abandoned because of a lack of knowledge of proper use, difficulties of maintenance, bad taste of water, and stress of collecting water. These options are not acceptable for local residents without the consideration of their social environments and support after introduction. However, it is difficult for local governmental organizations and NGOs to undertake activities with continuous supports after the installation from the middle and long term viewpoint because their activities are usually based on the project in several years funded by donors.

Arsenic contamination of drinking water should be considered as the disaster which is closely correlated between drinking water use and social environment. Safe drinking water is essential for human and water use is a part of daily life. Therefore it is important for installation of acceptable water supply options for local residents to recognize villages as a system and obtain their livelihood, not only to overview villages averagely based on macro statistical data (Sakamoto et al. 2007).

Although there are a lot of efforts and activities to mitigate this disaster, it is the important fact that there still remain local residents who suffer the arsenic contamination disaster. This chapter is to indicate the planning process for installation of water supply options based on viewpoints of local residents (Fukushima 2006).

Purpose of this study is classified into following two issues (Hagihara et al. 2008);

1. to obtain the local social environment and drinking water use situation and clarify the direction on planning process for installation of water supply options
2. to show the adaptive planning process which are selected by the technology assessment

14.2 Social Survey

14.2.1 Implementation of Social Survey

Social survey by interviews with local residents using the questionnaire was implemented in August and September in 2007 to obtain the social environment and conditions of drinking water use with the viewpoint to recognize the village as a social system. Questionnaire for the interview is prepared based on KJ method and ISM (Interpretive Structural Modeling) method. The questionnaire is composed of four major items of ‘Personal data’, ‘Drinking water’, ‘Sanitation’ and ‘Daily livelihood’.

Target village of the social survey was decided considering the situation of arsenic contamination and survey activities. In this study Basailbogh village in Srinagar upazila, Munshiganji district was selected. Figure 14.1 shows areas

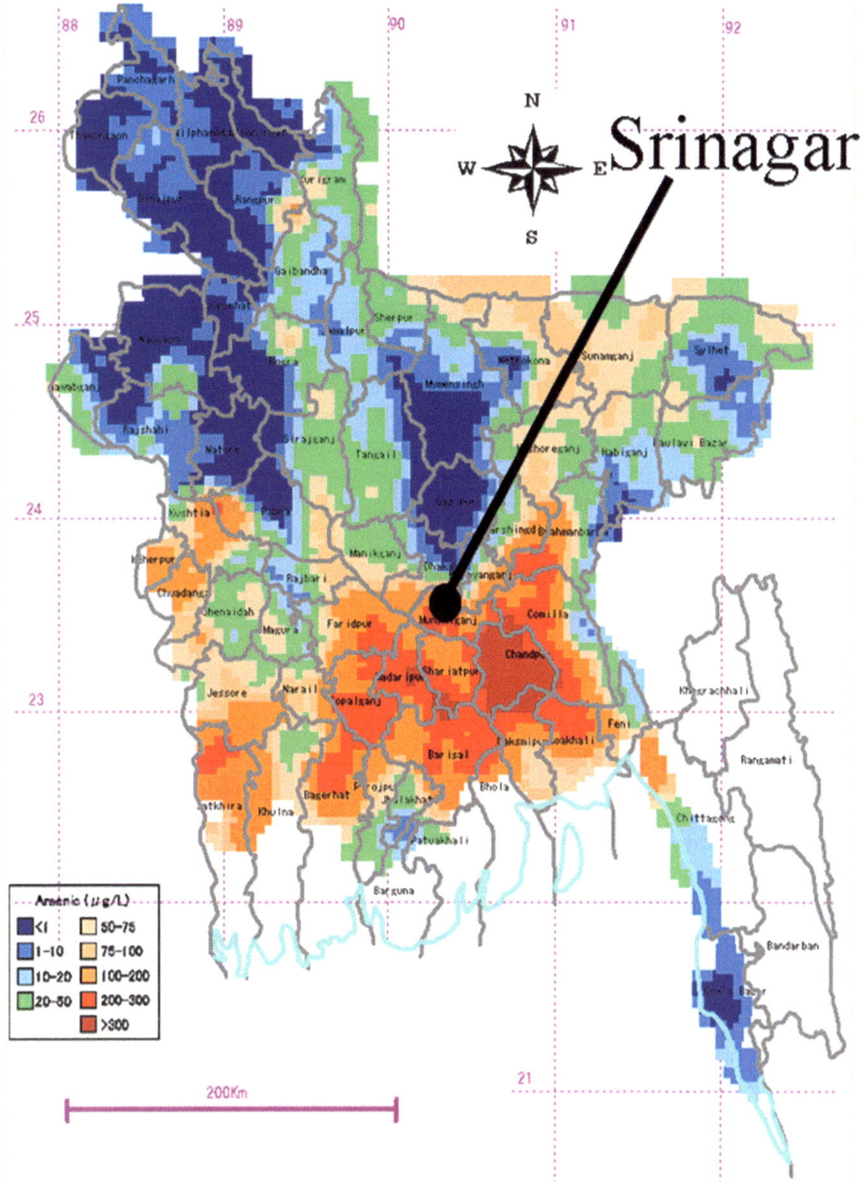

Fig. 14.1 Areas affected by arsenic and location of Srinagar (Based on Hossian 1996)

affected by arsenic in Bangladesh and location of Srinagar. Interviews with local residents were carried out with corporation of NGOs, JADE (Japan Association of Drainage and Environment) and SPACE (Society for People's Action in Change and Equity) and answers from 118 out of 123 families were obtained from the village.

14.2.2 Current Social Situation and Conditions of Drinking Water Use

Basailbogh is located approximately one hour from Dhaka by car. In the rainy season, the village is flooded and divided into some islands. Boats and bamboo bridges are used for the transportation. There are small community which is called "Bari" composed of several related families. In Basailbogh village, there were 21 Baris in 2007 (Table 14.1).

As for drinking water use conditions, almost 100 % of STWs were arsenic contaminated in the village and the main safe water options were deep tube-wells (DTWs). In December 2007, There were water supply options of three public DTWs, four private DTWs, and two private family size arsenic iron removal plants (AIRP) in the village. Although all families can use public options such as DTWs, some owners of private options prohibited other families to use them in the village.

Thirty families, 25 % of the total still had used arsenic contaminated STWs. Compared with other neighboring villages, situation of the drinking water use could be considered as developed in Basailbogh village. However, the fact that 25 % families still drink arsenic contaminated STWs in this developed village indicates the existence of residents left from the safe drinking water use development and it should not be avoided.

14.2.3 Cramer's Coefficient Analysis

For multi-variate analysis, selection of analysis items from the questionnaire is an important issue. In this chapter relation between items is clarified by Cramer's coefficient analysis. Cramer's coefficient is the value which indicates correlation between two categorical variables as shown below.

$$\text{Cramer's Coefficient} \quad \varphi = \left[\chi^2/\{N(k-1)\}\right]^{1/2}$$

then χ^2 is Chi-square Value, N is Number of Samples and k is Number of Category.

Based on the 5 % significance, coefficient over 0.25 is regarded as 'Related' considering number of categories. Related items are classified and summarized. Finally represented items which express features of social environment and conditions of drinking water use in the village are selected for multi-variate analysis in next step.

Table 14.1 Features of Bari and water source in Basailbogh village in 2007

Bari	Number	Water source	Description
Kerani	10	AIRP	As there are some migrant workers, economic condition of this Bari is better than others. Although two private AIRP had been installed in March of 2007, they were already abandoned in 2008 and a new private DTW was installed.
Chairman	4	DTW near Mosque	As former chairman of Union committee lives, this Bari is called as "Chairman Bari".
Bhunnya	3	DTW near Mosque	Economic condition is quite good and their houses are made of concrete, not zinc roof. Families have their business and educational level is also higher.
New Dhali	5	STW	As some families have worked on abroad, some houses are vacant. There is an big house with 3rd floors, but owner of the house lives in Dhaka. STWs are used for drinking water due to the long distance to DTWs.
Bhandari	14	Private DTW STW	This Bari can be divided into two groups, rich families at the northern part and poor families at the southern part. Most of poor families emigrated from outside and constructed houses at vacant area. An rich family installed a DTW by his own expense, he allowed only his relatives and some families to use it.
Baburshi	16	Private DTW	A DTW at Sarang Bari is used. Relation with Sarang Bari was bad due to the usage of boats and bamboo bridges. There are difference of economic conditions in Bari.
Sarang	10	Private DTW	There is an private DTW in Bari. Although families in this Bari answered that this DTW is private one, it was installed by the American NGO and was not limited to use.
Gazi	5	Roadside DTW	It is located at the cross road of the main road and road to southern west part and is flooded in the rainy season. Some families have small shops. As a family closes the road connected to the main road, other families must use a bamboo bridge.
Began	2	Roadside DTW	It is located next to the roadside DTW. Although the main road is a lot of traffic, this Bari is surrounded by the block wall. It was confirmed that women often enjoyed the conversation inside the Bari.
Member	3	Roadside DTW	It is flooded in the rainy season. A storage house was damaged by NARGIS. Economic conditions are not good.
Bhuiya	3	Roadside DTW	As a woman of committee member lives, motivation for the safe water and sanitation is high. They answered that they hope to install their own DTW in Bari.
Pokir	4	Roadside DTW	It is partially flooded in the rainy season. Economic conditions are bad and there are unmarried adult women. Although they do not have certain knowledge of arsenic problems, roadside DTW is used.

(continued)

Table 14.1 (continued)

Bari	Number	Water source	Description
Munshi	5	STW	It is located at the southern east part of the village. Economic conditions are bad. Although they know the arsenic contamination in STWs, they have no other choice for drinking water due to the long distance to DTWs.
Mosque	2	DTW near school	There is the Mosque. There are usually men enjoy conversation.
Mia	5	DTW near school	They have wide land and houses are relatively wider than other Baris. Some relatives have the business of construction works.
Member house	1	STW	It is totally flooded in the rainy season and a boat is used for the transpiration. A man of committee member lives and has the agriculture. DTW near school is not so far, but a STW is used for drinking water.
Ripon Miar	7	DTW near school STW	It is totally flooded in the rainy season and boats are used for the transpiration. An rich family installed a DTW by his own expense, he allowed only his relatives to use it.
Khaleque Miar	3	STW	It is totally flooded in the rainy season and boats are used for the transpiration. As the DTW near school is far and water taste is bad, they don't want to use.
Sarjon Alir	6	DTW near school STW	It is totally flooded in the rainy season and boats are used for the transpiration. Two (2) families use the DTW near school and others use STWs. Families who use the DTW have higher interests in arsenic problems.
Bhuiya2	5	STW	It is totally flooded in the rainy season and boats are used for the transpiration. DTW near school is not so far, but STWs are used for drinking water. Relation between families is not good and there are some problems in boats use and breeding of ducks.
Sheikh	4	DTW near school Roadside DTW	It is totally flooded in the rainy season and boats are used for the transpiration. Although an family installed a DTW by his own expense, he does not allow other families to use. These families are relatives, however there are no communication due to the land use dispute.

14.3 Direction on Planning Process of Water Supply Options

Direction on planning process of water supply options is prepared by using Factor Analysis and Covariance Structure Analysis with represented items obtained in the interview survey. Firstly, common factors are abstracted from these items by Factor Analysis. These factors are expected to be summarized information of local situation. Secondly, relations between these factors are structuralized by Covariance Structure Analysis.

14.3.1 Factor Analysis

Factor Analysis was implemented against nine represented items which express the local situation of Basailbogh village. Samples with lack of answers are removed from the analysis. Factor abstraction is Maximum-Likelihood Estimation Method. Orthogonal rotation of factor axis is Standard Varimax Rotation Method. Number of factors are judged from Scree Plot Diagram. RMSEA and p value are used for testing of the model. Results of Factor Analysis are summarized in Table 14.2.

As results, Common Factors of "Stress of Collecting Water", "Knowledge", "Interests in Arsenic Problems" and "Difficulties in Economy" were obtained. Reviewing these Common Factors based on current situation of the village, families whose house are located far from DTWs use STWs due to the burden of collecting water stress. If they don't have the knowledge of water supply options, DTWs are not used. Even if their houses are located near DTWs, they use STWs due to the lack of interests in arsenic contamination problems. In addition, families with difficulties in economy don't have the affordability to consider safe drinking water options.

Regarding "Knowledge" and "Interests in Arsenic Problems", activities to improve these issues are often dealt with simultaneously in actual. Therefore "Interests in Arsenic Problems" was applied for the direction because this factor is composed of questionnaire items which shows use of safe drinking water and expected to be more important from the viewpoint of motivation improvement of STWs users.

14.3.2 Covariance Structure Analysis

Latent Variables of [Stress of Collecting Water], [Interests in Arsenic Problems] and [Difficulties in Economy] are prepared for Covariance Structure Analysis to

Table 14.2 Summary of factor analysis

Outline	Factor name (contribution ratio %)		Composed items (factor loading)
Basailbogh (118 samples) p=0.981 RMSEA=0 Accumulated factor loading = 45.7%	1	Stress of collecting water (16.7%)	Time of collecting water (0.991)
			Physical stress of collecting water (0.594)
	2	Knowledge (12.3%)	Knowledge of options (0.999)
	3	Interest in arsenic problems (8.9%)	Interest in health (0.607)
			Safety of drinking water (0.522)
	4	Difficulties in economy (7.8%)	Interest in social issues (0.712)

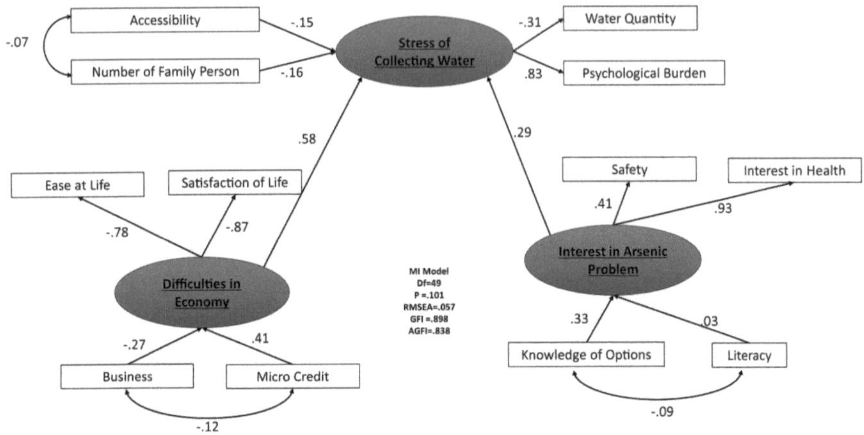

Fig. 14.2 Summary of covariance structure analysis

clarify the structure and relation in social environment and conditions of drinking water uses in Basailbogh village.

In this chapter, firstly Latent Variables are organized by multiple indicator multiple cause (MIMIC) model which consists of two kind of observation variables, exogenous variable and endogenous variable. Then, relation between Latent Variables are structuralized by Multiple Indicator model. p value, RMSEA, GFI and AGFI are used for testing of the model.

Finally, Multiple Indicator model with three Latent Variables are obtained as shown in Fig. 14.2.

Regarding coefficients between Latent Variables, coefficients with [Stress of Collecting Water] show positive values of 0.29 with [Interests in Arsenic Problems] and 0.58 with [Difficulties in Economy] respectively. It is explained that improvement of interests in arsenic problems leads families who use STWs to obtain appropriate knowledge of problems and to accept stress of collecting water. Families who have difficulties in economy are usually busy for their daily livelihood and feel more stress. Although it is difficult to improve their poverty directly by installation of water supply options, it is necessary to focus poor residents and to consider these situation in the planning process.

As a conclusion, reduction of [Stress of Collecting Water] are expected to be the acceptable direction on planning process for installation of water supply options.

14.4 Modeling and Evaluation of Stress of Collecting Water

Previous studies clarified that collecting water is a big burden for local residents and reduction of this is acceptable processes for them. In this section, the burden of collecting water is modeled and evaluated. Stress of collecting water is defined as two types, physical and psychological. In Bangladesh, collecting water is mainly shouldered by women. They carry 18–20 l(liters) water at once by using a chamber. They do not like to be exposed to men's eye except their relatives for religious reasons. Physical stress is defined as the amount of workload. Psychological stress is defined as time to collect water (time to be exposed to someone's eye). As mentioned above, land surface is flooded in Basailbogh village in the rainy season and routes of water collecting is changed by seasons. Although they can go anywhere through the road in the dry season, some residents must use boats or bamboo bridges to move in the rainy season. Stress of collecting water is modeled in both seasons and measured for all families in the village.

14.4.1 Modelling of Stress of Collecting Water

It is thought that the feeling of stress is changed by location in the village. For example, bamboo bridges and slopes are unstable. Considering the characteristic of traffic and movement, Basailbogh village is divided into eight sections. Section 0 is their own bari. Section 1 is main road with a lot of traffic. Section 2 is main road. Section 3 is other road. Section 4 is slope. Section 5 is boat area. Section 6 bamboo bridges. Section 7 is mosque area. Transportation distance in these sections are defined as $d_i(i = 0\cdots 7)$. To simplify measurement, all families are assumed to depart from the centre of each bari.

Physical Stress is modelled as follows. It shows workload of collecting water from their houses to available options.

$$\text{Physical stress} \quad \begin{aligned} P &= workload \\ &= n \times m \times \sum_{i=o}^{7} \alpha_i d_i \end{aligned}$$

n is number of family persons, m is weight of drinking water in a day per person, α_i is coefficient of physical stress in section i.

Psychological stress is modelled as follows. It shows time of collecting water from their houses to available options. In other words, it shows time to be exposed to men's eye.

$$\text{Psychological stress} \quad \begin{aligned} M &= time \\ &= \sum_{i=o}^{7} {}^{\beta_i d_i}\!/_{v_i} + T\beta_w + \sum_{i=o}^{7} {}^{\beta_i d_i}\!/_{\gamma_i v_i} \end{aligned}$$

β_i is coefficient of psychological stress in section i, v_i is velocity of walking in section i, T is time to pump water, β_w is coefficient of psychological stress in section w placed on options, γ_i is the change rate of velocity after filling water.

14.4.2 Estimation of Stress of Collecting Water

Figure 14.3 shows the measurement of water collecting stress and the map of Basailbogh village. Regarding families who use arsenic contaminated STWs, options which minimize stress in available options are measured. From the results of the measurements, all families who drink safe water choose available options in which stress is minimized.

14.4.3 Evaluation of Stress of Collecting Water

Public DTWs in Basailbogh village are installed at the accessible place such as the primary school and roadside. For the development of the water supply options, installation to reduce the stress for the village averagely can be considered in the initial phase. In fact, NGOs and local government generally aims to install options for the Community basis. However, there are some families who don't use these options installed for the Community and still use arsenic contaminated STWs.

Regarding the difference of the estimation between both types of stress, as answers of questionnaire survey show that physical stress is more burden than psychological in Basailbogh village, reduction of physical stress is focused in this study. However it is noted that this assumption can't be applied to other villages where gender problems are more severe and psychological stress is bigger burden

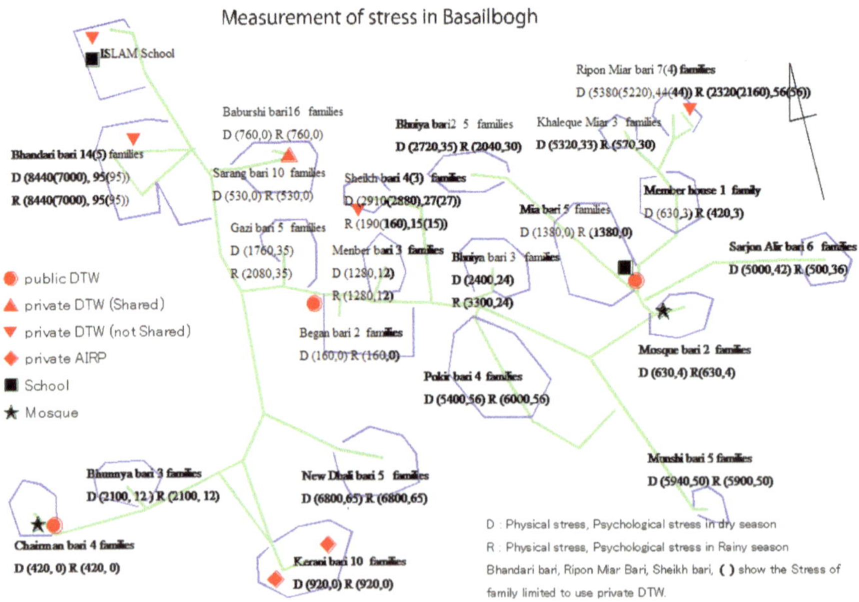

Fig. 14.3 Result of measurement of stress of collecting water in Basailbogh village

for women. It is important to select appropriate one based on actual situation of the target area.

14.5 Assessment of Water Supply Options by Multi Criteria Analysis

It had been confirmed in the site visits that there were many abandon water supply options due to lack of acceptability and support after installation. It is also important for installation of water supply options to select acceptable ones for local residents.

14.5.1 Specification and Feature of Water Supply Options

In Basailbogh village, DTWs, RWHs (Rain Water Harvesting), RWH for Baris, DW (Dug-Well), AIRPs, AIRP for Baris are available. PSF (Pond Sand Filter) is not available because there are few ponds and which is already used.

As water supply options have capacity of supply, assessments should focus on their capacity. Table 14.3 shows that options are classified according to the capacity of water supply. Although DTW and DW are community level options, they are considered as Bari level options in this analysis. Actually Bari introduced DTW by themselves in Bangladesh and they are very useful for local residents due to maintenance free and easiness to use.

14.5.2 Assessment of Water Supply Options by Multi Criteria Analysis

Multi Criteria Analysis are implemented to assess and to select acceptable water supply options for local residents. To select acceptable options for local residents, it is necessary to consider criteria from various viewpoints such as cost to install options, maintenance, etc. As performance of water supply options has merits and demerits respectively, assessment methodologies are preferred to be possible to evaluate both aspects. Therefore Concordance Analysis is applied which evaluate

Table 14.3 Supply capacity of options

Capacity	Options
1 family	RWH and AIRP
Bari (2–10)	DTW, RWH, RWH for Baris, DW, AIRP for Baris
Community (10–60)	DTW, RWH for Community, DW, AIRP for Community

relative preference between alternatives by Concordance degree and Discordance degree.

Multi-criteria are set for concordance analysis. (1) cost of install of the option, (2) maintenance, (3) cost of maintenance (year), (4) stability of water quantity, (5) sanitation, (6) taste of water are set for criteria. In the case of actual projects, the initial survey and after-support can be added. In this study degree of evaluation is set for one family, that is, it is important for local residents to imagine the actual water use situation. Considering the Basailbogh situation, capacity of "Some families" is set four (4) families and "Community" is set 15 families.

As concerned with outranking, the weights judgment is a difficult issue in concordance analysis. In this analysis, it is thought that each criteria is important for local residents and each weight is set as the same value. In the case of project side (1) is usually regarded as important. Results of Concordance analysis are shown Tables 14.4, 14.5, and 14.6.

By concordance analysis, RWH (Family), DTW (Bari), DTW (Community) are selected as acceptable water supply options for local residents. DTW is outranked at some families and community level. DTW advantage is maintenance free, stable water quantity through a year. Other options have disadvantages in these criteria. In a local observation, DTW are also used without problems in various areas.

Table 14.4 Result of concordance analysis (community)

Option	Concordance	Rank	Discordance	Rank	Integrated rank
DTW	0.714	1	−1.000	1	1
RWH	−0.571	3	0.143	2	2
DW	0.571	2	0.500	4	3
AIRP	**−0.714**	**4**	**0.357**	**3**	**4**

Table 14.5 Result of concordance analysis (Bari)

Option	Concordance	Rank	Discordance	Rank	Integrated rank
DTW	0.286	1	−0.576	1	1
RWH	0.143	2	0.008	4	3
RWH (for Bari)	−0.143	5	0.171	5	5
DW	0.000	4	1.394	6	5
AIRP	−0.429	6	−0.497	3	4
AIRP (for Bari)	**0.143**	**2**	**−0.500**	**2**	**2**

Table 14.6 Result of concordance analysis (family)

Option	Concordance	rank	Discordance	rank	Integrated rank
RWH	0.430	1	1.000	1	1
AIRP	0.290	2	1.000	1	2

14.6 Adaptive Planning Process for Installation of Water Supply Options

In this section, adaptive planning process for installation of water supply options are studied based on the direction of "Reduction of Stress of Collecting Water" by using acceptable options. In this study, the fact that there are still suffering residents against arsenic contaminated disaster in the village where there has already been installed some options for the Community and Baris is focused and the planning process to minimize the burden of them is studied.

Here, it is necessary for actual planning to consider various conditions such as cost, procurement of options, association of NGO and contractors. However if these conditions are prepared, applicable processes will be extremely constrained. Therefore to show what is the most acceptable process for local residents without these constraint conditions is expected to be the clarification of goals on the practical activities.

14.6.1 Assumption of Analysis

Figure 14.4 shows flow-chart of adaptive process to install water supply options. Although activities of local residents for collecting water is varied, assumptions of this action about the resident's drinking water use are set as follows:

1. Water supply options are public.
2. Options are installed at center of Bari to simplify measurement.
3. If there are options in their Bari, water collecting stress is 0.
4. Each families use the option which minimizes physical stress from available ones.
5. If there are options in their Bari, they always use safe water.
6. As local situation is harder than rainy season, stress is considered in dry season.

Based on these assumptions, stress of local residents are estimated.

14.6.2 Change of Drinking Water Use in Basailbogh village

Drinking water use in 2005 and 2007 in Basailbogh is clarified and compared to consider the actual change of water use. In August of 2005, six DTWs had been installed in the village. In December of 2007, two private DTWs and two AIRP were newly installed by themselves. As a result, number of families who use STWs had decrease from 45 to 30.

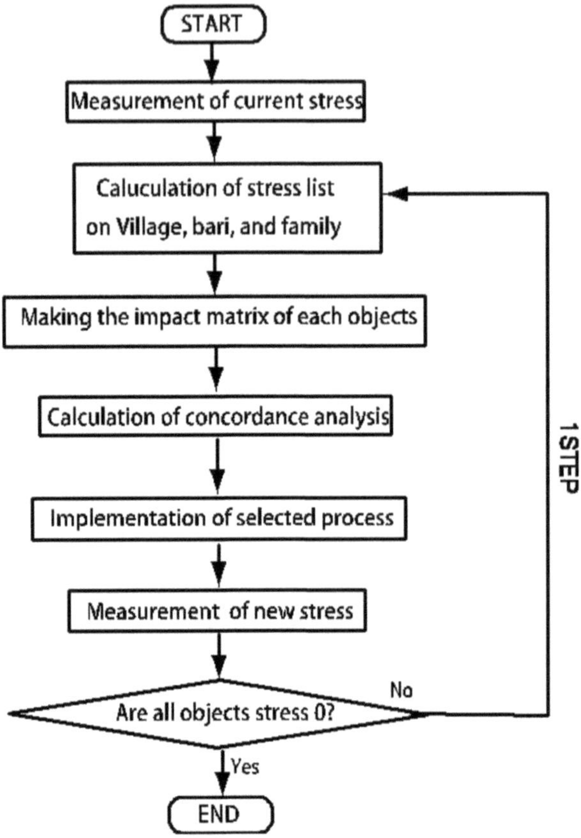

Fig. 14.4 Flow-chart of adaptive process

Basailbogh village was divided into four Community of drinking water use by public DTWs and private DTWs in 2005. If installation of options for community is aimed, it will be same as the installation for Baris in 2005. Namely, it is considered that installation for community basis had been done as of 2005. Therefore, Bari and Family are targeted in further study.

Rank of the stress of Bari in 2005 and 2007 are shown in Table 14.7. Bhandari, Kerani, Ripon Miyar Bari scored higher value of both stress than other Baris in 2005. Public DTWs were far from these Baris and stress of these Baris was equivalent to 50 % of total in Basailbogh village. It is clarified that the stress is drastically reduced by the installation of DTWs to these Baris.

Meanwhile Baris of Munshi, New Dhali and some families of Bhandari and Ripon Miar still use STWs. Stress of these Baris are almost same as above three Baris. However these Baris are relatively poor and could not install options by themselves.

Table 14.7 Summary of stress estimation of Bari in 2005 and 2007

2005						2007					
Physical	Family	Score	Psychological	Family	Score	Physical	Family	Score	Psychological	Family	Score
Bhandari	14	20440	Bhandari	14	266	Bhandari (1)	5	7000	Bhandari (1)	5	95
Kerani	15	14220	Kerani	15	124	New Dhali	5	6800	New Dhali	5	65
Ripon Miar	10	12620	Ripon Miar	10	94	Munshi	5	5940	Pokir	4	56
Munshi	5	5940	Pokir	4	52	Pokir	4	5400	Munshi	5	50
Pokir	4	5400	Munshi	5	50	Khaleque Miar	3	5320	Ripon Miar (1)	4	44
New Dhali	6	5000	Sheker	6	42	Ripon Miar (1)	4	5220	Sheker	6	42
Sheker	5	5000	New Dhali	5	41	Sheker	6	5000	Gazi	5	35
Sheikh	3	2880	Gazi	5	35	Sheikh (1)	3	2880	Khaleque Miar	3	33
Khaleque Miar	3	2850	Sheikh	3	31	Bhuiya2	5	2720	Bhuiya2	5	30
Bhuiya2	5	2720	Bhuiya2	5	30	Bhuiya	3	2400	Sheikh (1)	3	27
Bhuiya	3	2400	Bhuiya	3	24	Bhunnya	3	2100	Bhuiya	3	24
Bhunnya	3	2100	Khaleque Miar	3	16	Gazi	5	1760	Bhunnya	3	12
Gazi	5	1760	Bhunnya	3	12	Member	3	1280	Member	3	12
Member	3	1280	Member	3	12	Mosque	2	630	Mosque	2	4
Mosque	2	630	Mosque	2	4	Member house	1	630	Member house	1	3
Member house	1	630	Member house	1	3						

14.6.3 Adaptive Planning Process for Installation of Water Supply Options

Adaptive planning process for installation of water supply options, DTWs and RWH selected by Multi Criteria Analysis are studied in 2005 and 2007 to confirm actual transition by local residents in the village. It is effective to clarify the issues to be considered in actual planning.

14.6.3.1 Planning Process in 2005

Accumulated charts on the reduction of stress in Baris and Family are shown in Figs. 14.5 and 14.6 respectively.

Reduction of stress for Bari is 55 % of total in Step 1 and 80 % in Step 2. After Step 2, reduction amount become milder and it is considered that installation by Step 2 is effective in 2005.

As for Family, reduction of stress is 38 % of total in Step 1, 68 % in Step 2 and 87 % in Step 3.

Compared with this planning process and actual change, a private DTW had been installed to Bhandari and Ripon Miar respectively and two private AIRPs had been installed to Kerani Bari in actual. As for Kerani Bari, although it was

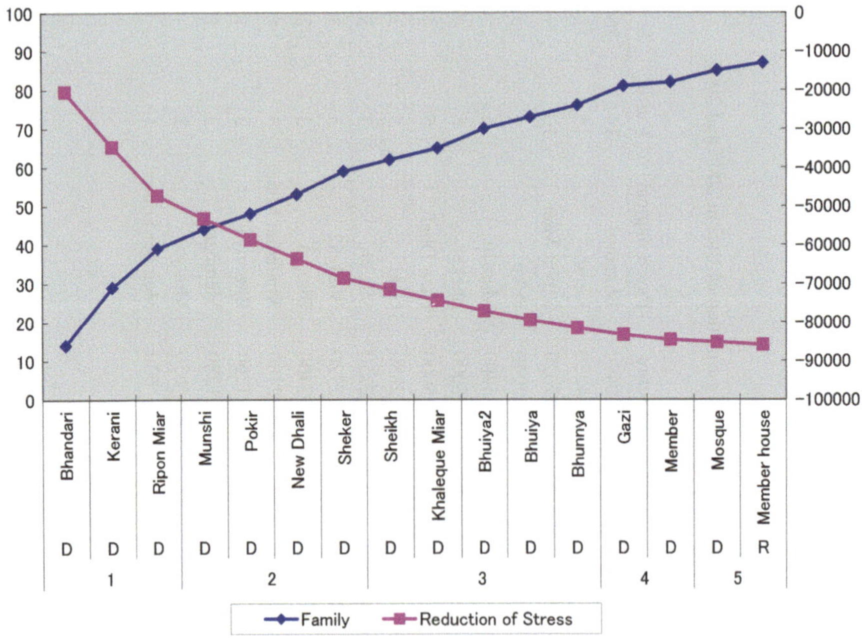

Fig. 14.5 Stress of Bari in 2005

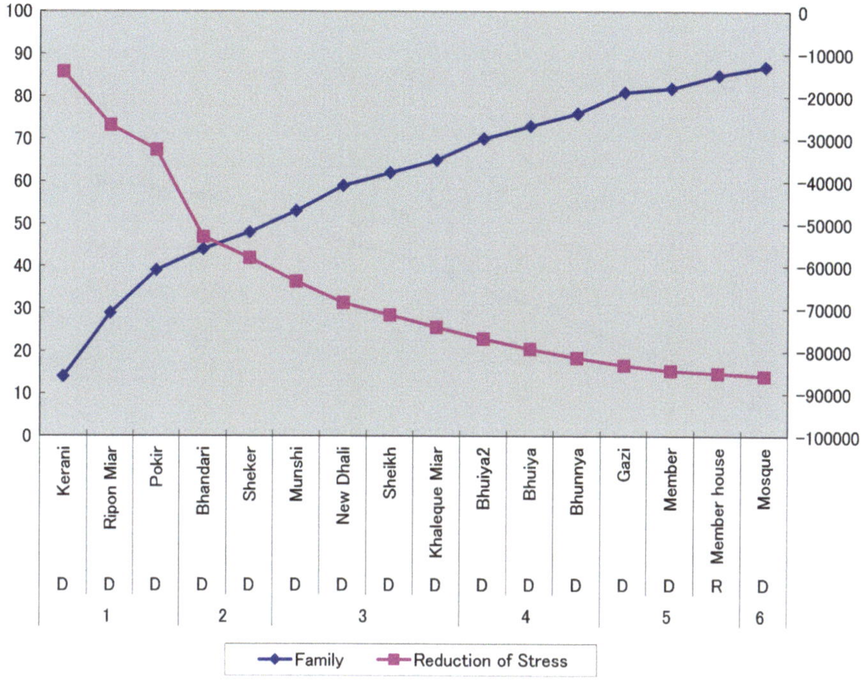

Fig. 14.6 Stress of family in 2005

explained to the owner that capacity of this AIRP is for a family, the owner could not refuse requests from his relatives and share two AIRPs in ten families. In 2008, these AIRPs had been removed due to the exceeding of removal capability of arsenic and DTW was newly installed.

As a result, planning process of Step 1 follows same installation process as actual transition of water use. However DTWs in Bhandari and Ripon Miar were private ones and owners refused some families to use DTWs. Therefore these refused families still use STWs.

14.6.3.2 Planning Process in 2007

Accumulated charts on the reduction of stress in Baris and Family are shown in Figs. 14.7 and 14.8 respectively.

Reduction of stress for Bari is 25 % of total in Step 1, 60 % in Step 2 and 69 % in Step 3. This chart is transited linearly until Step 3 and after that amount of reduction is decreased.

As for Family, reduction of stress for Bari is 25 % of total in Step 1, 59 % in Step 2 and 74 % in Step 3. Compared between Baris and Family, although target Baris are different, difference of reduction of stress is little.

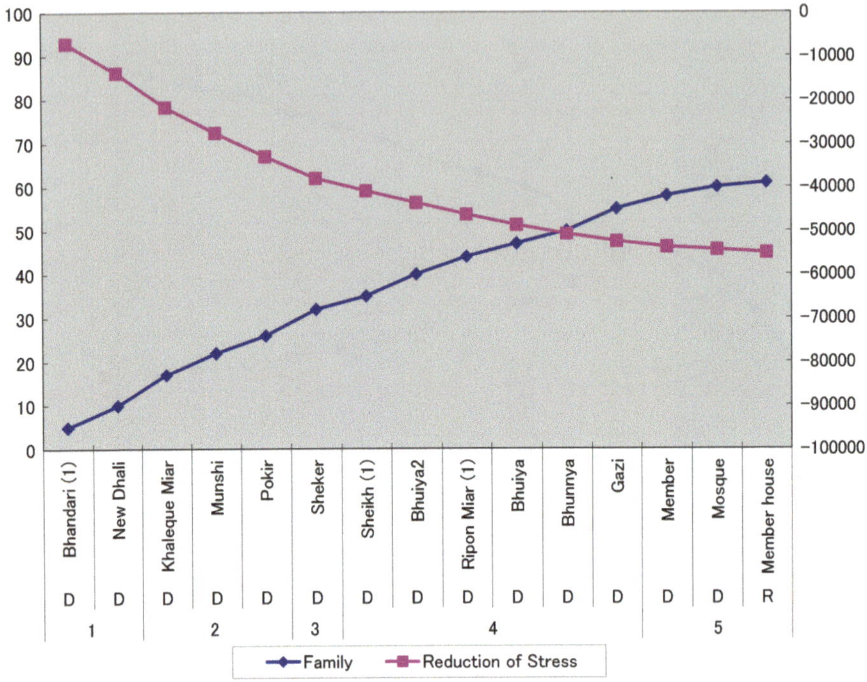

Fig. 14.7 Stress of Bari in 2007

Compared with the process in 2005, amount of reduction is drastically decreased in Step 1 because options had been already installed to Bhandari, Kerani, Ripon Miar Baris which had large stress. In 2007, number of families in targeted Baris are relatively less than other Baris and there was no large difference between process of Bari and Family. These fact indicates that Baris who needed options had already installed options by themselves that is as same as Bari level planning process.

14.6.4 Consideration of Planning Process for Installation of Water Supply Options

Based on the results of this analysis, planning process for the installation of water supply options can be divided into three phases. Phase 1 is targeted to the Community level when there are less available options in the village, Phase 2 is targeted to the Baris level when some Baris don't have acceptable options due to the stress of collecting water, and finally Phase 3 is targeted to the Family when some families are left in the development of Phase 1 and Phase 2. Development by the local government and NGOs are usually aimed to the Phase 1. In Basailbogh village

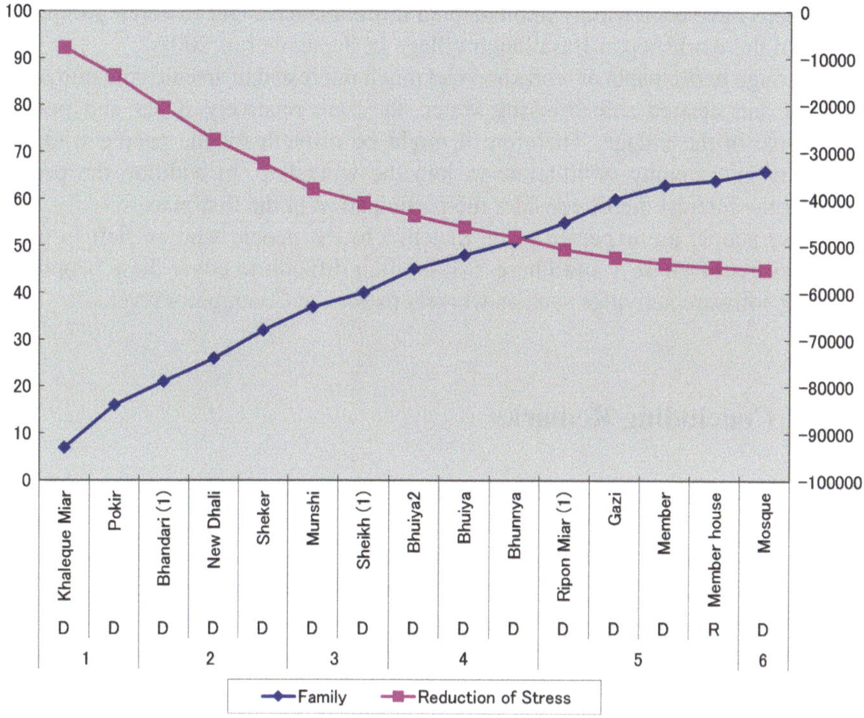

Fig. 14.8 Stress of family in 2007

it had been considered as Phase 2 in 2005 and transited from Phase 2 to Phase 3 by their own development.

Basailbogh village is located near Dhaka and logistic condition is good. Compared with other area, Basailbogh village is considered as the advanced village for the arsenic contamination disaster. However even in this advanced village, social situation and conditions of drinking water are different between each family and 25 % families still use arsenic contaminated STWs for drinking water. This fact indicated that development in Phase 1 and Phase 2 is still inadequate and is necessary to cover Phase 3. Activities for Family basis is also important to mitigate arsenic contaminated disaster.

Meanwhile, planning process in 2007 shows to reduce the stress step by step in Basailbogh by installation of 14 DTWs and one RWH. Considering practical activities for installation, it is obviously impossible for local government and NGOs to install these options in one village. Therefore, to plan more practical process, it should be shifted at any step from the installation of options to the improvement of the motivation for the safe drinking water as shown in the Sect. 14.3. That is to say, it is necessary to shift activities from the hardware aspect such as the installation of DTWs to the software aspect. This kind of educational campaigns such as workshops have been widely implemented in Bangladesh. Our research group also had held the workshop at Basailbogh village in September of 2007.

Although participants of workshop had much interested in arsenic contamination disaster and desired safe drinking water, they are relatively richer and powerful people in the village. Therefore it might be difficult for the people who had difficulty in economy or illiterate to join the workshop. In addition the people without the interest didn't consider the participation in the first place.

These people are expected to be matched to the people who are left in planning process of Phase 1 and Phase 2. It is often difficult to cover these people by existing software activities such as workshops for the Community level.

14.7 Concluding Remarks

In this chapter adaptive planning process for installation of water supply options were shown. The social environment and conditions of drinking water use in Basailbogh village was clarified by social survey. As a result it was confirmed that reduction of stress of collecting water is the acceptable direction on the planning process for local residents. Stress of collecting water was modeled and evaluated quantitatively. It showed that there were three objects for planning process, Community, Bari, and Family from local water use situation. Water supply options were assessed by concordance analysis and RWH for family and DTW were selected as acceptable options for local residents. Finally, the acceptable planning processes were shown and mentioned that planning process can be divided into three Phases to aim Community, Bari and Family.

Acknowledgements In this study, the social survey and local activity were supported by JADE, SPACE, AAN (Asia Arsenic Network), Maiko SAKAMOTO, associate professor of Tokyo University. We are grateful to them and local residents for their great contribution.

References

Fukushima, Y. (2006) A planning process on disaster mitigation for arsenic contamination drinking water in Bangladesh. *Master thesis*, Kyoto University.

Hagihara, Y., Shibata, S., Hagihara, K., Fukushima, Y., Sakai, A., & Takahashi, K. (2008). A Choice behavior analysis on safety drinking water in Bangladesh. *Disaster Prevention Research Institute Annals, 51*(B), 695–715. Kyoto University.

Hossian, M. (1996). *British geological survey technical report, Graphosman world atlas*. Dhaka: Graphosman.

Sakamoto, M., Fukushima, Y., & Hagihara, Y. (2007). Socio systems approach to disaster mitigation of arsenic contaminated drinking water in Bangladesh. *Journal of Japan Society of Hydrology and Water Resources, 20*(5), 432–449.

Chapter 15
Study on Planning Scheme to Improve the Living Environment Through Safe Water Supply and Sanitation in a Rural Village of Bangladesh

Akira Sakai, Tofayel Ahmed, and Maiko Sakamoto

Please note the Erratum to this chapter at the end of the book

Abstract To cope with groundwater arsenic contamination, which is widespread throughout Bangladesh, securing alternative drinking water sources is an urgent issue. However, the current prevalence of inappropriate sanitary practices in many parts of the country has resulted in extensive surface water pollution, which limits the safe use of such water for drinking. To facilitate improved water quality and water conservation, water supply and sanitation improvements should be integrated. In this chapter, a project aimed at improving the living environment through water supply safety and sanitation upgrades will be introduced. This chapter focuses on one of the two study-area villages discussed in Chap. 7 that are located in arsenic affected areas. However, in this village, due to the limited support provided previously, most villagers were forced to subsist on arsenic-contaminated groundwater, and were mostly unwilling to take steps that would improve the situation. Herein, the authors will present a scheme aimed at overcoming the problems that had resulted from the previous failed projects, and discuss locally appropriate technology options that consider continuous community-based management practices. This scheme focuses on pond sand filter (PSF) systems and ecological sanitation, as it was expected that use of PSF systems would increase villager awareness regarding their drinking water source, the quality of which must be maintained by sound ecological sanitation measures. Finally, after describing the current situation, issues related to the community-based management of the installed facilities will be discussed.

A. Sakai (✉)
University of Marketing and Distribution Sciences, Kobe 651-2188, Japan
e-mail: akira_sakai@red.umds.ac.jp

T. Ahmed
Japan Association of Drainage and Environment Bangladesh Office, Niketon, Gulshan-01, Dhaka 1212, Bangladesh
e-mail: tofaeljade@dhaka.net

M. Sakamoto
University of Tokyo, Kashiwa, Chiba 277-8653, Japan
e-mail: m-sakamoto@k.u-tokyo.ac.jp

K. Hagihara, C. Asahi (eds.), *Coping with Regional Vulnerability*, New Frontiers in Regional Science: Asian Perspectives 4, DOI 10.1007/978-4-431-55169-0_15

Keywords Planning scheme • Safe water supply • Community involvement • Community based management

15.1 Introduction

In many areas of Bangladesh, people continue to face serious problems related to water supply and sanitation. One urgent issue related to water supply is widespread groundwater arsenic contamination. However, due to water pollution caused by the prevalence of inappropriate sanitation and the lack of human excreta management systems, the availability of surface water that is safe for drinking is restricted. Therefore, efforts aimed at mitigating arsenic related health risks through the use of surface water as an alternative arsenic-free water source have not been effective. In order to prevent surface water pollution, water supply safety and improved sanitation should be integrated as part of efforts to improve the living environment.

In this chapter, a project aimed at improving the living environment by implementing a safe water supply and sanitation will be discussed. In the project area, all shallow tube-wells (STWs) are contaminated with arsenic. However, due to limited outside support, there were no alternative drinking water sources in the village, and most villagers subsisted on the arsenic-contaminated water. Furthermore, even though the villagers were aware of the arsenic-related health risks prior to the start of the project, there was little willingness to take steps that would improve their hazardous situation, primarily due to a lack of realistic prospects for obtaining a safe water supply.

Therefore, in the sections that follow, the authors will discuss locally appropriate technology options while considering continuous community-based management practices, and a scheme that aims at overcoming factors that have caused previous project failures will be presented. Then, based on the results of a questionnaire survey that was conducted after facility installation, the authors will show how the villagers have increased their awareness of water and sanitation issues, and have demonstrated increased willingness to improve their living environment through involvement in the applied planning scheme. Finally, after describing the current situation, subjects related to independent community-based management practices will also be discussed.

15.2 The Necessity of Integrating Water Supply and Sanitation

In rural areas of Bangladesh, health and environmental risks related to water supply and sanitation are interrelated (see Fig. 6.5 in Chap. 6), so it is necessary to integrate water supply and sanitation improvements with appropriate management practices. Inappropriate human excreta and wastewater management practices result in unhygienic environments, as well as pollution of surface water and shallow groundwater. While most areas of the country have switched to groundwater sources for drinking water, numerous villagers are still using surface water for various domestic purposes such as bathing, cooking and laundry. Hence, polluted surface water still poses significant health risks. In arsenic affected areas, if surface water cannot be used for drinking, it is difficult to obtain alternative drinking water sources, and thereby reduce health risks related to arsenic contamination. Therefore, it is clear that improved sanitation combined with appropriate human excreta management will enhance surface water availability and reduce both environmental and health risks.

In this study, appropriate human excreta management is the primary focus of the contemplated sanitation improvements. It is required to start with the idea that human excreta (urine and faeces) are valuable resources that can be used to regenerate depleted soil through the addition of nutrients and organic materials. Therefore, applying urine and faeces to agricultural land is both a form of human excreta management and a way to restore the natural nutrient cycle between excreta and food production. In contrast, improper management, such as the unregulated discharge of human excreta, not only precludes its proper use as fertilizer, it also pollutes the surrounding environment. It is also important to note that because it may be difficult to ensure a steady supply of commercial fertilizers in the future, utilizing human excreta will contribute to sustainable agriculture and food security. Maintaining agricultural productivity is also an important component in efforts to alleviate malnutrition, which is a significant factor for those suffering from arsenicosis and waterborne diseases, and thus contributes to health risk reductions.

The proper integration of water supply and sanitation improvements will result in various other benefits as well. These include, for example, surface water conservation, soil environment improvements, and reduced burdens on women tasked with collecting and transporting water – as well as corresponding improvements in their living conditions. Furthermore, by ensuring that villagers understand that a supply of safe drinking water cannot be guaranteed if surface water sources are not kept clean, it is possible to awaken concerns regarding other aspects of their living environment, and hence, the necessary consciousness to conserve their overall living environment can be expected to flourish.

It should also be noted that, in rural areas of Bangladesh, improvements in water supply safety and sanitation have the potential to reduce burdens on family finances through various benefits that include reducing medical expenditures, increasing agricultural yields and reducing fertilizer expenses, as well as increasing income generation opportunities by reducing disease occurrences (Sakai et al. 2010).

15.3 Study Area

15.3.1 Study Area Description

The study area in this chapter is Bansbaria Village in Keshobpur Upazila, Jessore District, which is one of the villages discussed in Chap. 7. At the time of the study, the water supply and sanitation conditions for the village were far inferior to those for the other village discussed in the same chapter. Therefore, the villagers of Bansbaria were far more vulnerable to water-supply and sanitation-related risks.

Bansbaria is a typical agricultural village surrounded by paddy fields. The village consists of about 120 households and farming is the primary occupation of about half of them, even though about 40 % of the farmers are landless. Approximately 40 % of the households generate their income by supplying day labour for agricultural or construction work. Additionally, 45 % of households have members that have worked as labourers in other areas. The overall literacy rate is 38 %.

This village is located in a flood-prone area that suffers from water logging, even after the flood season, due to accumulated soil that interrupts flood run-off. This can have a significant effect on agricultural land use and productivity. There are many ponds in the village, some of which are used for income generation activities, including raising fish and refining jute fibre. Such ponds are unsuitable for use as drinking water sources or for other domestic water needs. The description in this paragraph were compiled in March 2009, before the project started, at which time the authors conducted a questionnaire survey on all of the families in this village.

15.3.2 Water and Sanitation Status in the Study Area

Prior to the questionnaire survey, the authors conducted an arsenic concentration survey of the STWs in this village in 2008. The results showed that all of the STWs (which had depths of between 20 and 70 m) had arsenic concentrations exceeding the drinking water standard in Bangladesh (0.05 mg/L). Thus, even though arsenic-free deep-tube wells (DTWs) were available in some of the neighbouring villages, Bansbaria villagers seldom availed themselves of such water due to the physical and psychological burdens that this entailed. As a result, more than 90 % of the villagers subsisted on contaminated STW water. Despite this, at the time of this questionnaire survey, there were no arsenicosis patients in this village.

As for sanitation, 40 % of the villagers had no access to a toilet and normally defecated in open spaces. As for the toilets themselves, all but one, a hanging latrine, were pit latrines, and more than half of the pit latrines were unsealed. Furthermore, the frequency of disease occurrence, and thus medical expenses, were high enough to strain the budgets of poor families.

Despite the above, the villagers in Bansbaria did not express intense dissatisfaction, even though they knew about the arsenic contamination and were somewhat anxious due to its related health risks. However, they also expressed little in the way of willingness to improve their living environment. The reason for this was thought to be a form of cognitive dissonance resulting from limited experience with support from either local government officials or local non-government organizations (NGOs). This, as a consequence, meant that they saw no conceivable way to improve their living environment, even though they felt anxiety that they were exposed to high health risks related to arsenic contamination.

15.4 Selecting Technical Options

When technical options are applied to local communities, they need to be thought out rationally in order to ensure they meet the requirements of the local society and the geographical features of the area, which includes considerations such as annual rainfall amounts and flood influence. They must also be within the limits of available materials, capital, and local human resources. These requirements are also the preconditions for community-based management practices. It is also necessary to assess the technical options in consideration of local social barriers, such as traditional customs and religious prohibitions, and to consider post-installation social impacts in order to ensure long-term social acceptance.

15.4.1 Alternative Safe Water Sources

The technical options that can be used to mitigate arsenic contamination can be divided into two categories: alternative water sources and arsenic removal. The alternative safe water sources include DTWs, surface water (ponds, rivers), dug wells, and rainwater harvesting. When selecting a water source, it is necessary to consider water quality characteristics including iron content, salinity, and sensory factors such as taste and odour. Quantitative stability throughout a year must also be assessed. Additionally, since surface and dug well water must be treated and disinfected prior to consumption, it is vital to prevent contamination from the various pollutant sources that exist in living environments. When selecting rainwater harvesting, it is important to ensure sufficient tank volume availability and to set daily usage limitations in order to ensure sufficient water is available throughout the year.

As for arsenic removal, even though a number of approaches have been applied at various levels – from individual family to community use – a widely applicable technology does not yet exist in Bangladesh. The applicability of the technology is also dependent on the quality of the groundwater. In short, the primary disadvantages of the method are the uncertain reliability of arsenic removal, the complicated maintenance requirements of the devices themselves, and the lack of feasible ways to dispose of the removed arsenic.

15.4.2 Appropriate Sanitation Options

Regarding technical options for sanitation improvements in developing countries, a number of approaches have been discussed (Tilly et al. 2008). First, when selecting sanitation options, it is necessary to consider initial costs, energy and resource consumption, environmental impacts, operation and maintenance difficulties, useful equipment service life, and the effects of flooding. In rural areas of Bangladesh, sanitation facilities should be constructed for household level use, as most public toilets, including school toilets, eventually decay into disuse due to lack of ownership and negligent maintenance practices. However, collection-based sanitation systems, which consist of toilets, sewer lines, and off-site treatment facilities, are not feasible in much of the country for the following reasons; (i) pipe laying is costly, (ii) there is insufficient pipe laying capacity in this region, (iii) an abundant supply of flushing water is necessary, and (iv) treating collected wastewater usually requires significant resource expenditures.

There are various types of on-site sanitation options available, including toilets that are equipped with human excreta treatment functions and small-scale wastewater treatment systems for household use. However, since it was also necessary to prevent surface water pollution and preserve and enhance soil fertility, the applicable technical options available were limited. The process of recovering human excreta resources while simultaneously providing improved sanitation is called ecological sanitation (EcoSan), and it has been shown that urine diverted dry toilet (UDDT) type EcoSan toilets are technically and socially acceptable sanitation options in rural areas of Bangladesh. Therefore, because they have less impact on surface water quality, EcoSan toilets were selected the as sanitation option for the project.

15.4.3 Safe Water Supply Selection Criteria

Generally speaking, initial setup costs, operating costs and operational difficulties, water quantity stability, water quality, and equity among users are the primary criteria considered when selecting a water supply option. In this project, priorities were also placed on continuous community-based management feasibility, equity among beneficiaries, and the arousal of consciousness and concern for living environments.

Based on the results of a geological survey of the Bansbaria Village area, it was determined that the water quantity from dug wells would be insufficient to supply water over the course of an entire year due to the limited catchment area available, and because there are no rivers located near this village to replenish the water table. As for ponds, since a clay layer exists under the shallow groundwater table, it was determined that it would be possible to supply water throughout the year by excavating this clay layer in order to enhance and maintain the necessary storage capacity. From the viewpoint of equity, rainwater harvesting was considered suboptimal for household use because the roof materials of some households were unsuitable for catchment purposes.

15.4.4 Comparison Between DTW and PSF

Table 15.1 shows comparisons between DTWs, which are the most popular alternative drinking water sources, and PSF systems, which consist of holding ponds and sand filters. In this system, the drinking water source is pond water, which is utilized after slow-sand-filter purification. As can be seen in the table, although the initial DTW costs are lower, the iron content and salinity level of groundwater in this area are high. In addition, it is impossible to ensure that arsenic will never penetrate the water supply. Furthermore, while both facility types must be monitored and maintained, since a DTW does not require daily maintenance, local communities seldom pay sufficient attention to its management. In fact, even though the potential for arsenic contamination is undeniable, water quality is seldom monitored. Moreover, since no organization is normally willing to take responsibility for DTW maintenance, even small faults can sideline such wells for long periods of time. In contrast, the PSF system that has been selected for this project requires ongoing community monitoring efforts to ensure the pond water quality is kept high, and the sand filter facility requires communal maintenance labour. To preserve pond water quality, the pathogens that originate in human excreta are controlled by the use of

Table 15.1 DTW and PSF comparison

Technical options	DTW	PSF
Construction cost	50,000 Tk	50,000–100,000 Tk (Sand filter only)
No. of families	–100	30–120
Features	Risk of arsenic contamination High iron and salinity content (In this area)	Pond water quality and pond volume must be maintained Excavation to ensure adequate pond capacity is necessary Sand filter maintenance is necessary
Community-based management	Arsenic concentration is seldom monitored Maintenance efforts are haphazard	The community is responsible for preserving the pond water quality and maintaining the sand filter
Relationships with EcoSan	No relation	EcoSan is necessary to preserve pond water quality

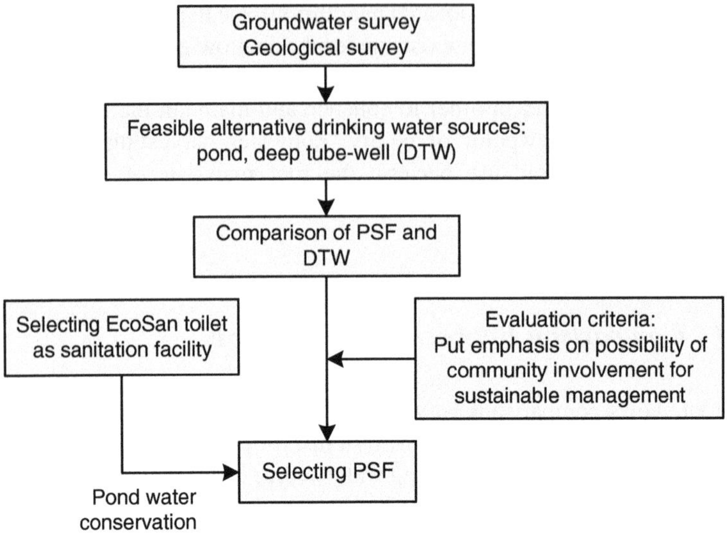

Fig. 15.1 Procedure used to select technical option

EcoSan toilets. Figure 15.1 shows an overview of the procedure used to select the technical option, along with the results.

15.4.5 Facility Plan

The integration of PSF and EcoSan ensures that EcoSan plays a significant role in maintaining the quality of drinking water. In addition, it was anticipated that the increased attention villagers would pay to pond water quality would stimulate their awareness of the need to conserve other aspects of their living environment. Consequently, the entire local community will eventually come to recognize that the ponds and sand filters are resources held in common, and thus must be managed communally. In addition to pond water conservation, the EcoSan toilets will facilitate sanitary improvements and assist in the recovery of human excreta derived resources, thus assuring sustainable agriculture and food security.

Figure 15.2 shows the safe water supply facility installation process, including the establishment of community-based organizations (CBOs), in Bansbaria Village. After considering the attitudes of pond owners in this area, pond capacities, original pond usage purposes, water quality, and water collection site accessibility, two ponds were selected for use. The selection process was not straightforward. In fact, since there were a limited number of candidate ponds, it was necessary to estimate their required holding capacities in terms of water demand, the presumed number of user families, the average number of family members, along with the amount of rainfall and evaporation. The unit demand of water was set to 9 l per person per day,

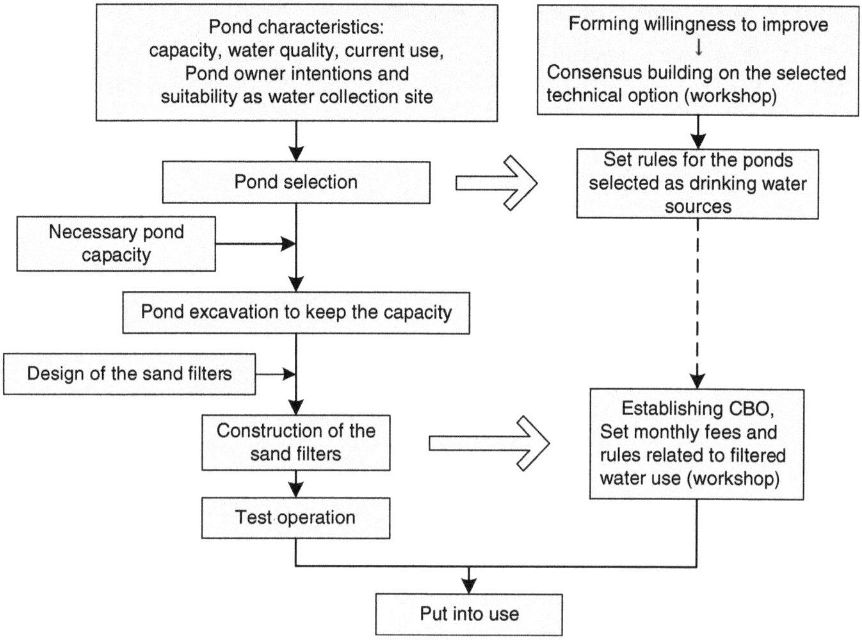

Fig. 15.2 Process used to install safe water supply facility, including establishing CBO

by limiting the use of PSF water to drinking and cooking. As existing pond capacities were insufficient, it was necessary to excavate the clay layer of the chosen ponds in order to ensure adequate capacities. Sand filter capacities were set based on the amount of water that had previously been calculated from the pond capacity estimates. To determine its scale, it was assumed that the daily water collection duration would be set at 10 h, and that the sand filtration flow rate would be 0.1 m/h.

Prior to the installation of the new facilities, arsenic-contaminated STWs were the sole source of water. Therefore, it was necessary for the villagers to travel to one of the neighbouring villages if arsenic-free drinking water was desired. Following the installation of the PSF systems, safe drinking water accessibility improved remarkably. Currently, the longest distance travelled by any of the villagers is 400 m and the collection time for water is less than 25 min. As for the EcoSan toilets, the numbers have gradually been increasing as donor funds become available, and have now reached 62. Since a number of toilets are shared among related households, 71 of the 120 households in the village have access to EcoSan toilets. Table 15.2 and Fig. 15.3 show an overview of the features and locations of water supply facilities in Bansbaria Village.

Table 15.2 Water supply facilities in Bansbaria village

PSF-1:	Pond capacity 1280 m^3 (before excavation: 630 m^3)	Sand filter: 45 households
PSF-2:	Pond capacity 3600 m^3 (before excavation: 1210 m^3)	Sand filter: 100 households

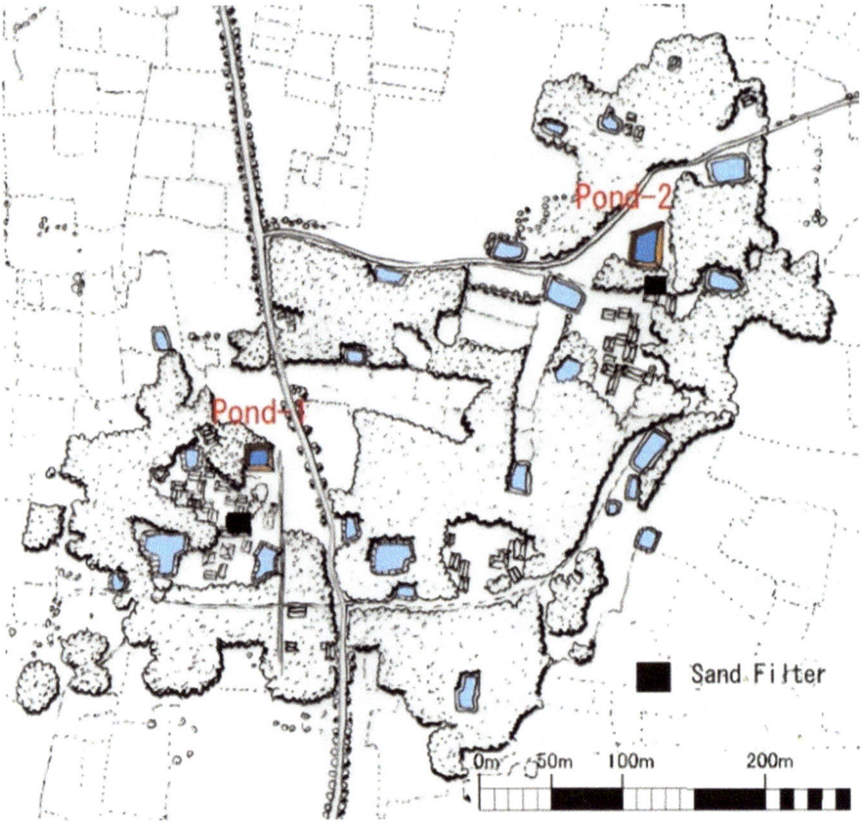

Fig. 15.3 Location of water supply facilities in Bansbaria village

15.5 Implementation Process and Changing Awareness

15.5.1 Implementation Process

Figure 15.4 shows the planning scheme implementation process. This process aimed at overcoming the problems that have led to previous project failures. Therefore, the following steps were included: (1) raising awareness on issues related to water supply and sanitation, (2) appropriate selection of technical options and precise information transfer, (3) local community involvement and adequate communication, and (4) establishing a responsible management organization.

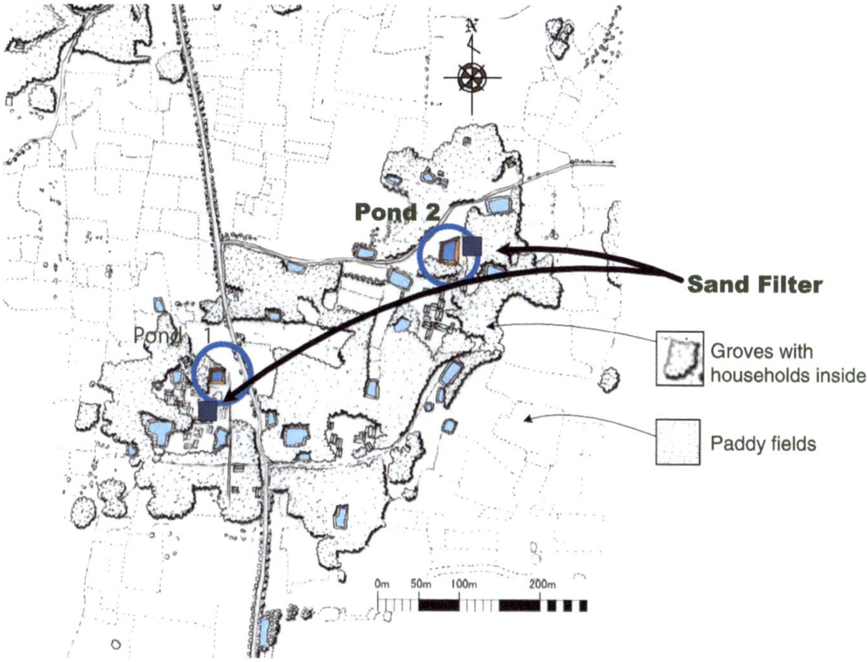

Fig. 15.4 Planning scheme implementation process

The plan was initiated through general information sharing regarding arsenic-related health risks, the necessity of improved sanitation, and the state of actual arsenic contamination in this village. This information dissemination was aimed at raising awareness about arsenic contamination and unhygienic environments among villagers. Then, by disseminating precise information about the technology adopted, the villagers were encouraged to adopt perspectives towards improving their current situation. As most villagers had no knowledge regarding arsenic mitigation options other than DTWs when the authors conducted the questionnaire survey, a number of influential villagers were taken to a site where they could observe a well-functioning PSF. As the result, they came to understand the technology. Furthermore, to spread understanding on the necessity of community involvement in pond water quality conservation efforts, an awareness program was carried out in household courtyards using visual materials over a period of 2 months.

If cognitive dissonances exist, the use of risk cognition and problem solving exercises can encourage people to eliminate them, and thereby stimulate increased willingness to solve the problems. Figure 15.5 shows how firm risk cognition relates to solving problem. In the case of local community water supply and sanitation related problems, continuous risk reduction through firm risk cognition is highly expected.

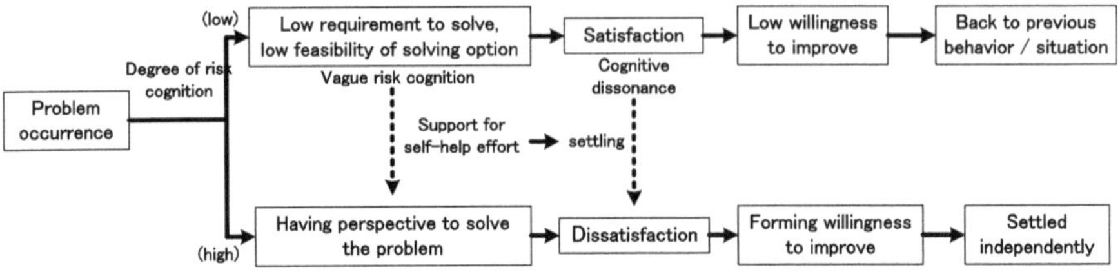

Fig. 15.5 Risk cognition and solving problem

Prior to constructing the facilities and excavating the ponds in order to make them suitable as drinking water sources, it was first necessary to obtain the consent of the landowners. In Bansbaria Village, preservation societies were established after selection of ponds as drinking water sources. These societies played a role in conserving the ponds and supervising the required excavations. After facility construction plans were finalized, a workshop was held and CBOs were established to manage the water supply related facilities. As continuous management is required, CBO members were selected from the educated younger generation, and the rules necessary to manage the ponds were decided at the management workshop. For example, it was decided that the treated water should not be used for purposes other than drinking and cooking, and that every PSF user would be required to pay fees, which would be saved by the CBO and used for future maintenance and overhaul expenses.

15.5.2 Changes Among Local Villagers

What kinds of changes have local villagers accepted through their involvement in the implementation process? A questionnaire survey targeting all of the households in the village was carried out in March 2010, before the villagers started to use water from the PSF. At this time, the total number of households was 118 and the number of installed EcoSan toilets was 19. Figure 15.6 shows the responses to the survey question, "What is the biggest change for you?" Most villagers answered that their awareness had been raised and that their hygienic practices had improved. However, there were also significant differences in the responses of EcoSan toilet users and non-users. For EcoSan toilet users, even though the number of installed toilets was small at that time, changes were observed in knowledge, awareness, and water-related practices as well as sanitation.

Since the survey was conducted before the PSF systems started to supply water, EcoSan non-users had not actually recognized any changes, although they changed their sanitary practices to some degree. When asked about the activities that caused them to change, more than 70 % of the respondents said that their awareness had been increased by the workshops in which numerous people had participated, with lower percentages being assigned to the household courtyard meetings or their

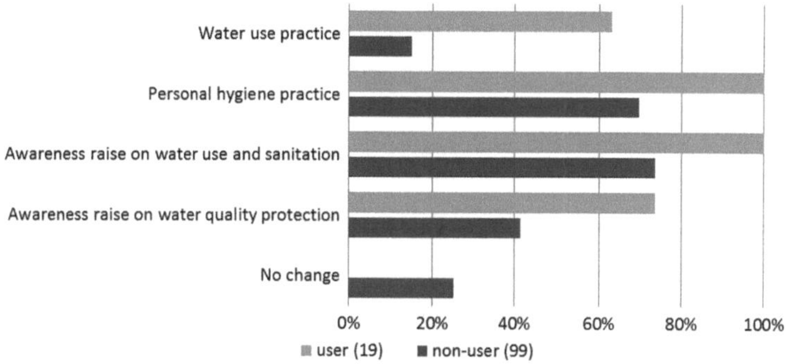

Fig. 15.6 Changes among villagers through involvement in the implementation process. *Figures in *parentheses* are the numbers of EcoSan toilet user and non-user families

personal observations of pond excavations and facilities construction, as shown in Fig. 15.7. Additionally, more than 70 % of the respondents said that personal communication had also influenced their changes. This answer indicates that it had previously been unusual for the local community to work together with outsiders to solve social problems.

Throughout the planning scheme, the effectiveness and advantages of the PSF facilities were explained repeatedly. Despite this, however, the percentage of villager respondents who said they would use the PSF did not exceed 80 %, and one fifth of those surveyed stated that they would not use the new facilities. As can be seen in Fig. 15.8, significant differences in understanding regarding the PSF effects – primarily the advantages of arsenic-free drinking water – were noted between the abovementioned two groups of respondents. It was this lack of understanding that caused the low percentage of respondents willing to use the PSF. Since the question was asked before the PSF was put into practical use, it was difficult for respondents to visualize the benefits that would accompany the resultant burden reduction. Another breakdown in communication was related to the responsibility of users to help maintain the PSF facility, which is something neither group fully recognized, in spite of the stipulated rules and fees that had been decided during the workshop. These survey results indicated that the awareness program alone was insufficient to generate the level of recognition required to keep the PSF systems well maintained independently.

15.5.3 *Willingness to Improve the Living Environment*

The result of the March 2009 questionnaire did not show significant amounts of willingness to improve their living environment, in spite of the serious water supply and sanitation situations (see Chap. 7, Fig. 7.13). However, as can be seen in Fig. 15.9, as improvements to the water supply and sanitation situations became

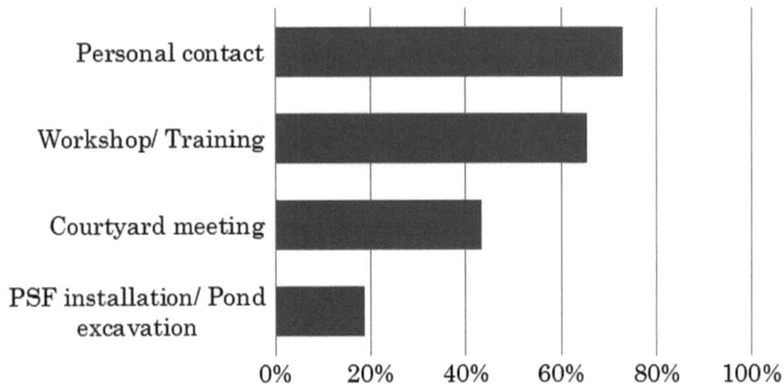

Fig. 15.7 Causes of villager attitude changes (multiple answers were allowed)

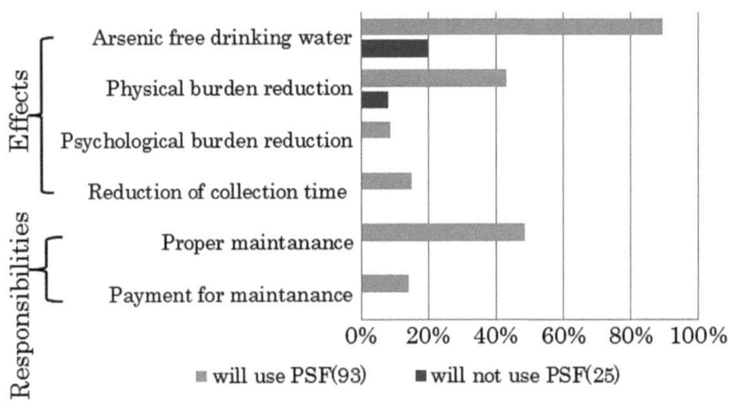

Fig. 15.8 Villager understandings related to PSF (multiple answers were allowed) *Figures in *parentheses* show the numbers of respondents who state they will use (or will not) use the PSF

more noticeable, this willingness increased. As for the low level of willingness observed before initiation of the planning scheme, it seemed that a cognitive dissonance existed due to a lack of realistic perspectives towards solving the problems. In such conditions, the villagers attempted to resolve this dissonance by thinking, "there is no need to improve" the situation. In contrast, after 1 year of having the technical options that would solve their problems presented to them, the dissonance was dissolved by the prospects of an improved living environment. Therefore, it can be said that presenting appropriate solutions through proper technology transfer contributes significantly to encouraging a willingness to improve, as shown in Fig. 15.5.

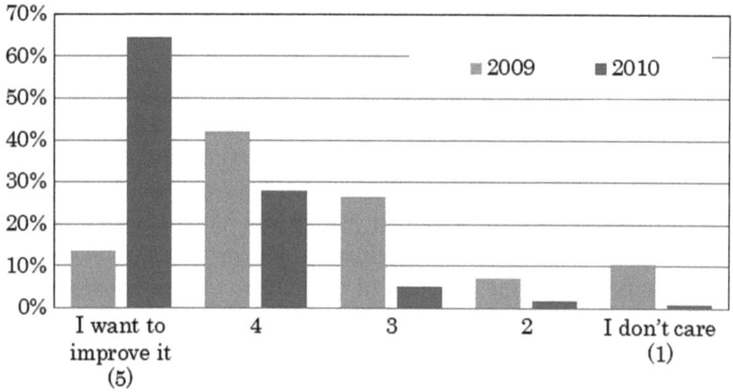

Fig. 15.9 Change in willingness to improve the living environment

15.6 Current Situation in Bansbaria Village

After conducting test operations and confirming that faecal coli-form bacteria were not present in the filtered water, PSF system water supply to the villagers began in June 2010. In addition to performing operation and maintenance, the CBO began collecting maintenance fees from the users. However, the villagers voiced some complaints. For example, some pond owner family members continued to enter the pond for bathing or washing, and domestic ducks and dogs sometimes invaded the ponds through broken fences. As the result, the number of villagers who said they would not drink PSF water, and simply use it for cooking instead has increased because the reduced iron content improved food taste. Furthermore, even after the facilities were handed over to the CBO, many villagers continue to depend on the outsiders that installed them to resolve differences, and a reliable relationship between CBO members and the villagers has not been achieved. This has resulted in suspicions and mistrust between them, such as questions as to whether the collected money has been used improperly.

Yet, it must be stressed that this was the first time the community has possessed resources in common, for which communal management is required. Therefore, it will be necessary to continue observations for a period in order to determine whether they can develop the capacity to solve such problems by themselves.

Furthermore, in addition to creating targets towards improving their living environments and forming responsible organizations, outsider support has also been provided to motivate the local community towards tackling a number of other unsolved issues, and it is expected that mutual trust between the CBO members and the villagers will improve if the CBO acquires experience in solving problems like those mentioned above, or newly identified problems, with the consent and assistance of the villagers.

15.7 Conclusions

In this chapter, while placing emphasis on the criteria necessary for independent and sustainable management practices, the authors discussed the technical options needed to integrate water supply and sanitation improvements, as well as a planning scheme that aimed at overcoming the factors that have resulted in previous project failures. The projects implemented at Bansbaria Village using this planning scheme, where serious arsenic contamination and improper sanitation practices had previously been endemic, resulted in substantive changes among villagers. It was especially notable that, as community members gained awareness about potential solutions related to water supply and sanitation through their own involvement in the planning scheme, they showed increased levels of willingness to improve their living environment.

During application of the planned scheme, the authors aimed at changing the perspectives of the villagers, who are the main persons concerned, specifically by emphasizing the need for increased awareness of water supply and sanitation issues. In addition to selecting locally appropriate technical options and obtaining villager consensus regarding them, the present scheme also provided villagers with enhanced perspectives towards self-improvement, and is therefore applicable to other local communities facing similar water supply and sanitation related problems.

It is felt that outsiders need to ensure proper communication with local society members at every step during the planning scheme, and must work diligently to encourage locals to participate fully in the scheme. More specifically, the responsibility of outsiders that provide assistance is not limited to facility installation; it is also necessary for them to assist in the creation of a CBO that is capable of independent facility management.

Acknowledgements This study was supported by the Japan Association of Drainage and Environment (JADE). The authors would like to express their gratitude to the villagers of Bansbaria for their cooperation, and to Mitsui Co. & Ltd. for its financial support in constructing the facilities necessary for safe water supply and sanitation in Bansbaria Village.

References

Sakai, A., Takahashi, K., & Azad-uz-zaman, Q. (2010). An example of benefit estimation related with ecological sanitation in rural areas of Bangladesh. *International Water Association, Decentralized Wastewater Treatment Solutions in Developing Countries*, Surabaya, 8 pp (CD-ROM).

Tilly, E., et al. (2008). Compendium of sanitation systems and technologies. *Swiss Federal Institute of Aquatic Science and Technology* (Eawag), Dübendorf, Switzerland.

Chapter 16
Waterside Environmental Management Incorporating Sustainability and Survivability

Kiyoko Hagihara, Yoshimi Hagihara, and Masanori Kawano

Abstract This chapter aims to demonstrate an adaptive waterside environmental management process that incorporates sustainability, survivability and participation. The concept of sustainability has give rise to questions about what it is supposed to mean: the sustainability of what, for whom, for how long, and why? For instance, there is a case where the sustainability in the same city and same region as a whole is achieved, but on the other hand, some people in a part of city and region may be in danger of not only risking their sustainability but also their very lives due to floods and ecological destruction at the waterside. In this chapter, we focus on differences among residents both in same area and in different areas and consider waterside management in urban area taking into account sustainability, survivability and participation along the adaptive waterside environmental management process. First, cost-benefit analysis is reviewed critically from the viewpoint of sustainability and survivability. It is suggested that the adaptive waterside environmental management process which uses a systems analysis methodology is a promising method to aid management decisions. The methodology is applied in three areas along Kamo River in Kyoto city, two of which are located in the upper river area and one in the downstream area. Based on a social survey of residents, environmental characteristics of each area are determined. The environmental valuation function for each area is then defined, and possible priorities for waterside environmental management are presented. Waterside environmental management incorporating sustainability and survivability is thus demonstrated.

K. Hagihara (✉)
School of Sociology, Bukkyo University, Kita, Kyoto 603-8301, Japan
e-mail: hagihara@bukkyo-u.ac.jp

Y. Hagihara
Kyoto University, 264 Suemarucho, Nakagyou, Kyoto 604-0901, Japan
e-mail: y-hagi@gf7.so-net.ne.jp

M. Kawano
Ministry of Land, Infrastructure, Transport and Tourism, 2-1-3 Kasumigaseki, Chiyoda-ku, Tokyo 100-8918, Japan
e-mail: kawano-m2y6@mlit.go.jp

K. Hagihara, C. Asahi (eds.), *Coping with Regional Vulnerability*, New Frontiers in Regional Science: Asian Perspectives 4, DOI 10.1007/978-4-431-55169-0_16

Keywords Waterside environmental management • Adaptive waterside planning methodology • Sustainability • Survivability • Participation

16.1 Introduction

In the last decade, and following the publication of Brundtland's report on sustainable development (WCED 1987), the number of books and papers with the word 'sustainable' in their title has grown enormously. However, the word is used in many different ways (Jordan 2008). The elasticity of the concept has given rise to questions about what it is supposed to mean: the sustainability of what, for whom, for how long, and why (O'Neill et al. 2008; O'Neill 2009; Adger and Jordan 2009)? For instance, consider the case of a region which is composed of both urban and rural areas. The rural areas may play important roles including supplying food as well as conserving national land, the head sources of streams and the natural environment. However, the rural areas have lost population, and it has become difficult for the community to fulfil these roles. The remaining rural residents may be in danger of not only losing their way of life, but also their very lives, due to floods and ecological destruction along the river. The urban areas are likely to be affected too. Therefore, as Hagihara and Hagihara (1991) have suggested, the depopulation problem in rural areas must be regarded as a problem for the sustainability of the whole region including both urban and rural areas. Sustainability is not a matter of simply maintaining some aggregate level of human and natural capital. It requires the maintenance of specific resources and goods that meet different human needs and capacities (O'Neill 2009).

Furthermore, public participation is widely and sometimes rather uncritically embraced as a 'good thing', but more is needed to be known about how best to incorporate it into the management of sustainability (Jordan 2008), and ultimately, what it actually delivers in terms of residents' and their communities survivability 'on the ground'. By survivability we mean, for individuals, continued life, and for communities their ability to continue to function as a community.

Cost-Benefit Analysis (CBA) has been widely used as an evaluation tool for public policymaking. However, although CBA is adequate to evaluate the efficiency of the policy in question, it does not take into account the equity issues and sustainability aspects of that policy. Therefore, multicriteria methods which include participatory multicriteria methods have been proposed to evaluate certain policies.

In this chapter, we focus on differences among residents both in same area and in different areas and consider waterside management in urban area taking into account sustainability, survivability and participation along the adaptive waterside environmental management process (Hagihara 2008; Hagihara and Sakamoto 2006) which is briefly shown by the following process: the differences both areas are clarified by the social survey, then an evaluation of these results with the aid of multicriteria analysis is conducted and if there are conflicts between areas conflicts analysis is executed. A social survey of valuation of waterside environment and daily life is firstly conducted among the residents in both areas, i.e., upper river area and down river area, along Kamo river in Kyoto city. Accordingly, residents in both areas can participate in providing information for the planning process by the social survey. By analyzing the results elements of both sustainability and survivability for both areas are derived. Thus we will be able to take into account the sustainability and survivability which include different kinds of elements in the waterside environmental management.

16.2 Cost-Benefit Analysis and Sustainability

16.2.1 Limitations of Cost-Benefit Analysis

Cost-benefit Analysis (CBA) is a common method used to aid decision-making (e.g., Nas 1996; Hanley and Barbier 2009). It has been widely used as an evaluation tool for public policymaking by governments in Japan and around the world.

The main strength of CBA as an appraisal technique is as a test for economic efficiency in resource allocation. The main criticism of CBA is the acceptability of the ethical framework underpinning it. CBA is primarily based on welfare economics. One of the welfare criteria is Pareto Optimality, which requires that no one is made worse off and at least one individual believes he/she is better off after a policy decision. The problem with this is that most policy changes make some people better off and some people worse off simultaneously. So, modern welfare economics is now based upon the Kaldor-Hicks principle of potential compensation, called the potential Pareto principle. That is, if the gainers from a policy gain sufficiently to compensate the losers, the policy is an improvement regardless of whether compensation is actually paid.

In applying the potential Pareto principle, however, it is possible that a policy could actually lower the sum of utilities if people with different levels of wealth have different marginal utilities of money. If the low-wealth person's marginal utility of money is higher than that of the high-wealth person, then it is possible that the utility loss of the low-wealth person could outweigh the utility gain of the high-wealth person. Thus, although the Pareto principle allows us to avoid interpersonal utility comparisons by guaranteeing increases in aggregate utility for policies with positive net benefits, the potential Pareto principle does not do so. The potential compensation criterion is useful in separating efficiency and equity, but has meant that discussions of actual compensation have been avoided on the grounds that equity issues are outside the economists' realm. (see, for example, Hanley and Spash 1993; Hanley 2001; Hanley and Barbier 2009).

16.2.2 From Cost-Benefit Analysis to Participatory Multicriteria Analysis

According to the economic literature, sustainable development is economic and social development that maintains a certain minimum level of human welfare for present and future generation of humans, in the sense of either maximizing welfare over time or meeting the demands of distributional justice between generations. However, if sustainability is also about equity in distribution over generations, then it raises the same question as to what it is we are supposed to be distributing equally. The question 'Equity of what?' is directly related to the common question among environmental activists and advocates, 'Sustainability of what?' (O'Neill et al. 2008; Munda 2005, 2010). Much of the environmental valuation literature is concerned with the issue of sustainability; however, they ignore income distribution (see, for example, Hanley and Spash 1993; Hanley 2001; Hanley and Barbier 2009). Yet a sense of fairness is a critical factor in economic decisions (Gowdy 2004).

Although CBA is adequate to evaluate the efficiency of the policy in question, it does not take into account the equity issues and sustainability aspects of the policy. Therefore, multicriteria analysis (MCA), which includes participatory multicriteria analysis, has been proposed to evaluate policies (Munda 2005; Kallis et al. 2006). Sustainability raises a set of issues based on the civil rights of current and future generations as well as respect for ecological systems (Messner 2006). MCA takes into account a much wider variety of methods than CBA (see, for example, Nijkamp 1977; Figueira et al. 2005; Getzner et al. 2005; Vincke 1992). An essential characteristic of MCA is the consideration of various evaluation criteria, which are weighted in the course of the analysis. In MCA approaches, diverse quantitative, qualitative, and fuzzy criteria can be defined to reflect different kinds of effects as well as trade-offs and synergies. Even CBA results can be included in MCA.

Recently, there has been a push for greater public participation and the inclusion of non-governmental stakeholders in project appraisal. There are different methods for including participation, such as scenario workshops, mediated modeling, and social multicriteria evaluation (Munda 2005). However, some difficulties have been identified for participatory MCA. Several methodological issues and questions are still subject to debate, such as: What MCA method and which participatory approach should be selected for a particular evaluation problem? Who should determine the criteria? Who decides on the weightings? Who is to be included in the participation process? How can objective results be attained (Messner 2006)?

With regard to river environmental management in Japan, there have been changes to the legal framework in recent years. Amendments to the first article of the River Law in 1997 expanded the aims of river management to include river environment management and conservation as well as flood control and water use. Further, the 16th article indicates that river improvement planning systems should be designed to incorporate the opinions of local residents. This is similar to the EU Water Framework Directive in 2000, which requires the 'active involvement' of interested parties in developing water resources and environment management plans (Article 14) (Burgess and Clark 2009; Klauer et al. 2006; Messner 2006).

There are challenges in meeting these requirements. For instance, the Yodo River watershed committee provided a report to a director of the Kinki area bureau of the Ministry of Land, Infrastructure and Transport, on methods for incorporating the opinions of local residents. The committee examined past attempts to incorporate the views of local residents and recommended two procedures be used. The first was a procedure which consisted of the provision of written options to local residents together with an explanation at a meeting. The second was a process that consisted of a dialogue with local residents through a series of meetings (the Yodo River watershed committee 2007). However, they tried both approaches and could not decide which is to be the best so they are still seeking the best method for incorporating the opinions of local residents.

In the above report, the committee pointed that only small portion of the participants at the meetings had stated their view, so it was difficult to be sure of the strength of various opinions. This indicates that there is a need to develop a more objective approach for collecting local residents' opinions. In the next section, it is shown that one approach is to use a questionnaire survey together with a systems analysis.

16.3 Waterside Environmental Management

16.3.1 Adaptive Waterside Planning Methodology

We have previously proposed an adaptive waterside environmental management process which uses a systems analysis methodology (Hagihara et al. 1995; Hagihara et al. 1998; See also Fig. 16.1). Systems analysis is a systematic methodology used to solve complicated problems. It is designed to aid decision maker(s) choose the best actions from a range of alternatives. The analysis is done through a series of processes, which include clarifying the decision maker(s)' purpose, evaluating comparative alternatives systematically, and developing new alternatives, if necessary. The adaptive waterside environmental management process we proposed consists of the following processes: (i) problem identification; (ii) field research; (iii) analysis to clarify the key issues; (iv) modelling alternatives; (v) evaluation which includes, for example, CBA, MCA and Social Impact Assessment; and (vi) conflict management if there are conflicts among stakeholders. This process provides a methodology for decision aiding, not decision making. If insufficient evaluation results are obtained and/or there is no possibility of compromise, we would return to the process of modelling alternatives and problem identification. Thus a particular feature of the systems analysis here is that the whole process is in fact a nested (Hagihara 2008). In this adaptive waterside environmental management methodology, each process requires information, for example, the physical

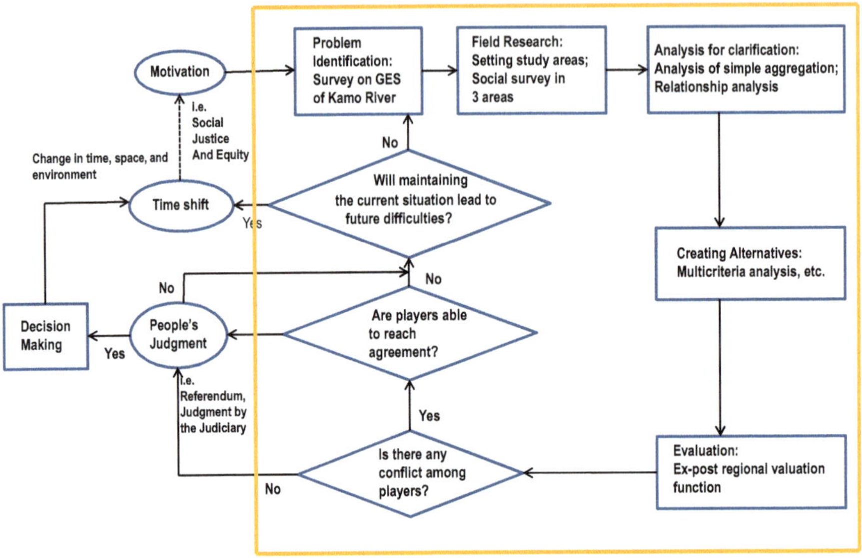

Fig. 16.1 Adaptive waterside planning methodology: inside the orange frame

features of the waterside environment; and what residents are concerned about and want. As a result, public participation for decision aiding is realized in each process by taking advantage of a common set of information. This information, derived from a survey of residents, using a well-designed questionnaire, can provide a lot of the information, and thus can meet the requirements for public participation. Participation by residents with diverse opinions is secured by statistical sampling.

16.3.2 Outline of Survey Watershed Areas

There are three layers of environment relevant in this study (Fig. 16.2): the Geo-environment where the law of geophysics governs, e.g. water quality, quality of landscape, slope of bank, risk of flood, risk of land slide; the Eco-environment where the law of ecology governs, e.g. number of trees, quantity of fish, quantity of birds; and the Socio-environment where rules of society govern, e.g. number of people, quality of daily life, economic stability, existence of toilets, quantity of waste, and the existence of play areas and festivals (Hagihara et al. 1998; Hagihara and Hagihara 2004). If the geological space where the Geo-environment exists is the entire earth, the Eco-environment cannot exist without the Geo-environment, and the Socio-environment without the Geo-and Eco-environments. Humans are members of the Geo-, Eco- and Socio-environments.

The features of physical Geo-, Eco- and Socio environment in case study areas have been already reported (Hagihara et al. 2009a, b). The followings are brief

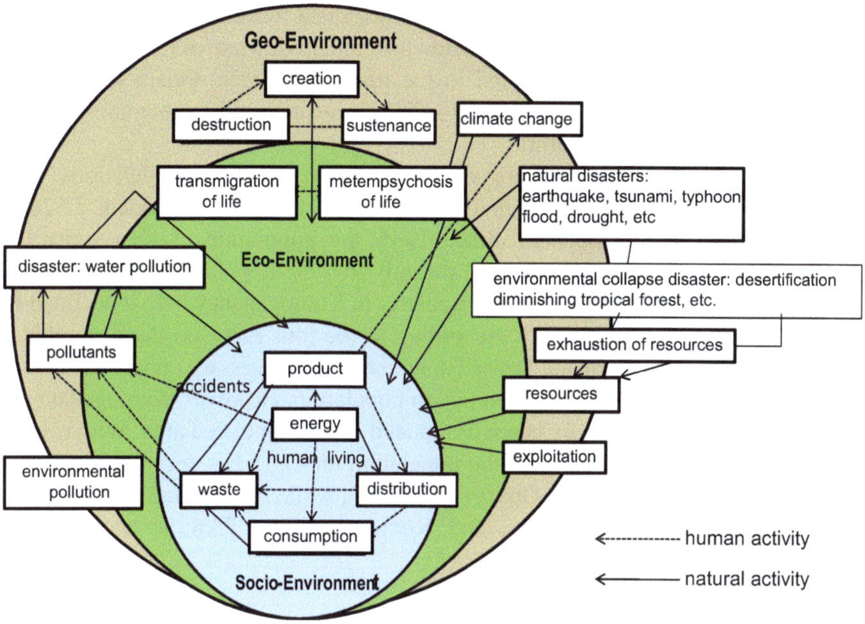

Fig. 16.2 GES (Geo-, Eco- and Socio-) environment

description of survey items on three environments. First, as for Geo-environment, a meteorological and hydrological condition, i.e., annual precipitation, river flow rate, water quality and ground water level were researched.

Seismic, flood and landslide risks were also investigated. Second, as for Eco-environment, vegetation of the Kamo River basin and distribution of animals in the area were surveyed. Then, with regard to Socio-environment, first, water utilization networks of water supply, sewage pipes, and rain water drain pipes were investigated. Second, network of roads were also investigated.

16.3.3 Selection of Case Study Research Areas

In order to select research areas each of which has a different environment with regard to GES environment along the Kamo River flows through Kyoto city, regional analysis was conducted with the aid of a principle component analysis in Kyoto city. Items used in the principle component analysis are as follows: population density, component distribution ratio of population by male and female age group including 0–14 years old; 15–29 years old; 30–64 years old and over 65 years old, component distribution ratio of the working population, vegetation areas including area of urban district; area of an artificial forest, area of a natural coniferous and broad-leaved forest, area of agricultural land.

Three areas, i.e. Suemaru, Ohara and Kumogahata, were selected as a result of the above analysis. Suemaru is located downstream, in downtown Kyoto city, while Ohara and Kumogahata are located in the upper river area. Ohara is a famous historical sightseeing area, while Kumogahata is located in the mountains and has no significant sightseeing features.

Demographically the areas are very distinct. In 2005 the populations were 1,475,000 in Kyoto city, 2,792 in Dohda where Suemaru is included, 2,526 in Ohara and 218 in Kumogahata. Since 1985, the population of Kyoto city has changed little and has even increased recently, but the population of two areas in the upper river area has decreased, particularly in Kumogahata where it declined by 30.1 % between 1985 and 2005. According to the four Laws on the depopulated areas since 1970, Kumogahata has satisfied one of the necessary conditions for the depopulated areas with regard to decline in population. However, Kumogahata is a part of Kyoto city and the city is not designated as a depopulated area. With respect to the structure of the population, the population under 15 has decreased in Kyoto city as a whole, particularly in Ohara (by 14.1 %), while the population over 65 has significantly increased in Ohara and Kumogahata, by 23.6 % and 18.2 %, respectively.

16.3.4 The Social Survey

Preliminary field research was undertaken in each area consisting of interviews of a range of residents about not only the waterside environment, but also their living circumstances. For example, with regard to their living circumstances, residents of Kumogahata, were very concerned about the sustainability of their living conditions. In particular they complained about poor medical services, inconvenient shopping, and the very poor bus services to the centre of the city where a hospital and a shopping centre are located. They were even worried about whether the area will continue to exist in the future.

Following this, to quantify residents' concerns and wishes for the waterside environment, a survey was conducted of residents of the three areas. The questionnaire was composed of the following types of questions: 'How are you aware of the waterside?', 'What is your impression of it?', and 'What actions do you want to see concerning it?'. Each questionnaire took into account the differences of the Geo-, Eco-, and Socio-environment in the three areas with the aid of the earlier field surveys, interviews with residents, the KJ method and the ISM (Interpretive Structuring Modeling) method. Specifically, the questionnaire is divided into sections relating to the Geo-, Eco-, and Socio-environment. The questionnaire also included common items in order to compare results among areas (See Hagihara et al. 2009a).

We conducted a mail survey of all households in Suemaru in 2006 and in the historical sightseeing communities of Ohara and Kumogahata in 2007. The percentage of respondents (number of respondents) were 44 %(61), 28 %(61), and 63 %(45), respectively (Fig. 16.3).

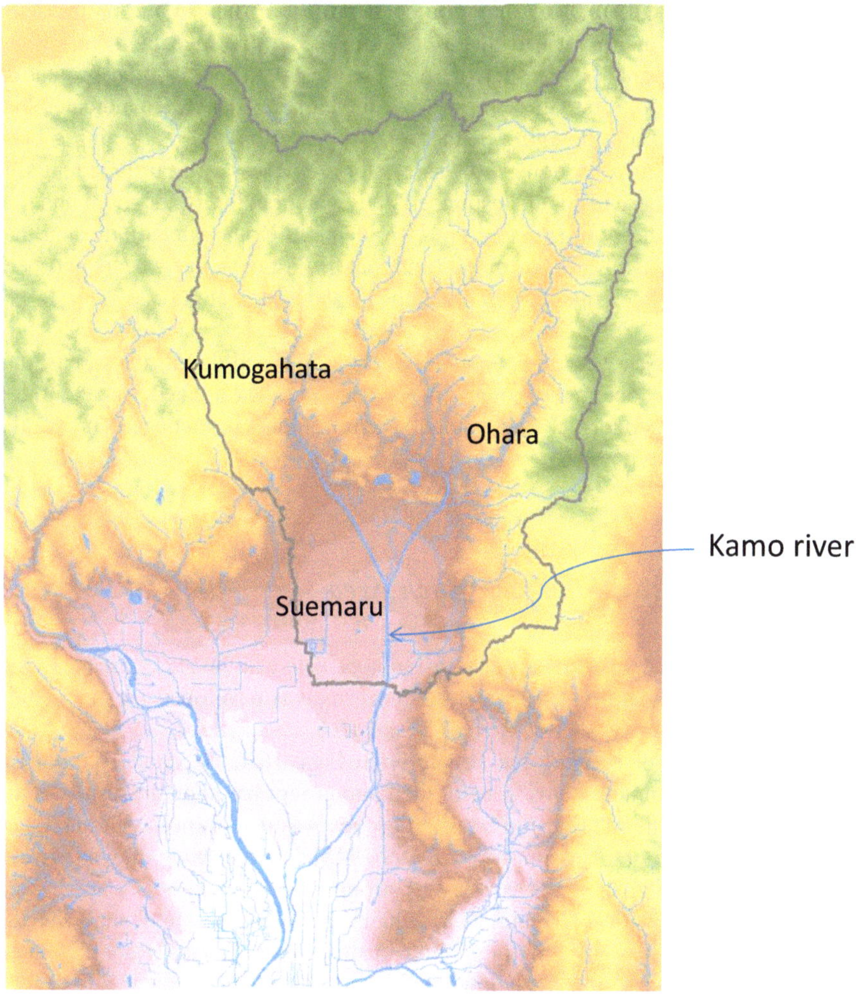

Fig. 16.3 Study areas in Kyoto city

16.4 Results of a Social Survey

16.4.1 Results of Simple Aggregation

The majority of residents in all areas responded that they are happy to live along the river. In Suemaru in the downstream area, residents responded that the 'river terrace is sufficiently wide', they 'enjoy the spring season (cherry blossoms)', and they are 'happy to see Daimonji-Okuribi (which is a famous religious event on the mountain over the river)', and so on. They enjoy playing at the waterside of the Kamo River and seeing cherry blossoms.

On the other hand, in both upper river areas, more than 70 % of residents responded that ‘industrial waste disposal facility is a nuisance to them’, ‘wild monkeys are a nuisance’, ‘insufficient forest is preserved’, and ‘flood protection is insufficient’. In addition to these responses, in Ohara, residents also indicated that ‘they are anxious about the aging of the area’s population and its declining birth rate’ and that ‘measures for domestic wastewater are necessary’. Further, a significant proportion of these residents responded that ‘water quality is not good’ and that ‘open space along the river is necessary’. Residents in Kumogahata indicated that ‘they feel at risk of landslide and flood’ and that ‘wild boars, deer, bears and bees are a nuisance’, ‘water quality is good’ and ‘open space along the river is not necessary’.

Different responses were thus obtained in the upper and down river areas, and even in the same upper river area. Given this, there is a need to take these differences into account in the waterside environmental management of the Kamo River.

16.4.2 GES Environment in Each Area

In order to systematically analyze the differences between the study areas, a GES environmental structure of the residents’ concerns was developed for each study area. They are depicted in Figs. 16.4, 16.5 and 16.6.

The procedure for creating the structures in Figs. 16.4, 16.5 and 16.6 was as follows: (i) Questionnaire items were divided into groups based on the results of the social survey; (ii) the relationship between groups was determined with the aid of Cramer’s Coefficient of Relationship (Cramer 1946) and lines were drawn between groups if their relationship was assessed as strong.

Cramer’s coefficient of relationship which use χ^2 -value and analyzes the relationship between two items is applied in this study. Cramer’s coefficient of relationship, ∅, is described as below.

$$\varnothing = \left[\chi^2/\{N(k-1)\}\right]^{1/2}$$

where χ^2 is χ^2 -value, N is sample number, k is the number of less category, or option in two items. The range of the value ϕ is between 0 and 1.0. With regard to the intensity of relationship, if $\phi \geq 0.3$, it is interpreted as strong relationship between two items.

In the figures, the names of the groups are written in bold characters. Based on these structures, there are a number of observations:

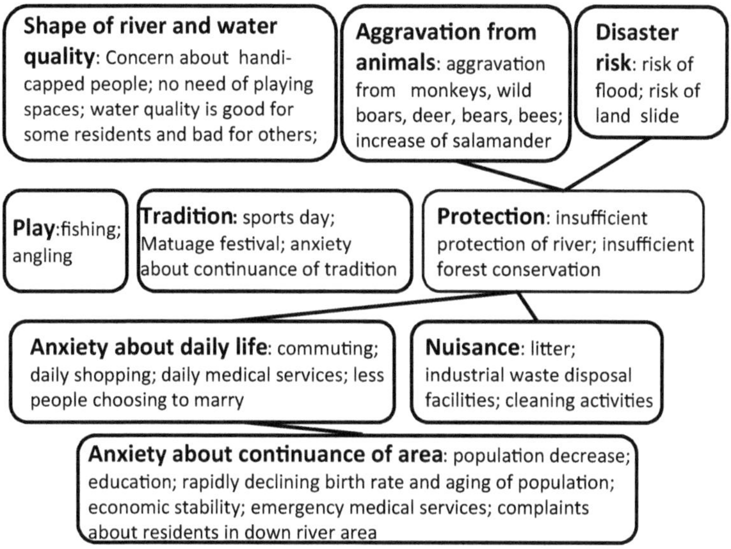

Fig. 16.4 GES environmental structure of residents' concerns in the upper river area of Kumogahata

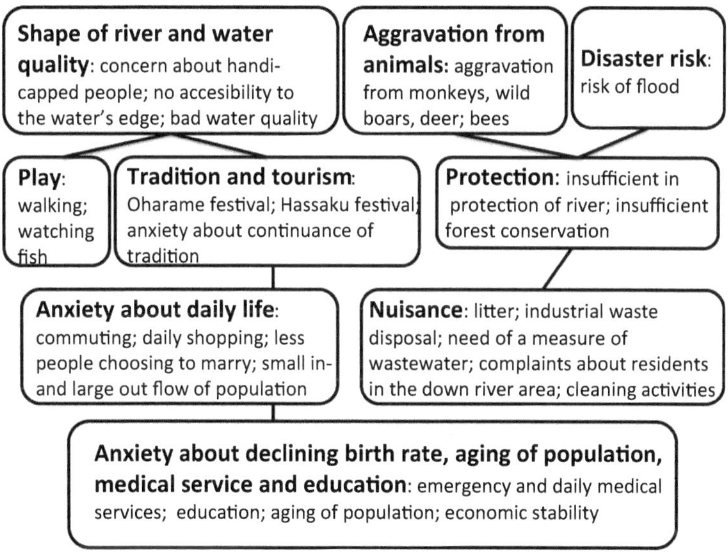

Fig. 16.5 GES environmental structure of residents' concerns in the upper river areas of Ohara

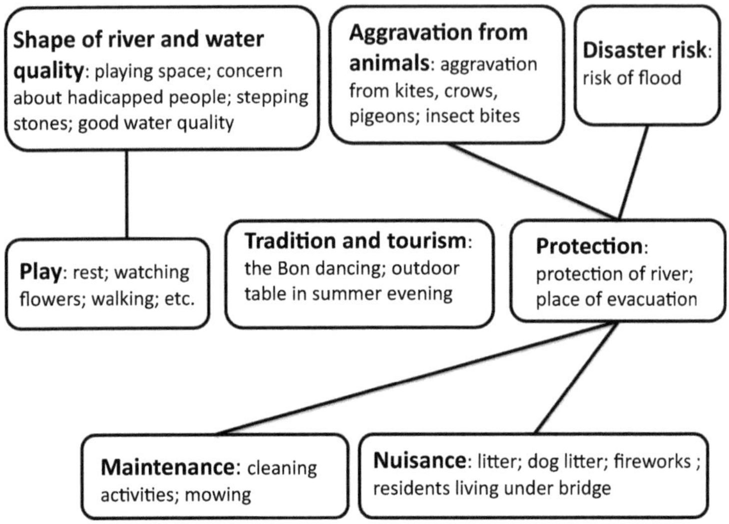

Fig. 16.6 GES environmental structure of residents' concerns in the down river area of Suemaru

16.4.2.1 Difference of Concerns Between Areas

Although the concerns are labeled with the same group names, what they refer to differs between areas. For example, 'disaster risk' appears in all figures, but in Kumogahata 'risk of landslide' is a concern in addition to 'risk of flood'.

Moreover, 'shape of river and water quality' appears in all figures and what it refers to is almost the same for each area, but it is viewed quite differently: with open space being sufficient in the down river area, while it is not sufficient in Ohara and it is not necessary in Kumogahata. Regarding 'loss of animals', the type of animal referred to differs; that is, the loss of kites and other birds are a concern in the down river area, while residents of the upper river areas are concerned about the welfare of mammals, such as deer. Further, the loss of the Japanese giant salamander, which is the world's largest amphibian and a national protected animal in Japan, is a concern in Kumogahata, but not in other areas.

In Kumogahata in particular, residents were anxious about daily life (i.e., commuting, medical services, and fewer people choosing to marry) as well as the continuance of the area (i.e., population decrease, economic stability, emergency medical services, and so on). On the other hand, residents in Ohara were not anxious about the continuance of the area in spite of being in the same upper river area. Residents in Suemaru were not anxious about the above issues at all.

16.4.2.2 Difference Among Groups Which Have a Strong Relationship with 'Shape of River and Water Quality'

The strong connection between 'play' and 'shape of river and water quality' implies that a substantial plan incorporating "play" is needed in the down river area. In Ohara there are strong concerns with 'tradition and sightseeing', because a lot of residents earn their living through tourism. Thus, in addition to improving the open space for playing and walking at the waterside, there is a need to take tourism into account in Ohara. On the other hand in Kumogahata residents think that there is no need to improve the waterside. What improvement at the waterside entails differs between areas, implying that these differences would need to be considered in waterside management.

16.4.2.3 The Relationship Between Items in the Geo- and Eco-environment and the Socio-environment

It is noted that groups of 'disaster risk (Geo-)' and 'loss caused by animals (Eco-)' are connected to 'nuisance (Socio-)' and 'anxiety about daily life (Socio-)'. This implies that measures for dealing with disasters should include, as well as physical ones, measures for reducing nuisance activity, cleaning activity, and other items of the Socio-environment. Consequently, 'protection of river' should be considered from the viewpoint of the Socio-environment, as well as that of the Geo- and Eco-environment. For example, in the case of dam construction (very common in regional development initiatives in Japan), not only should the effectiveness of flood control in the down river area be a consideration, but also the improvement or maintenance of the quality of daily life in the upper river areas should be adopted as criteria in weighing up alternative proposals.

16.5 Evaluation of Each Waterside Environment

To more precisely calibrate residents' perceptions of their waterside environment, a GES environmental valuation function is created for each area. We will be able to clarify "what" elements of GES environment and "how" they are evaluated by people with the aid of the GES environmental valuation function.

16.5.1 Creating Valuation Functions

The development of a GES environmental valuation function for each area was undertaken as follows: Numbers of questionnaire items are approximately more

Table 16.1 The results of the Hayashi's Quantification Theory III in Suemaru

j	Items	$x_{s,1j}$	$x_{s,2j}$	$x_{s,3j}$
1	River terrace is adequate	−0.037	0.012	−0.007
2	Recognize risk of flood	0.019	0.159	−0.036
3	Fond of cherry blossoms	−0.023	−0.024	0.013
4	Eliminate introduced species	−0.042	0.004	0.021
5	Knowledge about loss by kite	0.103	0.022	0.026
6	Rest	0.085	−0.025	0.133
7	Watching flowers	0.070	−0.015	−0.083
8	Dog litter	0.040	−0.047	−0.054
9	Enjoy watching fireflies	−0.084	−0.058	−0.057
10	Enjoy Okuribi	0.003	0.011	0.011
11	Participate in cleaning activities	−0.014	0.004	0.003
12	Appreciate lack of grass	−0.047	−0.010	0.024

than 30. First, we are trying to get useful information on the GES environment in each area from these many questionnaire items. Twelve representative variables were selected for explaining the GES environment in each area based on a following criteria, that is, they include items which are elements of each Geo-, Eco- and Socio-environment, and items to which many people respond (see the second column of Tables 3–5; Hagihara et al. 2009b). Then, using these items, valuation axes were obtained with the aid of the Hayashi's Quantification Theory III (Yasuda and Unno 1977). Finally, the GES environmental valuation function in each area is derived.

Tables 16.1, 16.2 and 16.3 show the results. The first subscript identifies the area, that is, s means Suemaru, o means Ohara and k means Kumogahata, and the second subscript identifies j item at the relevant axis. For example, $x_{s,1j}$ means j item's category score on the first axis in Suemaru.

Three valuation axes were obtained in each area from the results of the Hayashi's Quantification Theory III (Hagihara et al. 2009b). Numbers of valuation axes are decided by a criterion, that is, axes' cumulative contribution ratio is more than 50 %. This value, 50 %, is derived from the idea that we should use information as much as possible, but it is difficult, so we should use at least a half of the information which is given by people. The interpretation of these axes are shown in Table 16.4. In the upper river area there are axes which indicate concern about pleasure at the waterside and difficulties in daily life, while in the down river area there are axes which indicate anxiety about the waterside as a user and anxiety about flooding. In Table 16.4, the values in parenthesis are the cumulative contribution ratio and the values in each box are the contribution ratio of each axis.

A GES environmental valuation function is defined with sample scores and weights as follows:

Table 16.2 The results of the Hayashi's Quantification Theory III in Ohara

j	Items	$x_{o,1j}$	$x_{o,2j}$	$x_{o,3j}$
1	Water quality is bad	−0.037	0.038	−0.036
2	Easy access to the water's edge	0.200	0.039	0.020
3	Recognition of flood risk	0.026	0.028	−0.041
4	Increase the amount of the areas of coppice	0.038	0.005	0.001
5	Fond of pale chub (fish)	−0.032	−0.053	−0.033
6	Walking	0.002	−0.064	0.100
7	Illegal waste disposal	−0.037	−0.052	0.105
8	Enjoy the Hassaku dance	0.023	−0.058	−0.043
9	Aging of population	−0.022	−0.010	−0.015
10	Daily medical service	−0.067	0.057	−0.022
11	Daily shopping	−0.004	0.100	0.070
12	Continuance of tradition	0.025	−0.055	−0.042

Table 16.3 The results of the Hayashi's Quantification Theory III in Kumogahata

j	Items	$x_{k,1j}$	$x_{k,2j}$	$x_{k,3j}$
1	Water quality is good	−0.026	0.088	0.057
2	Easy access to the water's edge	−0.067	0.122	−0.008
3	Recognition of flood risk	−0.032	−0.012	−0.028
4	Increase the amount of the area of coppice	−0.020	−0.065	0.179
5	Fond of rhododendron	−0.034	−0.086	−0.007
6	Fond of Japanese sculpin	−0.052	−0.052	0.032
7	Deer are annoying	−0.037	0.019	−0.045
8	Enjoy Matuage festival	0.003	0.004	−0.034
9	Aging of population	0.054	0.024	0.062
10	Daily medical services	0.122	−0.004	−0.038
11	Daily shopping	0.085	−0.049	−0.030
12	Emergency medical services	0.103	0.053	−0.002

Table 16.4 An interpretation of axes

	Suemaru (53.3 %)	Ohara (51.6 %)	Kumogahata (50.2 %)
1st axis	Anxiety about the waterside as a user 24.8 %	Pleasure at the water-side: 21.3 %	Difficulties in daily life: 20.2 %
2nd axis	Anxiety about flood 16.1 %	Difficulties in daily life: 18.9 %	Pleasure at the water-side: 17.5 %
3rd axis	Feeling of the season 12.4 %	Complaint about residents in down river areas: 11.4 %	Complaint about the area: 12.6 %

$$D_i = \sum_r w_r \sum_j \frac{\delta_i(j) x_{rj}}{l_i},$$

where

$$\delta_i(j) = \begin{cases} 1 & \text{if sample i responds to item j} \\ 0 & \textit{otherwise} \end{cases},$$

w_r is weight at each axis, l_i is the number of items which sample i responds to 12 items, x_{rj} is category scores of j items at axis r.

Although it is not easy to decide weights in multicriteria analysis, contributing ratio on each axis is used as the weight in this chapter. Thus, the GES environmental valuation function for each area is as follows:

In Suemaru:

$$D_{si} =^{si} -\frac{24.8}{l_i}\sum_{j=1}^{12} \delta_i(j) x_{1j} - \frac{16.1}{l_i}\sum_{j=1}^{12} \delta_i(j) x_{2j} - \frac{12.4}{l_i} 12 \sum_{j=1}^{12} \delta_i(j) x_{3j}$$

In Ohara:

$$D_{oi} = \frac{21.3}{l_i}\sum_{j=1}^{12} \delta_i(j) x_{1j} - \frac{18.9}{l_i}\sum_{j=1}^{12} \delta_i(j) x_{2j} - \frac{11.4}{l_i}\sum_{j=1}^{12} \delta_i(j) x_{3j}$$

In Kumogahata:

$$D_{ki} = -\frac{20.2}{l_i}\sum_{j=1}^{12} \delta_i(j) x_{1j} + \frac{17.5}{l_i}\sum_{j=1}^{12} \delta_i(j) x_{2j} - \frac{12.6}{l_i}\sum_{j=1}^{12} \delta_i(j) x_{3j},$$

where subscripts s, o and k represent Suemaru, Ohara and Kumogahata, respectively.

These GES environmental valuation functions enable us to see how each item affects the sample i, i.e., people's GES environmental value and in what manner and by how much.

Further, the effect of item j on the GES environmental value in each area, x_j $(j = 1, .., 12)$, is defined as follows

$$x_j = \sum_r w_r x_{rj}$$

where x_{rj} is category scores of j items at axis r and w_r is the contributing ratio on each axis.

The value of x_j is the magnitude and the direction of effect by item j on the GES environmental valuation function. For example, if x_j is positive, the GES environmental value increases, and if it is negative, the GES environmental value decreases. The score in each area is shown as follows:

In Suemaru,

$$x_{sj} = -24.8x_{1j} - 16.1x_{2j} - 12.4x_{3j}$$

In Ohara,

$$x_{oj} = 21.3x_{1j} - 18.9x_{2j} - 11.4x_{3j}$$

In Kumogahata,

$$x_{kj} = -20.2x_{1j} + 17.5x_{2j} - 12.6x_{3j}$$

16.5.2 Identifying Priorities for Action

Based on the above equations, alternative actions can now be proposed for waterside management in each location.

16.5.2.1 Suemaru

Scores for Suemaru are shown in Table 16.5. From the total scores, the items 'rest', 'knowledge about the loss of kites', and 'recognize risk of flood' decrease GES environmental value in Suemaru. On the other hand, the items 'enjoy watching fireflies' and 'appreciate lack of grass' increase the GES environmental value. Consequently, important elements of waterside management should include actions that decrease the fear of the loss of kites and the risk of flood; conserve the ecological environment which will bring about the enjoyment of watching fireflies and the enjoyment of the season; as well as regular mowing and improving the area for walking.

16.5.2.2 Ohara

Scores for Ohara are shown in Table 16.6. The items 'daily shopping is inconvenient', 'daily medical service is inconvenient', 'water quality is bad', and 'illegal disposal of waste is annoying' decrease GES environmental value in Ohara. On the other hand, the items 'easy accessibility to the water edge' and 'enjoy the (traditional) Hassaku dance' increase the GES environmental value significantly. Consequently, introducing measures to improve drainage facilities for wastewater, thus increasing the water quality, as well as measures to improve accessibility to the

Table 16.5 Scores of Suemaru

j	Items	Anxiety and complaint in daily use	Anxiety about flood	Feeling of the season	Total scores
1	River terrace is adequate	0.918	−0.187	0.086	0.818
2	Recognize risk of flood	−0.465	−2.556	0.443	−2.578
3	Fond of cherry blossoms	0.559	0.393	−0.166	0.786
4	Extinguish introduced species	1.046	−0.063	−0.263	0.719
5	Knowledge about loss by kite	−2.553	−0.355	−0.321	−3.23
6	Rest	−2.114	0.405	−1.646	−3.355
7	Watching flowers	−1.746	0.244	1.034	−0.467
8	Dog litter	−0.981	0.76	0.67	0.448
9	Enjoy watching fireflies	−0.084	0.929	0.706	1.551
10	Enjoy Okuribi	0.351	−0.178	−0.137	0.036
11	Participate in cleaning activities	0.430	−0.065	−0.043	0.322
12	Appreciate lack of grass	1.163	0.158	−0.294	1.027

water's edge and to create substantial playing spaces should be a priority. Further, daily life should be improved to the minimum necessary for residents to continue to live in the community, and for the community to continue to function.

16.5.2.3 Kumogahata

Scores for Kumogahata are shown in Table 16.7. The items 'it is good to increase the amount of wooded areas', 'daily shopping is inconvenient' and 'daily medical service is inconvenient' decrease the GES environmental value in Kumogahata significantly. According to the interviews with residents, people who think that it is good to increase the amount of wooded areas also desire other changes in the area, in particular a transition from the timber industry to an alternative industry. On the other hand, the items 'it is easy to go the water's edge' and 'water quality is good' increase the GES environmental value in Kumogahata. Consequently, the improvement of the quality of daily life is the top priority, noting that it is not an easy task for governments to make changes to improve some aspects daily life.

Table 16.6 Scores of Ohara

j	items	Anxiety and complaint in daily use	Anxiety about flood	Feeling of the season	Total scores
1	River terrace is adequate	0.918	−0.187	0.086	0.818
2	Recognize risk of flood	−0.465	−2.556	0.443	−2.578
3	Fond of cherry blossoms	0.559	0.393	−0.166	0.786
4	Extinguish introduced species	1.046	−0.063	−0.263	0.719
5	Knowledge about loss by kite	−2.553	−0.355	−0.321	−3.23
6	Rest	−2.114	0.405	−1.646	−3.355
7	Watching flowers	−1.746	0.244	1.034	−0.467
8	Dog litter	−0.981	0.76	0.67	0.448
9	Enjoy watching fireflies	−0.084	0.929	0.706	1.551
10	Enjoy Okuribi	0.351	−0.178	−0.137	0.036
11	Participate in cleaning activities	0.430	−0.065	−0.043	0.322
12	Appreciate lack of grass	1.163	0.158	−0.294	1.027

Table 16.7 Scores of Kumogahata

j	items	Difficulties in daily life	Pleasure at the waterside	Complaint about the area	Total scores
1	Water quality is good	0.517	1.548	−0.718	1.347
2	Easy access to the water's edge	1.351	2.142	0.102	3.594
3	Recognition of flood risk	0.639	−0.217	0.358	0.781
4	Increase the amount of the area of coppice	0.396	−1.139	−2.253	−2.995
5	Fond of rhododendron	0.683	−1.498	0.090	−0.725
6	Fond of Japanese sculpin	1.052	−0.903	0.406	0.555
7	Deer are annoying	0.749	0.328	0.561	1.637
8	Enjoy Matuage festival	−0.067	0.073	0.432	0.438
9	Aging of population	−1.083	0.413	−0.781	−1.452
10	Daily medical services	−2.466	−0.078	0.479	−2.065
11	Daily shopping	−1.724	−0.862	0.372	−2.214
12	Emergency medical services	−2.073	0.933	0.025	−1.115

16.6 Concluding Remarks

The ambiguity in the definition of sustainability has led to many different interpretations. In this paper we first showed that the concept of "sustainability" should include both sustainability and survivability. We then explored an adaptive waterside environmental management process, which includes public participation, in case studies in the upper and down river areas of the Kamo River in Kyoto city in Japan.

In the adaptive waterside environmental management process, it has been shown that both upper and down river areas are able to be considered from the viewpoint of not only sustainability but also survivability. In particular, it has been shown that it is necessary to take into account differences among residents both in rural and urban areas in a region. Furthermore, the kind of measures which are needed to improve the riverside problems of each area can be identified through the use of an environmental valuation function based on a survey of residents. Thus it is shown the waterside environmental management process presented in this paper can take into account both sustainability and survivability for different areas and populations in a region.

Acknowledgement We are grateful to Dr. Wendy Jarvie for helpful comments and suggestions.

References

Adger, W. N., & Jordan, A. (2009). Sustainability: Exploring the processes and outcomes of governance. In W. N. Adger & A. Jordan (Eds.), *Governing sustainability* (pp. 3–31). Cambridge: Cambridge University Press.

Burgess, J., & Clark, J. (2009). Practitioner evaluation of participatory process in environmental decision making. In W. N. Adger & A. Jordan (Eds.), *Governing sustainability* (pp. 159–190). Cambridge: Cambridge University Press.

Cramer, H. (1946). *Mathematical methods of statistics*. Princeton: Princeton University Press.

Figueira, J., Greco, S., & Ehrgott, M. (Eds.). (2005). *Multiple criteria decision analysis: State of the art surveys*. New York: Springer.

Getzner, M., Spash, C. L., & Stagl, S. (2005). *Alternatives for environmental valuation*. Oxford: Routledge.

Gowdy, J. M. (2004). The revolution in welfare economics and its implications for environmental valuation and policy. *Land Economics, 80*(2), 239–257.

Hagihara, Y. (2008). *Kankyo to bousai no doboku keikakugaku* (*Adaptive system planning methodology for environment risk management*). Kyoto: Kyoto University Press.

Hagihara, K., & Hagihara, Y. (1991). The role of intergovernmental grants in underpopulated regions. *Regional Studies, 25*, 163–172. doi:10.1080/00343409112331346377.

Hagihara, K., & Hagihara, Y. (2004). The role of environmental valuation in public policymaking: The case of urban waterside area in Japan. *Environment and Planning C: Government and Policy, 22*, 3–13. doi:10.1068/c02101s.

Hagihara, Y., & Sakamoto, M. (2006). *Konfurikuto manejimento: mizusigen no sosharu risuku (Conflict management: Social risk in water resources' problems)*. Tokyo: Keiso Shobo

Hagihara, Y., Takahashi, K., & Hagihara, K. (1995). A methodology of spatial planning for waterside area. *Papers in Regional Science, 25*(2), 19–45.

Hagihara, Y., Hagihara, K., & Takahashi, K. (1998). *Toshi kankyo to mizube keikaku: sisutemuzu anarisisu ni yoru (Urban environment and waterside planning: With the aid of systems analysis)*. Tokyo: Keiso Shobo.

Hagihara, Y., Hagihara, K., Matsushima, T., & Kawano, M. (2009a). Kamogawa ryuiki no shakai-seitai kankyo no henka (Socio and eco environment changes of Kamo River basin). *Annals of Disaster Prevention Research Institute of Kyoto University*, 52(B), 831–850.

Hagihara, Y., Hagihara, K., & Kawano, M. (2009b). Jo-karyuiki no chiikitokusei ni motozuita mizube kankyo manejimento ni kansuru ichi-kousatu (A study on environmental management through characteristics of up and down stream regions). *Annals of Disaster Prevention Research Institute of Kyoto University* 52(B), 885–897.

Hanley, N. (2001). Cost-benefit analysis and environmental policymaking. *Environment and Planning C: Government and Policy, 19*, 103–118. doi:10.1068/c3s.

Hanley, N., & Barbier, E. B. (2009). *Pricing nature: Cost-benefit analysis and environmental policy*. Cheltenham: Edward Elgar.

Hanley, N., & Spash, C. L. (1993). *Cost-benefit analysis and the environment*. Cheltenham: Edward Elgar.

Jordan, A. (2008). The governance of sustainable development: Taking stock and looking forwards. *Environment and Planning C: Government and Policy, 26*, 17–33. doi:10.1068/cav6.

Kallis, G., Videira, N., Antunes, P., Pereira, A. G., Spash, C. L., Coccossis, H., Quintana, S. C., del Moral, L., Hatzilacou, D., Lobo, G., Mexa, A., Paneque, P., Mateos, B. P., & Santos, R. (2006). Participatory methods for water resources planning. *Environment and Planning C: Government and Policy, 24*, 215–234. doi:10.1068/c04102s.

Klauer, B., Drechsler, M., & Messner, F. (2006). Multicriteria analysis under uncertainty with IANUS – Method and empirical results. *Environment and Planning C: Government and Policy, 24*, 235–256. doi:10.1068/c03102s.

Messner, F. (2006). Applying participatory multicriteria methods to river basin management: Improving the implementation of the water framework directive. *Environment and Planning C: government and Policy, 24*, 159–167. doi:10.1068/c2402ed.

Munda, G. (2005). Multiple criteria decision analysis and sustainable development. In J. Figueira, S. Greco, & M. Ehrgott (Eds.), *Multiple criteria decision analysis: State of the art surveys* (pp. 953–986). New York: Springer.

Munda, G. (2010). *Social multi-criteria evaluation for a sustainable economy*. Berlin/Heidelberg: Springer.

Nas, T. F. (1996). *Cost-benefit analysis: Theory and application*. London: Sage Publications. Japanese edition: Nas, T. F. (2007). *Hiyou ben-eki bunseki: riron to ouyou* (trans: Hagihara, K. et al.) Tokyo: Keiso shobo.

Nijkamp, P. (1977) *Theory and application of environmental economics*. North-Holland, Amsterdam; Translated by Fujioka, A., Hagihara, K., & Kanesawa, T., et al. (1985). *Kankyo-keizaigaku no Riron to Ouyo*. Tokyo: Keiso Shobo.

O'Neill, J. H. (2009). Sustainability, welfare and value over time. In W. N. Adger & A. Jordan (Eds.), *Governing sustainability* (pp. 283–304). Cambridge: Cambridge University Press.

O'Neill, J. H., Holland, A., & Light, A. (2008). *Environmental values*. London: Routledge.

The Yodo River Watershed Committee. (2007). *Verdict: Toward the advanced people participation*. Osaka: Kinki bureau of the MLIT.

Vincke, P. (1992). *Multicriteria decision-aid*. Chichester: John Wiley & Sons.

World Commission on Environment and Development. (1987). *Our common future*. Oxford: Oxford University Press.

Yasuda, S., & Unno, M. (1977). Shakai Tokeigaku (*Social Statistics*). Maruzen, Tokyo.

Chapter 17
Third Party Intervention in Conflict Resolution: Dispute Between Bangladesh and India over Control of the Ganges River

Keith W. Hipel, Maiko Sakamoto, and Yoshimi Hagihara

Abstract To demonstrate the strategic influence a third party can have on negotiations, a formal approach to resolving a complex conflict is applied to an important international water resources controversy. Specifically, third party intervention is employed within the framework of the Graph Model for Conflict Resolution (GMCR) to systematically investigate the ongoing dispute between Bangladesh and India over the regulation of the Ganges River by India at the Farakka Barrage located just upstream on the Ganges River in India before it flows into Bangladesh. A general system of systems engineering approach to Third Party Intervention within the GMCR structure is designed to reflect a range of ways in which it can be implemented in practice. Having an insightful and powerful tool like Third Party GMCR permits one to ascertain how a Third Party can guide a serious conflict to a more reasonable resolution which may be mutually beneficial to all concerned parties.

Keywords Ganges river • Farakka barrage • Graph model for conflict resolution • International dispute • Third party intervention • Upstream-downstream conflicts

---

K.W. Hipel (✉)
Department of Systems Engineering, University of Waterloo, Waterloo, ON N2L 3G1, Canada
e-mail: kwhipel@uwaterloo.ca

M. Sakamoto
University of Tokyo, Kashiwa, Chiba 277-8653, Japan
e-mail: m-sakamoto@k.u-tokyo.ac.jp

Y. Hagihara
Kyoto University, 264 Suemarucho, Nakagyo-ku, Kyoto 604-0901, Japan
e-mail: y-hagi@gf7.so-net.ne.jp

K. Hagihara, C. Asahi (eds.), *Coping with Regional Vulnerability*, New Frontiers in Regional Science: Asian Perspectives 4, DOI 10.1007/978-4-431-55169-0_17

17.1 Introduction

Quite often parties in a given dispute become entrenched or locked-in on fixed positions and value systems, especially for a conflict which has been in play over a relatively long period of time. For example, because Iraq is located downstream of Syria on the Euphrates River originating in Turkey, serious conflicts have flared up over the years regarding how water in this river can be fairly shared among the nations. Because water is such a crucial resource to the survival and prosperity of these countries, a number of challenging conflicts have occurred over the years. In 1975, Iraq almost went to war with Syria when Syria failed to release sufficient water in the Euphrates River to downstream Iraq while Syria was filling its newly constructed Keban Reservoir located behind its newly constructed Thawra Dam situated upstream from Iraq. Fortunately, war was avoided by the timely intervention of a Third Party, in this case Saudi Arabia, which offered financial compensation to Syria if it released sufficient quantities of water to downstream Iraq. Hipel et al. (2014) carried out a strategic investigation of this potentially lethal 1975 controversy between Syria and Iraq as well as later disputes which flared up in 1990 and 1998 using the Graph Model for Conflict Resolution (GMCR) (Kilgour et al. 1987; Fang et al. 1993; Hipel et al. 2011b). This constitutes a dramatic illustration of how a timely intervention by a Third Party can produce an outcome which is more preferred by all concerned parties. Accordingly, appropriate assistance by a Third Party not directly involved in the dispute under study, can be extremely important in producing mutually beneficial, win/win or Pareto optimal outcomes. This overall process is commonly referred to as Third Party Intervention and is addressed in detail in this chapter. In fact, Hipel et al. (2014) employed Third Party ideas presented in this chapter and related publications in combination with GMCR to study the aforementioned three connected disputes between Syria and Iraq.

Upstream-downstream conflicts constitute an important class of water resource disputes in which external guidance may be required in order to reach an enhanced resolution. The situation existing in Asia between China and its downstream neighbouring countries constitutes another large-scale example for which Third Party intervention may be required at an increasing rate over time, especially in the face of climate change. Virtually all of the major rivers in South and Southeast Asia, as well as within China itself, trace their source to the massive glaciers largely located within the Himalayan Mountains and Tibetan Plateau Region of China. Hence, China possesses significant control over the hydrological destiny of its many downstream neighbouring nations. Therefore, India, for example, strongly opposes China diverting water from Tibet to the dry northeast region of China. At a bilateral level, Bangladesh and India are currently enmeshed in a bitter dispute over the control of the flow of the Ganges River by India to downstream Bangladesh. In fact, this serious conflict motivated the development of the Third Party Intervention methodology presented in this chapter within the framework of GMCR.

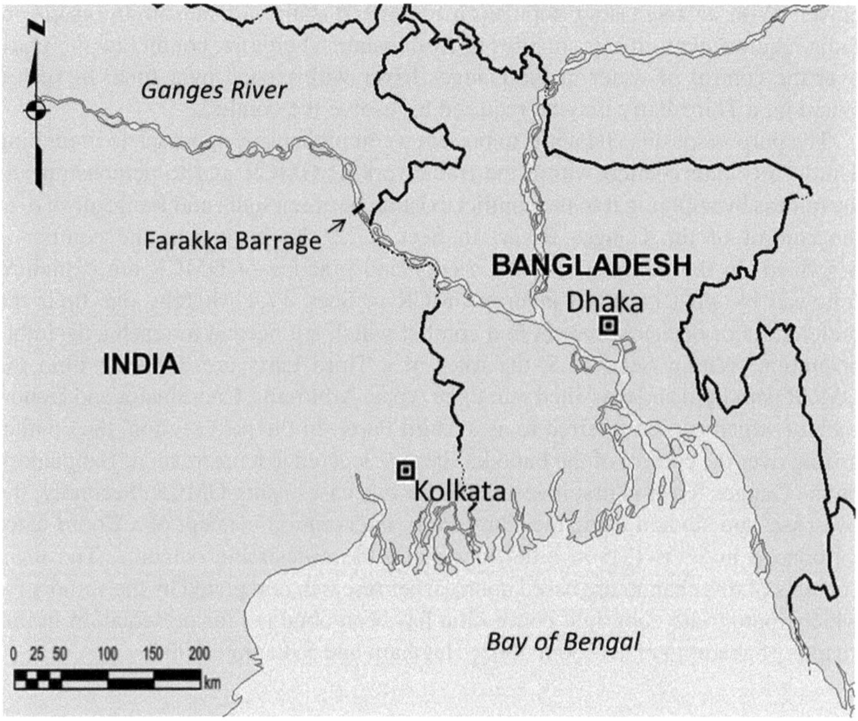

Fig. 17.1 Map of Bangladesh

As can be seen in the map in Fig. 17.1, India lies upstream of Bangladesh on the Ganges River, whose origin is located in the Tibetan Plateau. Because the river constitutes an important source of water for both countries within a region lacking sufficient supplies of fresh water, many conflicts have arisen over the utilization of this valuable water resource both between and within these nations. As the downstream country, Bangladesh is situated in a disadvantageous position. To satisfy water demand and avoid using groundwater polluted by natural arsenic, Bangladesh would like to use as much water as possible from the Ganges River, especially during the dry season. Overall, Bangladesh is in an inferior position, both economically and topographically, to India.

With a population of about 160 million people, Bangladesh is the eighth largest country in the world in terms of the number of people. However, with a land area of only 148,000 km^2, its population density is just over 1,000 people per km^2, which is the largest in the world for a nation having more than 1,000 km^2 of territory. In terms of land area, Bangladesh is ranked as the 87th largest nation. Having a per capita Gross Domestic Product (GDP) income of only 960 USD, places Bangladesh at a rank of 155 among the countries in the world (International Monetary Fund 2014). Poverty is rampant throughout the country. In contrast, India is one of the most rapidly developing nations in the world and is becoming a major economic power. With an increasing population in concert with vast industrial expansion, India requires ever greater supplies of fresh water. Therefore, conflict taking place over the control of water in the Ganges River will expand over time. In such a situation, a Third Party may be required to resolve the conflict.

The purpose of this chapter is to present a comprehensive approach to modelling Third Party intervention within the framework of GMCR and to demonstrate its usefulness by applying it to the conflict existing between India and Bangladesh over the control of the Ganges River. In Sect. 17.2, the history of the conflict is described. In the next section, the theory and practice of GMCR are explained followed by what is called inverse GMCR in Sect. 17.4 whereby one finds the preferences of decision makers in a conflict which are needed to reach a desirable resolution. Within Sect. 17.5, the roles of a Third Party are defined within the GMCR paradigm and classified into three types: Arbitrator, Coordinator and Donor, each of which can be referred to as a Third Party. In the next section, the conflict arising over the control of the Farakka Barrage located just upstream of Bangladesh on the Ganges River is first investigated for two cases using GMCR. Secondly, the two cases are studied using the Third Party intervention concept of a Coordinator introduced in Sect. 17.5 in order to reach a more desirable outcome. The main contents of this chapter are based upon earlier research completed by the authors for which appropriate copyright permission has been obtained for presentation in this chapter (Sakamoto et al. 2004, 2005; Hagihara and Sakamoto 2004).

17.2 Background to the Ganges River Controversy Between Bangladesh and India

As depicted in Fig. 17.1, Bangladesh is located in a vulnerable position downstream from India on the Ganges River. A detailed description of a serious conflict that took place between these two nations over control of the flow of the Ganges River is furnished by Kondo (1997). In 1975, India unilaterally built the Farakka Barrage across the Ganges River near its border with Bangladesh. At first, a temporary treaty was agreed upon by India and Bangladesh. This treaty stipulated that India could withdraw 310–350 m^3/s of water at the Farakka Barrage and discharge 1,245–1,400 m^3/s of water downstream from April 21 to May 30, when the Ganges River usually has the least amount of flow in a given year. A few months later, the treaty expired and India commenced withdrawing water without any agreement until the two nations concluded a second treaty in 1977. The withdrawals and releases allowed by India under the 1977 Treaty are listed in Table 17.1.

The 1977 treaty was concerned with water resources allocation of the Ganges River during the dry season. Initially, Bangladesh and India claimed different definitions for what constitutes the dry season. Bangladesh argued that it was from November to May, while India insisted that it was from March to May. Due to of the compromise that was reached, the dry season in this treaty was defined as

Table 17.1 1977 treaty between India and Bangladesh (m^3/s)

Month	Day	Flow 1948–73	Withdrawal by India	Release to Bangladesh
January	1–10	2,790	1,133	1,657
	11–20	2,542	1,090	1,452
	21–31	2,336	991	1,345
February	1–10	2,244	934	1,310
	11–20	2,096	892	1,204
	21–28	1,982	870	1,112
March	1–10	1,848	758	1,090
	11–20	1,798	722	1,076
	21–31	1,728	708	1,020
April	1–10	1,671	680	991
	11–20	1,572	588	984
	21–30	1,558	581	977
May	1–10	1,600	609	991
	11–20	1,678	680	998
	21–31	1,855	758	1,097

being from January to May. India obtained the right to withdraw more water than in the 1975 temporary treaty, while Bangladesh was obliged to concede more than India. According to the 1977 treaty, India had to withdraw water such that the rate of flow released to Bangladesh should not be less than 80 % of the water allocated to Bangladesh. This treaty was in effect until 1984.

Subsequent to 1984, there was no rule stipulating the utilization of the Ganges River water resources between the two nations until a new treaty was concluded in 1996. The general allocation rules were specified as shown in Table 17.2 for this treaty which is still in force.

Within the 1996 treaty, Bangladesh made even more concessions than in the treaty covering the years from 1977 to 1984, because in the 1996 treaty the amount of water which Bangladesh can acquire is sometimes less than 800 m^3/s. Table 17.3 clearly demonstrates that India has a certain amount of water that it keeps for itself to satisfy its needs, and the remaining water is discharged to Bangladesh.

For the time being, the 1996 treaty has apparently brought a settlement to the controversy between the two nations, but in reality the conflict is not completely resolved. Since the treaty is clearly advantageous to upstream India, downstream Bangladesh has many legitimate complaints about it. Specifically, the amount of water in the Ganges River flowing into Bangladesh is a result of the amount of water withdrawn by India at the Farakka Barrage. Accordingly, neighbourly relationships between Bangladesh and India are an important factor for reducing Bangladesh's vulnerability to droughts and floods. On the other hand, poor communication between the two nations increases the risk of water shortages and inundations.

Table 17.2 1996 treaty between India and Bangladesh (m^3/s)

Month	Day	Flow 1949–88	Withdrawal by India	Release to Bangladesh
January	1–10	3,045	1,133	1,912
	11–20	2,766	1,133	1,633
	21–31	2,553	1,133	1,420
February	1–10	2,445	1,133	1,312
	11–20	2,347	1,133	1,214
	21–28	2,240	1,133	1,107
March	1–10	2,108	1,116	991
	11–20	1,952	960	991
	21–31	1,832	991	841
April	1–10	1,789	798	991
	11–20	1,774	991	783
	21–30	1,727	736	991
May	1–10	1,907	991	916
	11–20	2,084	1,093	991
	21–31	2,318	1,133	1,185

Table 17.3 Rules of allocation in the 1996 treaty (m^3/s)

Availability at Farakka	India	Bangladesh
Less than 1,982	50 %	50 %
1,982– 2,124	Balance of flow	991
More than 2,124	1,133	Balance of flow

The zone surrounding the Farakka Barrage is heavily guarded. The unilateral construction of the dam and the signing of a treaty by India indicate the great importance that India places on the Farakka Barrage. As noted earlier, Bangladesh is one of the world's poorest countries and is located downstream of the Farakka Barrage. India, on the other hand, is a formidable economic power which is situated upstream. One wonders how Bangladesh can improve its position with respect to the utilization of water resources from the Ganges River when it is in such a relatively weak position from economic, political and geographical viewpoints. Subsequent to outlining GMCR and inverse GMCR in the next two sections as well as Third Party intervention in Sect. 17.5, it is demonstrated in Sect. 17.6 how Third Party intervention can enhance Bangladesh's situation.

17.3 Graph Model for Conflict Resolution: Theoretical Overview and Implementation

The Graph Model for Conflict Resolution (GMCR) (Kilgour et al. 1987; Fang et al. 1993) constitutes significant improvements over the *conflict analysis* procedure of Fraser and Hipel (1979, 1984), which in turn is an extension of *metagame analysis*

(Howard 1971). These three approaches to multiple participant-multiple objective decision making can be thought of as belonging to a branch of *game theory* that is quite different from the more traditional methods based on the classic work of von Neumann and Morgenstern (1953). Hipel (2009a, b) and Kilgour and Eden (2010) provide descriptions of a variety of useful conflict methods which can be applied to a range of different kinds of conflict problems. Hipel et al. (1993) furnish an overview of the use of game theoretic models in engineering decision making while Hipel and Bernath Walker (2011) explain how conflict techniques can be utilized in environmental management.

The systematic investigation of a given dispute using the GMCR methodology consists of two main stages: modelling and analysis, which can be supplemented by a range of follow-up analyses. A graph model is composed of a set of decision makers (DMs), N ($|N| \geq 2$), a set of feasible states, S, and each DM's relative preferences among the feasible states in the conflict under study. For $s, t \in S$, $s \succ_i t$ means that DM i prefers state s to t, while $s \sim_i t$ indicates that DM i is indifferent between s and t.

When contemplating what move to make from a given state in a conflict, DMs think like a chess player in terms of moves and countermoves when trying to decide what to do next. Hence, intuitively one may wish to capture the set of moves in one step available to each DM using what is called a directed graph. In particular, DM i's directed graph is denoted by $G_i\ (S, A_i)$ for which S is the set of all vertices and $A_i \subset S \times S$ constitutes the set of all arcs such that $(s, s) \notin A_i$ for all $s \in S$. A complete graph model consists of the set of DMs, the set of feasible states which includes the possible movements by each DM as encoded in that DM's directed graph, and the relative preferences of each DM between all pairs of states. If the states can be ranked from most to least preferred for a DM where ties are allowed, the preferences are labelled as being transitive or ordinal for that DM. In most disputes preferences tend to be transitive, especially when the conflict is well underway. However, GMCR can handle intransitive preferences whereby DM i prefers state s to t, state t to k but prefers state k to s. Another key advantage of the graph model is its flexibility in handling irreversible moves, whereby a DM can cause the conflict to move from one state to another, but not vice versa. In a graph model, irreversible moves are simply identified by unidirectional arcs.

In practice, movements under the control of a DM from a particular state are efficiently recorded in what is referred to as a reachable list, which contains the set of all states to which DM $i \in N$ can move in one step. DM i's reachable list from state $s \in S$ is the set $\{t \in S \mid (s, t) \in A_i\}$, denoted by $R_i\ (s) \subset S$. If a DM can unilaterally improve from one state to a more preferred state, this is called a unilateral improvement (UI). DM i's UI list from state s is the set $\{t \in R_i\ (s) \mid t \succ_i s\}$, denoted by $R_i^+ (s) \subset S$.

When participating in a conflict, DMs may behave in different ways. Accordingly, the graph model uses a rich variety of solution concepts for describing human behaviour when DMs interact strategically with one another as each DM attempts to do as well as possible within the social constraints of what others can do. Four insightful solution concepts which have been defined within a graph model context are Nash stability (Nash 1950, 1951), general metarationality (Howard 1971), symmetric metarationality (Howard 1971), and sequential stability (Fraser and Hipel 1979, 1984). Within this chapter, Nash stability and sequential stability are employed in analysing the Farakka Barrage dispute and are defined below for the general case of two or more DMs.

Nash Stability For DM $i \in \mathbf{N}$, a state $s \in \mathbf{S}$ is Nash stable for DM i if and only if $\mathbf{R}_i^+(s) = \varnothing$. Under Nash stability, the initial state s is therefore stable if and only if DM i cannot unilaterally move from s to any state it prefers more. A state which is Nash stable is also referred to as being rational.

Sequential Stability For DM i ∈ N, a state s ∈ S is sequentially stable if and only if for all x $\in R_i^+(s)$, there exists $t \in R_{N-i}^+(x)$ such that $t \leq s$. Therefore, a state s is sequentially stable for DM i if and only if he or she is deterred from making any UI away from s because subsequent UIs by the other DMs, N – i, could result in a state less preferred by i than the initial state s.

A state that is stable for every DM is referred to as an equilibrium or resolution. If a conflict evolves to a certain equilibrium, the conflict will remain at that equilibrium because no DM has an incentive to move away from it on his or her own. Nonetheless, it is possible that DMs could move to another equilibrium if a mutually beneficial coalition can be formed.

A follow-up analysis permits one to explore extensions to the GMCR methodology to ascertain if and how the equilibrium results and other strategic findings could vary. A regular stability analysis determines how well a given DM can do on his or her own. By carrying out a coalition study, one can check if a DM can fare even better by cooperating with others via joining a coalition in which all coalition members benefit (Kilgour et al. 2001; Inohara and Hipel 2008a, b). In a sensitivity analysis, one may wish to check if small meaningful changes in preference of one or more DMs affect the resolutions. The consequences of preference uncertainty in a stability analysis can also be determined using levels or strength of preference (Hamouda et al. 2004, 2006; Xu et al. 2009b), unknown preference (Li et al. 2004a), fuzzy preferences (Bashar et al. 2012, 2014; Hipel et al. 2011a) and Grey preferences (Kuang et al. 2015). One can also find out how attitudes can affect an outcome to a dispute since, for example, one may suspect that having a positive attitude will lead to a better conclusion (Inohara et al. 2007; Bernath Walker et al. 2012). One can employ what is called status quo analysis to ascertain if a desirable outcome can be reached from a specified starting or status quo state (Li et al. 2004b, 2005).

A decision support system (DSS) permits both practitioners and researchers to employ a given methodology for analysing problems (Sage 1991) such as those arising in the field of water resources (Hipel et al. 2008a, 2015). The DSS called GMCR II (Hipel et al. 1997; Fang et al. 2003a, b) can be employed for strategically analysing a real world conflict such as the Farakka Barrage conflict in Sect. 17.6. A new DSS called GMCR + (Kinsara et al. 2015b) can be utilized for not only executing a regular GMCR analysis but also for implementing what is called Inverse GMCR (Kinsara 2015a) described next.

17.4 Inverse Graph Model for Conflict Resolution (Inverse GMCR)

To reach a desirable outcome for the participants in a conflict, preferences, including entrenched preferences from one or more DMs in the dispute, may have to change. In a formal GMCR study the key information for calibrating a conflict model must be obtained: the DMs, each DM's options or courses of actions under its control and the relative preferences of each DM over the feasible states. As shown in the top position of Fig. 17.2, the engine in a DSS such as GMCR II calculates the stable states for each DM and state according to a range of solution concepts defining possible human behaviour under conflict. From this, the potential equilibria can be ascertained as the set of states which are stable for all DMs according to a given solution concept.

In a particular negotiation, a Third Party or a DM involved in the negotiations may wish to know how the preferences of one or more DMs should change in order to be able to reach an attractive final outcome. This process falls under a class of important methodologies called reverse engineering: given the output and system (stability engine in GMCR), what input is needed (preferences in GMCR)? When

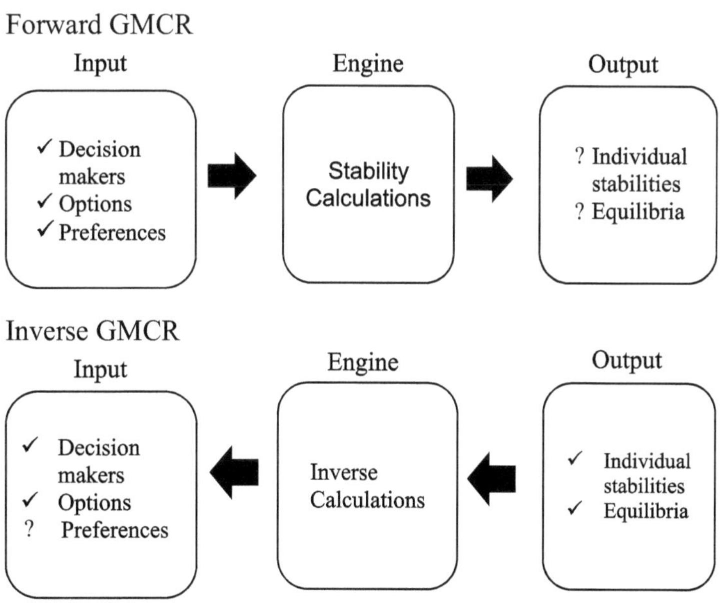

Fig. 17.2 (**a**) Forward and (**b**) Inverse GMCR (Based on the graph by Kinsara et al. 2015b)

employed within the paradigm of GMCR, this process is called Inverse GMCR (Kinsara et al. 2015a) and is shown in the lower portion of Fig. 17.2. Fortunately, Kinsara et al. (2015b) have developed algorithms for implementing Inverse GMCR in practice and have applied it to the aforementioned conflict over water in the Middle East. Their system, which is referred to as GMCR +, can handle both the Forward GMCR (top of Fig. 17.2) and Inverse GMCR (bottom of Fig. 17.2). Moreover, stability calculations within GMCR + can be executed using both the logical (Kilgour et al. 1987; Fang et al. 1993) and matrix (Xu et al. 2009a) definitions of GMCR. Inverse GMCR offers a powerful tool for employment by facilitators or negotiators participating in a Third Party Intervention, which is discussed next.

17.5 Third Party Intervention in Conflict Resolution

The purpose of this section is to present the procedure depicted in Fig. 17.3 as a viable approach for effectively addressing challenging conflict using a Third Party to enhance the situation. By understanding how a Third Party can bring about a more attractive resolution for all concerned parties or DMs interacting in a dispute, one can ascertain the best role that a Third Party can play. This overall approach is commonly referred to as Third Party Intervention and has been widely studied in the literature by authors such as Raiffa (1982), Raiffa et al. (2002), Bercovitch and Jackson (2009), Greig and Diehl (2012), and researchers who contributed to a set of encyclopaedia papers on the topic of conflict resolution (Hipel 2009a, b). The key contribution of the procedure devised in this section is the operationalization of Third Party Intervention within the paradigm of GMCR. This permits Third Party Intervention to be readily employed by practitioners and researchers for formally investigating real world conflict. Even though a conflict may initially involve more than two DMs, the terminology Third Party is still used for the person who intervenes in the dispute in order to bring about a satisfactory resolution. In practice, having two DMs initially involved in a controversy is quite common, such as in the upstream-downstream conflict between the two nations examined in Sect. 17.6, and this is the reason why the external person who is brought into the dispute to aid the two conflicting parties in reaching an attractive outcome is referred to as a Third Party.

Raiffa (1982) defines the functions of a Third Party, which is called an intervener in his book, as a facilitator, mediator, arbitrator, or rules manipulator. More specifically, "A facilitator is a person who arranges for the relevant parties [or DMs] to come to the negotiating table" (Raiffa 1982). The facilitator may decide not to become involved in the actual process of negotiation, but he or she might play a facilitating role to obtain an agreement. "A mediator is an impartial outsider who tries to help the negotiators in their quest to find a compromise agreement" (Raiffa 1982). The mediator can assist with the negotiation process, but he does not have the authority to impose a solution; rather, his purpose is to induce the negotiators to

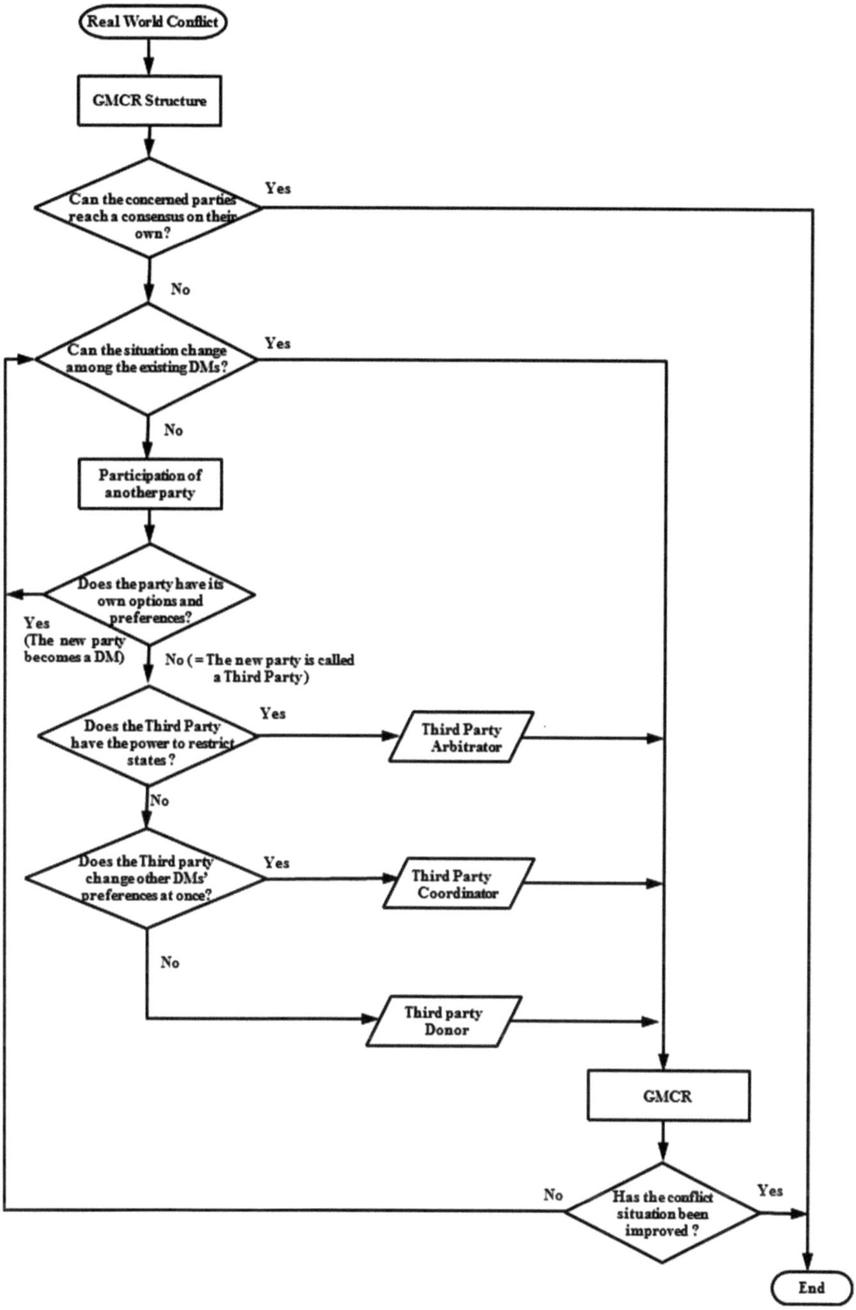

Fig. 17.3 Third party intervention within the Graph Model for Conflict Resolution

determine whether a compromise is preferred by each party to the no-agreement alternative. "An arbitrator, after hearing the arguments and proposals of all sides and after finding out 'the facts,' may also try to lead the negotiators to devise their own solutions or may suggest reasonable solutions" (Raiffa 1982). "A rules manipulator is given the authority to alter or constrain the process of negotiation – or, put another way, to modify the rules of the game" (Raiffa 1982). The rules manipulator can in fact not only propose such rules, but can also forbid the utilization of various moves that might lead to disastrous outcomes.

In a later book, Raiffa et al. (2002) define the roles of a Third Party, which is called an external helper in their book, as facilitation, mediation, arbitration, and rules manipulation. They believe that it is difficult to classify the functions of a Third Party. As an alternative, they introduce the ideas of evaluative and non-evaluative participation. Evaluative indicates that an external helper determines who receives what on the basis of his determination of which party is right and which participant has behaved more appropriately. For the situation of a non-evaluative external helper, it is not a question of who is right or wrong.

In this chapter, a Third Party is defined to be an entity that is not an actual stakeholder but is motivated to assist in bringing about a more desirable outcome, resolution or equilibrium. Figure 17.3 outlines how the Third Party procedure developed by the authors is implemented in practice. As shown in the flow chart, the approach starts at the top-left side with an actual dispute requiring resolution and ends at the bottom-right part of the graph. Moreover, if a Third Party is needed, it is categorized as being an Arbitrator, Coordinator or Donor as listed in the central portion of the figure.

Moving downwards from the start of the flow chart at the top left in Fig. 17.3, one can see that the real-world conflict being considered is formally investigated within the structure of the Graph Model for Conflict Resolutions (GMCR) outlined in Sect. 17.3, as indicated in the rectangle located second from the top on the left part of the flowchart. In practice, one could use a decision support system such as GMCR II or GMCR +, mentioned at the end of Sect. 17.3 and in Sect. 17.4 for the case of GMCR +, to apply the GMCR methodology to an actual dispute. For a small dispute, one could perform the stability calculations by hand using what is called Tableau Form designed by Fraser and Hipel (1979, 1984). Whatever the case, if a party has its own preferences and options it controls with respect to the conflict, it can be considered to be an actual DM in the dispute within GMCR. If, as written in the third enclosure from the top on the left in Fig. 17.3, the DMs involved in the conflict can reach a consensus on their own, the negotiations reach a resolution acceptable to all DMs and the dispute comes to an end. If not, a more in-depth conflict analysis study may be needed. If this study indicates that the situation among the DMs does not change and is deadlocked, another party may be needed. If this new party has its own options and preferences over the states, it becomes a new DM as shown in the middle-left portion of Fig. 17.3. On the other hand, if the new party does not have its own preferences over the possible states in the conflict, it is defined as a Third Party, which can be further categorized as Arbitrator, Coordinator or Donor. If a Third Party has the power to exclude certain states, and restrict

DMs from moving to other states, the Third Party is called an Arbitrator. The difference between Coordinator and Donor is whether the Third Party can change other DMs' preferences at the current moment or not. If it can change DMs' preferences immediately by offering some options to the DMs, the role of the Third Party is defined as being a Coordinator. If it cannot, the Third Party is referred to as a Donor. A Donor does not have a direct influence on the conflict situation at the present moment, but in the long term, it might help the DMs to devise a new option or change their preferences. Employing the Inverse GMCR decision technology mentioned in Sect. 17.4 provides an operational means by which the Donor can ascertain which preference changes are needed by one or more DMs to create a final desirable outcome.

Compared with Raiffa et al.'s (2002) definition, an Arbitrator is almost the same as a rules manipulator. Other definitions of Raiffa et al. (2002) are not similar because they presume that the structure of a conflict is fixed and also that a Third Party does not have the power to change the situation. In this study, however, a Third Party may alter the structure of a conflict implicitly or explicitly. As for the concepts of evaluative and non-evaluative participation, a Third Party in this procedure can perform both roles by considering the final desirable outcome from the perspective of either interpretation.

17.6 Farakka Barrage Conflict Case Study

In the next subsection, a detailed strategic analysis of the Farakka Barrage conflict using the GMCR methodology outlined in Sect. 17.3 reveals that it is difficult to reach a resolution which is acceptable for Bangladesh according to two different preference structures assumed for India (called Case A and Case B). However, when a Third Party is included in the study following the procedure explained in Sect. 17.5 and depicted in Fig. 17.3, a final win/win outcome can be achieved.

17.6.1 First Conflict Investigation: Current Situation Without a Third Party

The DMs and options in the Farakka Barrage conflict outlined in Sect. 17.2 are listed on the left in Table 17.4. As can be seen, Bangladesh controls the single option of agreeing to the current operation of the Farakka Barrage (called Agree, for short, in Table 17.4). India possess the two options of operating the Farakka Barrage according to the existing operating rules (Operate) and changing the present operating rules to benefit Bangladesh (Change).

On the right side in Table 17.4, the possible states or scenarios that could occur in the Farakka dispute are shown as columns of Ys and Ns. The selection of an

Table 17.4 Decision makers, options, and states in the Farakka barrage conflict

DMs and options		States							
Bangladesh									
Agree	Agree to the current operation of the Farakka barrage	N	Y	N	Y	N	Y	N	Y
India									
Operate	Operate the Farakka barrage according to the existing operating rules	N	N	Y	Y	N	N	Y	Y
Change	Change the present operating rules	N	N	N	N	Y	Y	Y	Y
Label		1	2	3	4	5	6	7	8

Table 17.5 Ranking of states for Bangladesh

DMs and options	States							
Bangladesh								
Agree	Y	Y	N	N	Y	N	N	Y
India								
Operate	Y	N	N	N	N	Y	Y	Y
Change	Y	Y	Y	N	N	Y	N	N
Label	8	6	5	1	2	7	3	4

option by a given DM is indicated by a Y opposite the option and not taking an option is marked using an N. When either a Y or N is written opposite all of the options of a given DM, this constitutes a strategy for the DM. A state is formed when each DM selects a strategy; thus, each column of Ys and Ns constitutes a state in the Farakka Barrage conflict. For instance, in the seventh state from the left in Table 17.4, the N for the Bangladesh option indicates that Bangladesh does not agree to the current operation of the Farakka Barrage. Accordingly, Bangladesh has selected the strategy (N). By operating the Farakka Barrage according to the existing operating rules, which are the changed rules, India has chosen the strategy (Y,Y). Combining Bangladesh's strategy (N) and India's strategy (Y,Y) creates state (N,Y,Y), when written horizontally in text rather than as a column as in Table 17.4. For convenience, state (N,Y,Y) is designated as state 7 in Table 17.4, with other numerical labels given to the remaining states.

The next step of the modelling process is to order the states in Table 17.4 to reflect the preferences of each DM. Table 17.5 displays the states in the conflict ranked from most preferred on the left to least preferred on the right according to Bangladesh's preferences. As shown, the most preferred situation for Bangladesh is state 8 in which India operates the Farakka Barrage according to the changed rules and Bangladesh agrees to this. The least preferred state is 4 in which India operates the dam following the original rules and Bangladesh accepts this.

For a relatively small dispute like the one investigated herein, one can rank the states by hand for each DM. When dealing with small, medium and larger games, a preference solicitation technique called Option Prioritization can be employed to

Table 17.6 Ranking of states for India

DMs and options	States							
Bangladesh								
Agree	Y	Y	N	N	Y	Y	N	N
India								
Operate	Y	Y	Y	Y	N	N	N	N
Change	N	Y	N	Y	N	Y	N	Y
Label	4	8	3	7	2	6	1	5

rank the states from most to least preferred where ties are allowed (Hipel et al. 1997; Fang et al. 2003a, b). In this approach, preference statements expressed in terms of options are listed from most to least important. Subsequently, under the assumption of transitive preferences, an algorithm is used in GMCR II and GMCR + to rank the states according to preference for the DM being considered.

Two different situations are now considered for India's preferences as explained below according to Cases A and B. For both situations, Option Prioritization is used to rank the states for India. In addition, stability analysis reveals that state 3 is an equilibrium for both cases.

Case A A DM often expresses its preferences in terms of which options it would like to see selected or avoided. India's highest priority is that it operates the Farakka Barrage according to the existing operating rules. Therefore, in Table 17.6 in which states are ranked from most preferred on the left to least preferred on the right for India, notice that the set of states having a Y opposite the option Operate (states 4, 8, 3, and 7) is positioned to the left of the other four states having an N beside the option Operate (states 2, 6, 1 and 5). Next, India would like to see Bangladesh agree to the current operation of the Farakka Barrage (option Agree). Therefore, for the set of four states on the left (states 4, 8, 3 and 7) in Table 17.6, one can see that states 4 and 8 are more preferred than states 3 and 7. Likewise, for the set of four states listed on the right in Table 17.6 (states 2, 6, 1 and 5), notice that states 2 and 6 are more preferred than states 1 and 5. Finally, India's third priority is not to select its option Change. The way in which this affects the final ordering of states in Table 17.6 can be clearly seen. When needed, conditional preference statements can also be taken into account using Option Prioritization, which satisfies all of the rules for what is called First Order Logic.

This ranking of states for Bangladesh and India shown in Tables 17.5 and 17.6, respectively, constitutes the key modelling information required as input for a stability analysis. As mentioned in Sect. 17.3, the DSSs called GMCR II and GMCR + can be utilized to calculate stability for each state and each DM according to a range of solution concepts describing possible human behaviour under conflict. Additionally, these DSSs can be used with small, medium and large conflicts. Because the conflict considered here is small in size, some of the stability calculations for Nash and sequential stability are now explained for Case A using what is called Tableau Form developed by Fraser and Hipel (1979, 1984).

Table 17.7 Stability analysis tableau

Bangladesh								
	E						E	
Stability	r	r	s	r	s	s	r	u
State ranking	8	6	5	1	2	7	3	4
UIs			6		1	8		3
India								
Stability	r	s	r	u	u	u	u	u
State ranking	4	8	3	7	2	6	1	5
UIs		4		3	4	4	3	3
					8	8	7	7
						2		1

When calculating stability by hand for a small conflict, the Tableau Form given in Table 17.7 is quite convenient to use. Notice that for Bangladesh and India in this table, the ranked states are listed according to the ordering shown at the bottom of Tables 17.5 and 17.6, respectively, using the state numbers to designate states. The stability of each state for each DM is indicated using the lettering *r*, *s* and *u* to stand for Nash stable, sequentially stable when sanctioning is required, and unstable, respectively.

Each number written under a given state in Table 17.7 is a unilateral improvement (UI). A UI is a more preferred state to which a particular DM can unilaterally move by a change in strategy, assuming that the other DM's strategy remains the same. Because a UI from a state is preferred by the DM under consideration, the UI appears to the left of that state in the ranking of states. For example, consider state 6 for India in Table 17.6. For this state, India has selected the strategy of not operating the Farakka Barrage according to the existing operating rules and changing the present operating rules. Hence, India has taken the strategy (N,Y), when written horizontally in text. If Bangladesh maintains the strategy (Y), India could unilaterally improve from state 6 to states 4, 8, or 2 by appropriately changing its option selections from (N,Y) to (Y,N), (Y,Y), or (N,N), respectively. Of these, state 4 is most preferred by India, and is therefore placed immediately under state 6 in the state ranking for India in the lower part of Table 17.7. State 2 is also preferred to 6, but it is the least preferred among the UIs, 6, 8, and 2. Therefore, state 2 is written at the bottom of the column.

Any state that does not have a UI written below it is Nash stable for the DM under consideration, since the DM cannot improve its position on its own. Therefore, an *r* is written above states 8, 6, 1 and 3 for Bangladesh and states 4 and 3 for India in Table 17.7 to indicate Nash stability. For a state to be sequentially stable (*s*) when UIs are present, all UIs below the state for a given DM must be sanctioned by the other DM. Consider, for instance, the stability of state 2 from the viewpoint of Bangladesh. From the upper portion of Table 17.7, one can see that Bangladesh has a UI from state 2 to 1. Accordingly, in state 2, in which Bangladesh agrees to the current operation and India does nothing, Bangladesh can improve its situation by deciding not to agree by unilaterally improving to state 1. Nonetheless, as shown in the lower portion of Table 17.7, India has a UI from state 1 to state 3 or 7. By moving to state 3, India follows the current operating rules. Because state 3 is less preferred to state 1 by Bangladesh (see the upper portion of Table 17.7), the UI by Bangladesh is effectively blocked. However, it is only necessary to find at least one sanction for a given UI in order for it to be sanctioned. Since all the possible UIs from state 2 are sanctioned (in this case there is only one UI), an *s* is written above state 2 in the state ordering for Bangladesh to indicate sequential stability.

For a state to be unstable for a DM, at least one UI must not be sanctioned. Consider, for instance, state 7 from the point of view of India. From the lower portion of Table 17.7, one can see that India has a UI to state 3. Because Bangladesh has no UI from state 3 (state 3 is Nash stable for Bangladesh), the UI by India cannot be blocked and therefore a *u* is written above state 7 for India to indicate that the state is unstable for India.

To form an equilibrium or possible compromise resolution, a state must be stable for all of the DMs. Notice in Table 17.7 that state 8 is rational (*r*) for Bangladesh and sequentially sanctioned (*s*) for India. Accordingly, state 8 is an equilibrium. The only other equilibrium state is state 3, which is rational for both DMs. In Table 17.7, an E is written above states 8 and 3 in Bangladesh's ordering of states to indicate that they constitute equilibria.

By referring to the Y-N option form notation used in Tables 17.5 and 17.6, one can understand that state 8 represents the desirable situation in which India operates the Farakka Barrage according to changed operating rules and Bangladesh agrees to this. In contrast, state 3 stands for the scenario in which India does not change the rules and operates the dam according to the existing rules without Bangladesh's consent. Nonetheless, as can be seen in these two tables, both DMs prefer state 8 over the status quo situation, state 3. Unfortunately, if either of the DMs independently changes its strategy selection to try to improve the situation, the resulting state is less preferred for that DM. This process is illustrated in Table 17.8. In particular, as shown in the third column from the right in Table 17.8, if Bangladesh changes its option selection from not agreeing to agreeing, the result is state 4 which is less preferred than state 3 by Bangladesh (see Table 17.5). Thus, this change of option choice is a unilateral disimprovement for Bangladesh. As depicted in the middle part of Table 17.8, if India unilaterally decides to change the

Table 17.8 Transitions from state 3

DMs and options	Bangladesh			India			Together		
Bangladesh									
Agree	N	→	Y	N		N	N	→	Y
India									
Operate	Y		Y	Y		Y	Y		Y
Change	N		N	N	→	Y	N	→	Y
Label	3	→	4	3	→	7	3	→	8
	Unilateral disimprovement			Unilateral disimprovement			Joint improvement		

Table 17.9 Ranking of states for India for case B

DMs and options	States							
Bangladesh								
Agree	Y	N	Y	N	Y	N	Y	N
India								
Operate	Y	Y	Y	Y	N	N	N	N
Change	N	N	Y	Y	N	N	Y	Y
Label	4	3	8	7	2	1	6	5

present operating rules, this creates a unilateral disimprovement for India from state 3 to 7.

In summary, neither nation on its own has the motivation to move from state 3 to 8, because such a movement constitutes a unilateral disimprovement. To reach state 8, which is more preferred over state 3 by both countries, Bangladesh and India must cooperate with one another and jointly move to state 8, as is illustrated in the right column in Table 17.8. Accordingly, a Third Party is needed to encourage communication and understanding between the two DMs to bring about this win/win resolution.

Case B India's highest priority is to operate the Farakka Barrage according to existing operating rules. Next in importance, India would not like to adopt the option Change. Finally, India's third priority is to see Bangladesh agree to the existing operating rules. The way in which these prioritized preference statements affects the final ordering of states according to preference can be clearly seen in Table 17.9, where states are ordered from most preferred on the left to least preferred on the right.

A stability analysis can be carried out using the preferences for Bangladesh and India in Tables 17.5 and 17.9 (Case B), respectively, in order to determine the stability of each state for each DM. The only state which is stable according to rational or sequential stability for both DMs is state 3, which is rational for each of them. Hence, state 3 is the equilibrium for Case B.

Interpretation For the first conflict analyzed in this subsection, in which there is no Third Party, state 8 is the most preferable potential resolution for Bangladesh, and this state can be realized when the state ranking for India is that given under Case A. As for Case B, if India prefers state 8 to 3, and 6 to 1 in Table 17.9, the state ranking for India can be changed to be the same one as in Case A in Table 17.6, where state 8 is achieved as an equilibrium. This change in preference can be brought about if India's third priority under Case B becomes its second priority, which is identical to Case A. Stated otherwise, India prefers that Bangladesh agrees to the current operation of the Farakka Barrage, more than India does not change the present operating rules. This change in priorities by India would cause state 8 to become more preferred than state 3, and state 6 to be more preferred than 1.

In the present circumstances, it may be difficult for Bangladesh and India on their own to reach state 8, since cooperation is required. Nevertheless, the participation of a Third Party may cause this to happen. Accordingly, in the next subsection a Third Party is brought into the study to ascertain if a breakthrough can be achieved.

17.6.2 Second Conflict Investigation: Inclusion of a Third Party

As explained in Sect. 17.5 and depicted in the central portion of Fig. 17.3, a Third Party can play one of three possible roles: Arbitrator, Coordinator or Donor. In this subsection, a Third Party is brought into the study of the Farakka Barrage conflict in order to encourage India to appropriately change its preferences to bring about a more desirable result. Therefore, the Third Party is acting as a Coordinator. The India/Bangladesh conflict is now strategically examined for the situation in which a Coordinator is present.

A Donor can be utilized to bring about significant changes in preferences and other model parameters over a long period of time. Sakamoto and Hagihara (2001) presented a decision model for conflict which allows for preferences in a conflict to change over time. Moreover, the overall preference changes that are required can be determined using the Inverse GMCR approach described in Sect. 17.4.

To control potential outcomes of a conflict, an Arbitrator can exclude certain states. For example, assume that Bangladesh and India have the ranking of states shown in Tables 17.5 and 17.9, respectively. This, of course, is Case B in the previous subsection in which only state 3 is an equilibrium. This undesirable situation could be improved if an Arbitrator excludes state 3 as a possible resolution. When state 3 is omitted from the conflict model, a stability analysis finds states 1 and 4 to be equilibria. If the Arbitrator does not allow states 1 and 3, or 3 and 4, then state 5 is achieved as an equilibrium. In addition, if the Arbitrator simultaneously excludes states 1, 3, and 4, then states 5 and 6 are found to be equilibria. From Bangladesh's point of view, state 6 is more preferred than state 3, as can be seen in Table 17.5.

The modelling and analyses given below are similar to the first conflict in Sect. 17.6.1, except that a Coordinator is now present. In the ensuing analyses, it is assumed that all of the states are equally preferred by the Coordinator, who must act in an unbiased manner to assist in bringing about an acceptable solution for both nations involved in the Farakka Barrage dispute.

Case A For the situation in which a Coordinator or Third Party is participating in the Farakka Barrage conflict, the DMs, options and states are as shown in Table 17.10. The Third Party has the option of taking some type of action which is called Act in the left column in Table 17.10 below the DM Third Party, which will encourage Bangladesh and India to reach a mutually attractive agreement. To be able to conveniently refer to the potential states given in Option Form in Table 17.10, the states are numbered from 1 to 16. The equivalent state numbers from the first conflict in Sect. 17.6.1 are given at the bottom of Table 17.10 under

Table 17.10 Decision makers, options, and states with a third party

DMs and options	States															
Bangladesh																
Agree	N	Y	N	Y	N	Y	N	Y	N	Y	N	Y	N	Y	N	Y
India																
Operate	N	N	Y	Y	N	N	Y	Y	N	N	Y	Y	N	N	Y	Y
Change	N	N	N	N	Y	Y	Y	Y	N	N	N	N	Y	Y	Y	Y
Third party																
Act	N	N	N	N	N	N	N	N	Y	Y	Y	Y	Y	Y	Y	Y
Label	1	2	3	4	5	6	7	8	9	10	11	12	13	14	15	16
State numbers in conflict 1	1	2	3	4	5	6	7	8	1	2	3	4	5	6	7	8

Table 17.11 Ranking of states for Bangladesh with a third party

DMs and options	States															
Bangladesh																
Agree	Y	Y	Y	Y	N	N	N	N	Y	Y	N	N	N	N	Y	Y
India																
Operate	Y	Y	N	N	N	N	N	N	N	N	Y	Y	Y	Y	Y	Y
Change	Y	Y	Y	Y	Y	Y	N	N	N	N	Y	Y	N	N	N	N
Third party																
Act	Y	N	Y	N	Y	N	N	Y	N	Y	Y	N	N	Y	N	Y
Label	16	8	14	6	13	5	1	9	2	10	15	7	3	11	4	12
State numbers in conflict 1	8	8	6	6	5	5	1	1	2	2	7	7	3	3	4	4

the assumption that the Third Party is ignored. However, unless stated otherwise, the state numbers for the conflict with a Third Party are utilized in the remainder of this subsection. The status quo situation is captured by state 3 in this second conflict.

To order the states in the second conflict according to preference for Bangladesh, it is assumed that Bangladesh prefers that the Third Party takes an action rather than not take any action. Except for that, the essential preferences of Bangladesh are the same as in the first conflict in Sect. 17.6.1. The ranking of states from most to least preferred for Bangladesh is displayed in Table 17.11. As can be seen at the bottom of this table, the state numbers from the first conflict are once again listed.

From the Indian perspective, it is assumed that India prefers that the Third Party takes an action rather than not take any action. Except for this preference assumption, the preferences for India are essentially identical to those given under Case A in Sect. 17.6.1 in Table 17.6. Therefore, India's highest priority is operating the Farakka Barrage according to the existing operating rules. Secondly, India would like Bangladesh to agree to the current operating rules. Thirdly, India would like the Third Party to act. Finally, India's fourth priority is not to change the current

Table 17.12 Ranking of states for India with a third party

DMs and options	States															
Bangladesh																
Agree	Y	Y	Y	Y	N	N	N	N	Y	Y	Y	Y	N	N	N	N
India																
Operate	Y	Y	Y	Y	Y	Y	Y	Y	N	N	N	N	N	N	N	N
Change	N	Y	N	Y	N	Y	N	Y	N	Y	N	Y	N	Y	N	Y
Third party																
Act	Y	Y	N	N	Y	Y	N	N	Y	Y	N	N	Y	Y	N	N
Label	12	16	4	8	11	15	3	7	10	14	2	6	9	13	1	5
State numbers in conflict 1	4	8	4	8	3	7	3	7	2	6	2	6	1	5	1	5

operating rules. Founded upon these prioritized preference statements, the state ranking for India is as displayed in Table 17.12.

Assuming the preferences displayed in Tables 17.11 and 17.12 for Bangladesh and India, respectively, a stability analysis is carried out to ascertain the stable states for each DM as well as the potential resolutions or equilibria. The four equilibria for Case A in the second conflict are states 3, 8, 11 and 16. The stability findings reveal that states 3 and 11 are rational for both DMs, while states 8 and 16 are rational for Bangladesh and sequentially sanctioned for India.

Among the four possible equilibria, state 16 is the most desirable resolution for both Bangladesh and India. The dynamic movement required to reach state 16 is depicted in Table 17.13. Specifically, as shown in the bottom right in Table 17.13, if Bangladesh, India, and the Third Party move together from state 3, they can jointly attain state 16, which is an improvement for all three parties. Nonetheless, Bangladesh may not trust India to implement its strategy, by selecting Change, to bring about state 16, and, likewise, India may not have confidence in Bangladesh to behave in a trustworthy manner. As illustrated in the top part of Table 17.13 on the left and centre, if either Bangladesh or India moves alone, the resulting state is a unilateral disimprovement. Accordingly, the assistance of the Third Party is required to help build confidence and trust between the two nations. From the top right part of Table 17.13, one can see that if Bangladesh and the Third Party move together from state 3, a unilateral disimprovement (state 12) occurs for Bangladesh. However, as demonstrated in the bottom left portion of Table 17.13, when India and the Third Party move together from state 3, the result is state 15, which is more preferred by India. It should be emphasized that state 15 is an improvement for India even though Bangladesh is not changing its position when going from state 3 to 15. Based upon this insight, the Third Party can encourage Bangladesh to believe that India and the Third Party can move together with Bangladesh. Hence, state 16 can be reached through this mutual trust. Even when the Third Party does not act and Bangladesh and India cooperate on their own, as shown in the middle lower column in Table 17.13, Bangladesh and India have a joint improvement from state 3 to 8, which is more preferred by both DMs. However, India prefers state 4 more

Table 17.13 Transition from state 3

DMs and options	Bangladesh			India			Bangladesh and third party		
Bangladesh									
Agree	N	→	Y	N		N	N	→	Y
India									
Operate	Y		Y	Y		Y	Y		Y
Change	N		N	N	→	Y	N		N
Third party									
Act	N		N	N		N	N	→	Y
Label	**3**	→	**4**	**3**	→	**7**	**3**	→	**12**
	Unilateral disimprovement			Unilateral disimprovement			Joint disimprovement		
DMs and options	India and third party			Bangladesh and India			All together		
Bangladesh									
Agree	N		N	N	→	Y	N	→	Y
India									
Operate	Y		Y	Y		Y	Y		Y
Change	N	→	Y	N	→	Y	N	→	Y
Third party									
Act	N	→	Y	N		N	N	→	Y
Label	**3**	→	**15**	**3**	→	**8**	**3**	→	**16**
	Joint improvement			Joint improvement			Joint improvement		

than state 8, which is achieved by a unilateral improvement by India. Therefore, the movement from state 3 to 16 is more likely to take place since the Third Party is more actively involved and state 16 is more preferred to state 8 by both Bangladesh and India.

Case B For this situation, the ranking of states for Bangladesh is assumed to be the same ordering provided for Case A in Table 17.11. From its viewpoint, India prefers states in which the Third Party acts more than when it does not act. India's highest priority is that it operates the Farakka Barrage following the existing operating rules. Secondly, India would like the Third Party to take an action. Thirdly, India would not like to choose its option Change. Fourthly, India would like Bangladesh to agree to the current operating rules. Based upon these preference statements, the ranking of states according to preference for India is as shown in Table 17.14.

Employing the ranking of states furnished in Tables 17.11 and 17.12, for Bangladesh and India, respectively, calculates stability analysis revealing that states 3 and 11 are the only equilibria. Each of states 3 and 11 is rational for both DMs. When compared to the first conflict, these states are similar to state 3, which represents the present situation. Nothing has changed by the participation of the

Table 17.14 Ranking of states for India with a third party

DMs and options	States															
Bangladesh																
Agree	Y	N	Y	N	Y	N	Y	N	Y	N	Y	N	Y	N	Y	N
India																
Operate	Y	Y	Y	Y	Y	Y	Y	Y	N	N	N	N	N	N	N	N
Change	N	N	Y	Y	N	N	Y	Y	N	N	Y	Y	N	N	Y	Y
Third party																
Act	Y	Y	Y	Y	N	N	N	N	Y	Y	Y	Y	N	N	N	N
Label	12	11	16	15	4	3	8	7	10	9	14	13	2	1	6	5
State numbers in conflict 1	4	3	8	7	4	3	8	7	2	1	6	5	2	1	6	5

Third Party. Accordingly, a further assumption is required to enhance the conflict situation.

One possible approach is to add an Arbitrator who has the power to dictate changes. The Arbitrator could, for example, remove state 11 from the conflict in Table 17.10 and maintain the state rankings of Bangladesh and India provided in Tables 17.11 and 17.14, respectively. This is executed in order to assign relevance to an action taken by the Third Party, since it will not have an available action if it cannot persuade India to change. Under this changed model, states 3, 12, and 15 are found to be potential equilibria. States 3 and 12 are rational for both DMs, while state 15 is sequential for Bangladesh and rational for India.

State 15 implies improvement of the conflict situation because both DMs prefer it more than state 3. Nonetheless, state 15 indicates that India operates the Farakka Barrage even though Bangladesh does not agree to this, which means that this state cannot be considered as a viable or robust resolution for the India-Bangladesh conflict. On the other hand, state 15 also reveals that India can change the present operating rules to benefit Bangladesh. This attitude change by India can be interpreted as an improvement of the conflict situation.

The long-lasting and resilient resolution for the conflict is represented by state 16. However, to realize state 16 as an equilibrium, state 12 should be eliminated, although this is a procedure that only an Arbitrator can implement.

For an ongoing dispute like the India/Bangladesh conflict, in which the DMs have been in conflict with each other for years and the situation has been tense, it will be difficult to resolve this conflict situation with a single measure because the DMs' preferences are fundamentally different. To improve the situation, some steps made within different roles by a Third Party will be required. The realistic and practical framework given in Sect. 17.5 regarding the involvement of a Third Party could be useful in clearly analysing the process of conflict resolution within a challenging and complex conflict environment and providing valuable strategic insights for reaching a win-win lasting final outcome

17.7 Concluding Remarks on Effective Governance

As demonstrated by the Farakka Barrage controversy between India and Bangladesh investigated in the previous section, Third Party intervention has a key role to play for fairly resolving a tough conflict situation. Moreover, the general framework for formally incorporating a Third Party into the paradigm of GMCR put forward in Sect. 17.5 and depicted in Fig. 17.3, provides a powerful tool for bringing a dispute to a successful conclusion. By employing the Inverse GMCR procedure outlined in Sect. 17.4, one can determine the preferences needed by disputing parties to reach an equitable result, which can be encouraged by a Third Party.

Effective governance is required by society to cope with a broad range of interconnected issues including regional vulnerability in which society would like to prevent and mitigate damage from disasters caused by humans and natural causes such as flooding, earthquakes and tsunamis. Within Bangladesh, factors pertaining to disasters can be classified as either natural or socio-environmental ones. Socio-environmental factors can be categorized more specifically as socio-environmental destruction, socio-environmental pollution, and socio-environmental cultural factors. The water resources conflict between Bangladesh and India studied in this chapter falls within the domain of socio-environmental cultural factors. Whatever the exact situation, conflict, which will almost inevitably arise, can be systematically addressed by employing the Third Party GMCR methodology put forward in this chapter.

As explained by authors such as Hipel et al. (2008a, b; 2009), effective governance can be achieved by designing and implementing it within the paradigm of a system of systems engineering structure which is integrative, adaptive and participatory in order to achieve desirable systems properties such as robustness, sustainability and fairness. Within the field of water resources, a comprehensive approach called Integrated Water Resources Management (GWP and INBO 2009) has been successful in managing water resources in a balanced and sustainable manner. All of these attractive and relatively recent approaches to sound governance have to somehow deal with conflicting interests and, hence, a conflict resolution mechanism will be needed to lead negotiations in a positive direction towards win/win resolutions encouraged by innovative policies. Certainly, the Third Party concepts to conflict resolution presented herein should prove to be useful.

Acknowledgements The authors would like to thank Mr. Conrad W. Hipel for his professional editing of this chapter.

References

Bashar, M. A., Kilgour, D. M., & Hipel, K. W. (2012). Fuzzy preferences in the graph model for conflict resolution. *IEEE Transactions on Fuzzy Systems, 20*(4), 760–770. doi:10.1109/TFUZZ.2012.2183603.

Bashar, M. A., Kilgour, D. M., & Hipel, K. W. (2014). Fuzzy option prioritization for the graph model for conflict resolution. *Fuzzy Sets and Systems, 246*, 34–48. doi:10.1016/j.fss.2014.02.11.

Bercovitch, J., & Jackson, R. D. W. (2009). *Conflict resolution in the twenty-first century: Principles, methods and approaches*. Ann Arbor: University of Michigan Press.

Bernath Walker, S. G., Hipel, K. W., & Inohara, T. (2012). Attitudes and preferences: Approaches to representing decision maker desires. *Applied Mathematics and Computation, 218*(12), 6637–6647. j.amc.2011.11.102.

Fang, L., Hipel, K. W., & Kilgour, D. M. (1993). *Interactive decision making: The Graph Model for Conflict Resolution*. New York: John Wiley & Sons.

Fang, L., Hipel, K. W., Kilgour, D. M., & Peng, X. (2003a). A decision support system for interactive decision making, part 1: Model formulation. *IEEE Transactions on Systems, Man and Cybernetics, Part C, SMC-33*(1), 42–55.

Fang, L., Hipel, K. W., Kilgour, D. M., & Peng, X. (2003b). A decision support system for interactive decision making, part 2: Analysis and output interpretation. *IEEE Transactions on Systems, Man and Cybernetics, Part C, SMC-33*(1), 56–66.

Fraser, N. M., & Hipel, K. W. (1979). Solving complex conflicts. *IEEE Transactions on Systems, Man, and Cybernetics, 9*(12), 805–816.

Fraser, N. M., & Hipel, K. W. (1984). *Conflict analysis: Models and resolutions*. New York: North-Holland.

Global Water Partnership (GWP), & International Network of Basin Organization (INBO). (2009). A handbook for integrated water resources management in basins, published by GWP and INBO, 2009. Available at www.gwpforum.org and www.inbo-news.org. Accessed on 31 Mar 2015.

Greig, J. M., & Diehl, P. F. (2012). *International mediation*. Cambridge: Polity Press.

Hagihara, Y., & Sakamoto, M. (2004). Conflict management on utilization of the Ganges water resources between Bangladesh and India. *Annals of the Disaster Prevention Research Institute, Kyoto University, 47*(B), 13.

Hamouda, L., Kilgour, D. M., & Hipel, K. W. (2004). Strength of preference in the Graph Model for Conflict Resolution. *Group Decision and Negotiation, 13*, 449–462.

Hamouda, L., Kilgour, D. M., & Hipel, K. W. (2006). Strength of preference in graph models for multiple decision-maker conflicts. *Applied Mathematics and Computation, 179*(1), 314–327. doi:10.1016/j.amc.2005.11.109.

Hipel, K. W. (Ed.). (2009a). *Conflict resolution* (Vol. 1). Oxford: Eolss Publishers.

Hipel, K. W. (Ed.). (2009b). *Conflict resolution* (Vol. 2). Oxford: Eolss Publishers.

Hipel, K. W., & Bernath Walker, S. (2011). Conflict analysis in environmental management. *Environmetrics, 22*, 279–293. doi:10.1002/env.1048.

Hipel, K. W., Fang, L., & Kilgour, D. M. (1993). Game theoretic models in engineering decision making. *Journal of the Japan Society of Civil Engineers, Infrastructure Planning and Management, 470*(IV-20), 1–16.

Hipel, K. W., Kilgour, D. M., Fang, L., & Peng, X. (1997). The decision support system GMCR in environmental conflict management. *Applied Mathematics and Computation, 83*(2 and 3), 117–152.

Hipel, K. W., Fang, L., & Kilgour, D. M. (2008a). Decision support systems in water resources and environmental management. *Journal of Hydrologic Engineering, 13*(9), 761–770.

Hipel, K. W., Obeidi, A., Fang, L., & Kilgour, D. M. (2008b). Adaptive systems thinking in integrated water resources management with insights into conflicts over water exports. *INFOR, 46*(1), 51–69.

Hipel, K. W., Obeidi, A., Fang, L., & Kilgour, D. M. (2009). Sustainable environmental management from a system of systems perspective. In M. Jamshidi(ed) (Ed.), *System of systems engineering: Innovations for the 21st century* (pp. 443–481). New York: Wiley. Chapter 18.

Hipel, K. W., Kilgour, D. M., & Bashar, M. A. (2011a). Fuzzy preferences in multiple participant decision making. *Scientia Iranica, Transactions D: Computer Science & Engineering and Electrical Engineering, 18*(3(D1)), 627–638.

Hipel, K. W., Kilgour, D. M., & Fang, L. (2011b). The Graph Model for Conflict Resolution. In J. J. Cochran, L. A. Cox, P. Keskinocak, J. P. Kharoufeh, & J. C. Smith (Eds.), *Wiley encyclopedia of operations research and management science* (pp. 2099–2111). New York: Wiley. Vol. 3 of 8.

Hipel, K. W., Kilgour, D. M., & Kinsara, R. A. (2014). Strategic investigations of water conflicts in the Middle East. *Group Decision and Negotiation, 23*(3), 355–376. doi:10.1007/s10726-012-9325-3.

Hipel, K. W., Fang, L., Cullmann, J., & Bristow, M. (Eds.). (2015). *Conflict resolution in water resources and environmental management* (p. 291). Heidelberg: Springer. doi:10.1007/978-3-319-14215-9. ISBN 978-3-319-14215-9.

Howard, N. (1971). *Paradoxes of rationality: Theory of metagames and political behaviour*. Cambridge: MIT Press.

Inohara, T., & Hipel, K. W. (2008a). Coalition analysis in the Graph Model for Conflict Resolution. *Systems Engineering, 11*(4), 343–359.

Inohara, T., & Hipel, K. W. (2008b). Interrelationships among noncooperative and coalition stability concepts. *Journal of Systems Science and Systems Engineering, 17*(1), 1–29.

Inohara, T., Hipel, K. W., & Walker, S. (2007). Conflict analysis approaches for investigating attitudes and misperceptions in the war of 1812. *Journal of Systems Science and Systems Engineering, 16*(2), 181–201.

International Monetary Fund (IMF). (2014). *World economic outlook data base*. Washington, DC: IMF.

Kilgour, D. M., & Eden, C. (Eds.). (2010). *Handbook of group decision and negotiation*. Dordrecht: Springer.

Kilgour, D. M., Hipel, K. W., & Fang, L. (1987). The graph model for conflicts. *Automatica, 23*(1), 41–55.

Kilgour, D. M., Hipel, K. W., Peng, X., & Fang, L. (2001). Coalition analysis in group decision support. *Group Decision and Negotiation, 10*, 159–175.

Kinsara, R. A., Kilgour, D. M., & Hipel, K. W. (2015a). Inverse approach to the Graph Model for Conflict Resolution. IEEE Transactions on System, Man, and Cybernetics: Systems, *45*(5), 734–742. doi:10.1109/TSMC.2014.2376473.

Kinsara, R. A., Petersons, O., Hipel, K. W, & Kilgour, D. M. (2015b). Advanced decision support system for the Graph Model for Conflict Resolution. Journal of Decision Systems, special issue on Integrated Decision Support Systems, *24*(2), 117–145. doi: 10.1080/12460125.2015.1046682.

Kondo, N. (ed) (1997). Gendai minami ajia no kokusai kankei (The present international relations in South Asia), Institute of Developing Economies, Japan External Trade Organization, Mihamaku, Chiba, pp. 111–138.

Kuang, H., Bashar, M. A., Hipel, K. W., & Kilgour, D. M. (2015). Grey-based preference in a Graph Model for Conflict Resolution with multiple decision makers. *IEEE Transactions on Systems, Man and Cybernetics: Systems*. doi:10.1109/TSMC.2014.2387096.

Li, K. W., Hipel, K. W., Kilgour, D. M., & Fang, L. (2004a). Preference uncertainty in the Graph Model for Conflict Resolution. *IEEE Transactions on Systems, Man, and Cybernetics, Part A, 34*(4), 507–520.

Li, K. W., Kilgour, D. M., & Hipel, K. W. (2004b). Status quo analysis of the Flathead River conflict. *Water Resources Research, 40*(5), 9. doi:10.1029/2003WR002596. W05S03.

Li, K. W., Kilgour, D. M., & Hipel, K. W. (2005). Status quo analysis in the Graph Model for Conflict Resolution. *Journal of the Operational Research Society, 56*, 699–707.

Nash, J. F. (1950). Equilibrium points in n-player games. *Proceedings: National Academy of Sciences, 36*, 48–49.
Nash, J. F. (1951). Non-cooperative games. *Annals of Mathematics, 54*, 286–295.
Raiffa, H. (1982). *The art and science of negotiation*. Massachusetts: Harvard University Press.
Raiffa, H., Richardson, J., & Metcalfe, D. (2002). *Negotiation analysis: The science and art of collaborative decision making*. Massachusetts: Belknap Press of Harvard University Press.
Sage, A. P. (1991). *Decision support systems engineering*. New York: Wiley.
Sakamoto, M., & Hagihara, Y. (2001). A study of the changing process of the conflict on a large-scale regional development project. *Studies in Regional Science, 31*(3), 177–190.
Sakamoto, M., Hagihara, Y., & Hipel, K. W. (2004). A study on the roles of a third party in the Ganges River water conflict between Bangladesh and India (in Japanese). In Proceedings of the environment research conference, sponsored by the committee on environmental systems of the Japan society of civil engineering, held in Tokyo, 3–4 Nov 2004, Vol. 32, pp. 29–36.
Sakamoto, M., Hagihara, Y., & Hipel, K. W. (2005). Coordination process by a third party in the conflict between Bangladesh and India over regulation of the Ganges River. In Proceedings of the 2005 I.E. international conference on systems, man and cybernetics, held in the Hilton Waikoloa Village, The Big Island of Hawaii, 9–12 Oct 2005, pp. 1119–1125.
von Neumann, J., & Morgenstern, O. (1953). *Theory of games and economic behaviour* (3rd ed.). Princeton: Princeton University Press.
Xu, H., Hipel, K. W., & Kilgour, D. M. (2009a). Matrix representation of solution concepts in multiple decision maker graph models. *IEEE Transactions on Systems, Man, and Cybernetics, Part A: Systems and Humans, 39*(1), 96–108.
Xu, H., Hipel, K. W., & Kilgour, D. M. (2009b). Multiple levels of preference in interactive strategic decisions. *Discrete Applied Mathematics, 57*, 3300–3313.

Erratum to: Study on Planning Scheme to Improve the Living Environment Through Safe Water Supply and Sanitation in a Rural Village of Bangladesh

Erratum to:
Chapter 15 in: K. Hagihara, C. Asahi (eds.), *Coping with Regional Vulnerability*, New Frontiers in Regional Science: Asian Perspectives 4, DOI 10.1007/978-4-431-55169-0_15

The online version of the original chapter can be found under 10.1007/978-4-431-55169-0_15

K. Hagihara, C. Asahi (eds.), *Coping with Regional Vulnerability*, New Frontiers in Regional Science: Asian Perspectives 4, DOI 10.1007/978-4-431-55169-0_18

In Chapter titled "*Study on Planning Scheme to Improve the Living Environment Through Safe Water Supply and Sanitation in a Rural Village of Bangladesh*" Figure 15.4 is incorrect. The correct version is given below:

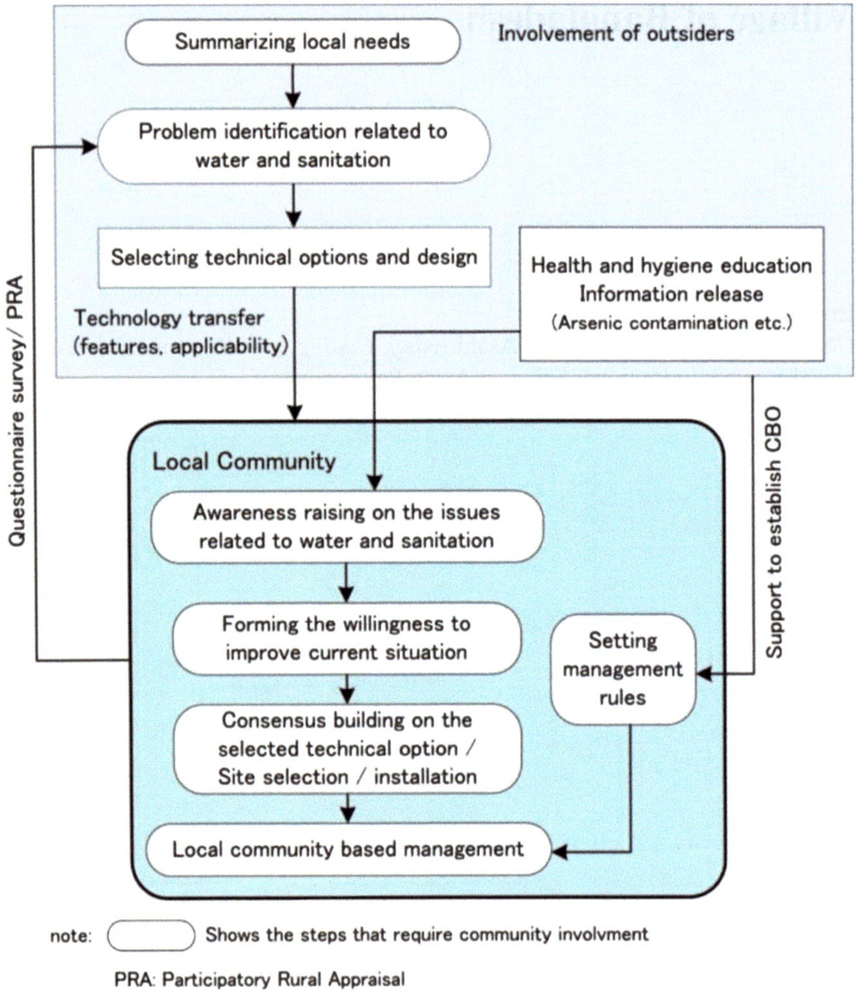

Fig. 15.4

Index

K. Hagihara, C. Asahi (eds.), *Coping with Regional Vulnerability*, New Frontiers in Regional Science: Asian Perspectives 4, DOI 10.1007/978-4-431-55169-0

MIX
Papier aus verantwortungsvollen Quellen
Paper from responsible sources
FSC® C105338

If you have any concerns about our products,
you can contact us on
ProductSafety@springernature.com

In case Publisher is established outside the EU,
the EU authorized representative is:
Springer Nature Customer Service Center GmbH
Europaplatz 3, 69115 Heidelberg, Germany

Printed by Libri Plureos GmbH
in Hamburg, Germany